Representations of Semisimple Lie Algebras in the BGG Category $\mathcal{O}$

GRADUATE STUDIES
IN MATHEMATICS **94**

Representations of Semisimple Lie Algebras in the BGG Category $\mathcal{O}$

James E. Humphreys

AMERICAN MATHEMATICAL SOCIETY
Providence, Rhode Island

2000 *Mathematics Subject Classification.* Primary 17B10; Secondary 20G05, 22E47.

For additional information and updates on this book, visit
www.ams.org/bookpages/gsm-94

Library of Congress Cataloging-in-Publication Data

Humphreys, James E.
  Representations of semisimple Lie algebras in the BGG category $\mathcal{O}$ / James E. Humphreys.
    p. cm. — (Graduate studies in mathematics ; v. 94)
  Includes bibliographical references and index.
  ISBN 978-0-8218-4678-0 (alk. paper)
  Softcover ISBN 978-1-4704-6326-7
  1. Representations of Lie algebras.   2. Categories (Mathematics)   I. Title.
QA252.3.H864   2008
512′.482—dc22

2008012667

To my future readers Zoë Humphreys, Asher Gerlis,
Emily Hunter, and Miranda Hunter

Contents

Preface

Representation theory plays a central role in Lie theory and has developed in numerous specialized directions over recent decades. Motivation comes from many areas of mathematics and physics, notably the Langlands program. The methods involved are also diverse, including fruitful interactions with "modern" algebraic geometry. Here we focus primarily on algebraic methods in the case of a semisimple Lie algebra $\mathfrak{g}$ over $\mathbb{C}$ with universal enveloping algebra $U(\mathfrak{g})$, where the prerequisites are relatively modest.

The category $\operatorname{Mod} U(\mathfrak{g})$ of all $U(\mathfrak{g})$-modules is much too large to be understood algebraically. Fortunately, many interesting Lie group representations can be studied effectively in terms of a more limited subcategory where modules are subjected to appropriate finiteness conditions: the *BGG category* $\mathcal{O}$ introduced in the early 1970s by Joseph Bernstein, Israel Gelfand, and Sergei Gelfand. Their papers, stimulated in part by Verma's 1966 thesis [**251**], have led to far-reaching work involving a growing list of researchers. In this book we discuss systematically the early work leading to the Kazhdan–Lusztig Conjecture and its proof around 1980. This is at the core of more recent developments, some of which we go on to introduce in the later chapters. Taken on its own, the study of category $\mathcal{O}$ offers a rewarding tour of the beautiful terrain that lies just beyond the classical Cartan–Weyl theory of finite dimensional representations of $\mathfrak{g}$.

Part I (comprising Chapters 1–8) is written in textbook style, at the level of a second year graduate course in a U.S. university. The emphasis here is on highest weight modules, starting with Verma modules and culminating in the determination of formal characters of simple highest weight modules in the setting of the Kazhdan–Lusztig Conjecture (1979). The proof of this conjecture requires sophisticated ideas from algebraic geometry which go well beyond the algebraic framework of earlier chapters. Thus Chapter 8 marks a shift toward the survey style used in the remainder of the book.

The chapters in Part II can to a large extent be read independently. They supplement the more unified theme of Part I in a variety of ways, often motivated by problems arising in Lie group representations. The book ends with an introduction to the influential work of Beilinson, Ginzburg, and Soergel on Koszul duality.

I have tried to keep prerequisites to a minimum. The reader needs to be comfortable with the basic structure theory of semisimple Lie algebras over $\mathbb{C}$ (summarized in Chapter 0) as well as with standard algebraic methods including elementary homological algebra.

Exercises are scattered throughout the text (mainly in Part I) where I thought they would do the most good. Some of the more straightforward ones are used later in the development. At any rate, the most important exercise for the reader is to engage actively with the ideas presented. Examples are also interspersed, though unfortunately it is difficult to gain much direct insight from low rank cases of the sort which can be done by hand. The deeper parts of the theory have required some imaginative leaps not based on examples alone.

The substantial reference list includes all source material cited, together with related books and survey articles. I have added a somewhat arbitrary sample of other research papers to point the reader in directions such as those sketched in Chapter 13. There is also a list of frequently used symbols, most of which are introduced early in the book. Anyone who consults the literature will encounter a wide array of notational choices; here I have tried to keep things simple and consistent to the extent possible.

The mathematics presented here is not original, though parts of the treatment may be. Many people have provided helpful feedback on earlier versions of the chapters, including Troels Agerholm, Henning Andersen, Brian Boe, Tom Braden, Jon Brundan, Walter Mazorchuk, Wolfgang Soergel, Catharina Stropple, and Geordie Williamson. I am especially indebted to Jens Carsten Jantzen for his detailed suggestions at many stages of the writing. His ideas have left a lasting imprint on the study of category $\mathcal{O}$. Naturally, the final choices made are my own responsibility. Corrections and suggestions from readers are welcome.

J. E. Humphreys
February 2008
`jeh@math.umass.edu`

Review of Semisimple Lie Algebras

The main technical prerequisite for this book is a good working knowledge of the structure of semisimple Lie algebras over an algebraically closed field of characteristic 0 such as $\mathbb{C}$. (In fact, it is enough to work over a splitting field of characteristic 0, as indicated below.) In this preliminary chapter we summarize results with which the reader should be familiar, coupled with some explicit references to the textbook literature. The basic structure theory through the classification by Dynkin diagrams is treated in a very large number of sources, of which we cite only a few here, e.g., Bourbaki [**45, 46**], Carter [**60**], Humphreys [**125, 129**], Jacobson [**143**].

The subject can be approached from a number of angles, including the traditional theory of Lie groups and the theory of linear algebraic groups; but group theory generally remains in the background here. Since notation varies considerably in the literature, the reader needs to be aware of our conventions (which are often closest to those in [**125**]); these are intended to steer something of a middle course among the available choices. Frequently used notations are listed at the end of the book.

0.1. Cartan Decomposition

The basic object of study here is a *semisimple Lie algebra* $\mathfrak{g}$ over a field of characteristic 0, having a *Cartan subalgebra* $\mathfrak{h}$ which is *split*: the eigenvalues of $\operatorname{ad} h$ are in the field, for all $h \in \mathfrak{h}$. Write $\ell := \dim \mathfrak{h}$. For convenience we normally take $\mathbb{C}$ for the field, unless the contrary is stated.

Denote by $\Phi \subset \mathfrak{h}^*$ the *root system* of $\mathfrak{g}$ relative to $\mathfrak{h}$ (which by convention does not contain 0), whose abstract properties will be recalled in the following section. To each root $\alpha \in \Phi$ corresponds a nonzero 1-dimensional subspace of $\mathfrak{g}$ called a *root space*:

$$\mathfrak{g}_\alpha = \{x \in \mathfrak{g} \mid [hx] = \alpha(h)x \text{ for all } h \in \mathfrak{h}\}.$$

Usually we fix a *simple system* $\Delta \subset \Phi$ having ℓ elements and corresponding *positive system* $\Phi^+ \subset \Phi$ having m elements. This defines a *Cartan decomposition* $\mathfrak{g} = \mathfrak{n}^- \oplus \mathfrak{h} \oplus \mathfrak{n}$, where $\mathfrak{n} := \bigoplus_{\alpha > 0} \mathfrak{g}_\alpha$ and $\mathfrak{n}^- := \bigoplus_{\alpha < 0} \mathfrak{g}_\alpha$.

The corresponding *standard Borel subalgebra* is $\mathfrak{b} := \mathfrak{h} \oplus \mathfrak{n}$, whose opposite Borel subalgebra is $\mathfrak{b}^- := \mathfrak{h} \oplus \mathfrak{n}^-$. These are maximal solvable subalgebras of $\mathfrak{g}$; more generally, any such subalgebra is called a Borel subalgebra. Any subalgebra $\mathfrak{p}$ containing a Borel subalgebra is called *parabolic*; those containing $\mathfrak{b}$ are *standard*. There are 2^ℓ of these, in natural bijection with subsets $\mathrm{I} \subset \Delta$ (where we interpret I as the Greek letter "iota" for notational consistency). A parabolic subalgebra decomposes as $\mathfrak{p} = \mathfrak{l} \oplus \mathfrak{u}$, where the *Levi factor* $\mathfrak{l}$ is *reductive* (the direct sum of its semisimple derived algebra and an abelian subalgebra) while $\mathfrak{u}$ is the largest nilpotent ideal of $\mathfrak{p}$. When $\mathfrak{p} = \mathfrak{p}_\mathrm{I} \supset \mathfrak{b}$, the center of $\mathfrak{l}_\mathrm{I}$ lies in $\mathfrak{h}$ and $\mathfrak{u}_\mathrm{I} \subset \mathfrak{n}$; here the root system $\Phi_\mathrm{I} = \Phi \cap \mathbb{Z}\mathrm{I}$ of (the derived algebra of) $\mathfrak{l}_\mathrm{I}$ lies in Φ and has I as a simple system, whereas $\mathfrak{u}_\mathrm{I} = \bigoplus \mathfrak{g}_\alpha$ with α running over those $\alpha \in \Phi^+$ not in Φ_I.

The Lie algebra $\mathfrak{g}$ acts on itself by derivations $\operatorname{ad} x$, where $(\operatorname{ad} x)(y) := [xy]$. Let G be the *adjoint group*, generated by all automorphisms $\exp(\operatorname{ad} x)$ with $x \in \mathfrak{g}$ nilpotent. All Cartan subalgebras of $\mathfrak{g}$ are conjugate under G; their common dimension ℓ is the *rank* of $\mathfrak{g}$. The *adjoint representation* $\operatorname{ad} : \mathfrak{g} \to \operatorname{Der} \mathfrak{g}$ is a first example of a representation of $\mathfrak{g}$. The associated *Killing form* $(x, y) := \operatorname{Tr}(\operatorname{ad} x \operatorname{ad} y)$ is nondegenerate. The algebra $\mathfrak{g}$ decomposes uniquely (up to order of summands) into the direct sum of simple ideals. In turn, the simple Lie algebras are uniquely determined by their (irreducible) root systems; these are classified explicitly as types $\mathrm{A}_\ell, \mathrm{B}_\ell, \mathrm{C}_\ell, \mathrm{D}_\ell, \mathrm{E}_6, \mathrm{E}_7, \mathrm{E}_8, \mathrm{F}_4, \mathrm{G}_2$. This often makes case-by-case proofs possible.

Subalgebras of type A_1 in $\mathfrak{g}$ play a special role. Such an algebra is isomorphic to $\mathfrak{sl}(2, \mathbb{C})$, which has a basis (h, x, y):

$$h := \begin{pmatrix} 1 & 0 \\ 0 & -1 \end{pmatrix}, \quad x := \begin{pmatrix} 0 & 1 \\ 0 & 0 \end{pmatrix}, \quad y := \begin{pmatrix} 0 & 0 \\ 1 & 0 \end{pmatrix}.$$

Each $\alpha \in \Phi^+$ determines such a subalgebra (call it $\mathfrak{s}_\alpha$) with basis denoted $(h_\alpha, x_\alpha, y_\alpha)$.

We always work with a *standard basis* of $\mathfrak{g}$, consisting of root vectors $x_\alpha \in \mathfrak{g}_\alpha$ and $y_\alpha \in \mathfrak{g}_{-\alpha}$ (for $\alpha > 0$) together with $h_\alpha = [x_\alpha y_\alpha]$ for $\alpha \in \Delta$ so that all $\alpha(h_\alpha) = 2$. There are many ways to choose such a basis. For example, a *Chevalley basis* has very simple structure constants, all in $\mathbb{Z}$, though it is not usually essential to make this kind of refined choice; see [**46**, VIII, 2.4] (where the convention is that $[x_\alpha y_\alpha] = -h_\alpha$), [**125**, §25].

0.2. Root Systems

The occurrence of root systems in the Lie algebra setting has led to a somewhat more widely applicable notion of *abstract root system* Φ in a vector space over $\mathbb{R}$. Ultimately the classification turns out to be the usual one in terms of Dynkin diagrams, but the axiomatic treatment suggests useful generalizations and is logically independent of the Lie algebra theory.

In the structure theory of $\mathfrak{g}$, the root system spans a $\mathbb{Q}$-form E_0 of the dual space $\mathfrak{h}^*$, where the Killing form is nondegenerate. Thus $E := \mathbb{R} \otimes_{\mathbb{Q}} E_0$ has a natural structure of euclidean space. Humphreys [**125**, §9] then defines an abstract root system in a finite dimensional euclidean space E over $\mathbb{R}$, with inner product denoted (λ, μ), to be a finite set Φ of vectors spanning E and not containing 0. It is required that for each $\alpha \in \Phi$, $\mathbb{R}\alpha \cap \Phi = \{\pm\alpha\}$. Moreover, the euclidean reflection s_α defined by $\lambda \mapsto \lambda - 2(\lambda, \alpha)/(\alpha, \alpha)\alpha$ sends Φ to itself. Further, the *Cartan invariant* $\langle \beta, \alpha^\vee \rangle := 2(\beta, \alpha)/(\alpha, \alpha)$ lies in $\mathbb{Z}$ for all $\alpha, \beta \in \Phi$. Here $\alpha^\vee := 2\alpha/(\alpha, \alpha)$ is the *coroot* of α. The $\mathbb{Z}$-span Λ_r of Φ in E is called the *root lattice*. (The notation in [**125**] is a little different.)

It is this formulation which we adopt, identifying E_0 with the $\mathbb{Q}$-span of the roots in $\mathfrak{h}^*$, which also contains the root lattice Λ_r.

In Bourbaki [**45**, VI, §1] (where the notation and initial viewpoint are different), one starts just with a finite subset Φ as above in a finite dimensional vector space E over $\mathbb{R}$. It is required that each $\alpha \in \Phi$ determine a unique *coroot* $\alpha^\vee$ in E^*, so that the "reflection" $s_{\alpha, \alpha^\vee}$ permutes Φ. Here $s_{\alpha, \alpha^\vee}(\lambda) = \lambda - \langle \lambda, \alpha^\vee \rangle \alpha$. Further, the values of each $\alpha^\vee$ on Φ must lie in $\mathbb{Z}$.

In either version one gets a *dual root system* $\Phi^\vee := \{\alpha^\vee \mid \alpha \in \Phi\}$ (living in the euclidean space E in our version, but in the dual space E^* in the Bourbaki version). This is a root system whose Dynkin diagram is dual to that of Φ; in particular, if Φ is of type B_ℓ, then $\Phi^\vee$ is of type C_ℓ. In the Killing form identification of $\mathfrak{h}$ and $\mathfrak{h}^*$, the coroot $\alpha^\vee$ corresponds to $h_\alpha \in \mathfrak{h}$:

$$\langle \beta, \alpha^\vee \rangle = \beta(h_\alpha) \text{ for all } \beta \in \Phi.$$

Each choice of *simple system* Δ in Φ defines a partition into subsets of positive and negative roots (denoted respectively Φ^+ and Φ^-). Here Δ forms a basis of E (or a $\mathbb{Z}$-basis of Λ_r), while each $\beta \in \Phi^+$ can be written uniquely as $\beta = \sum_{\alpha \in \Delta} c_\alpha \alpha$ with $c_\alpha \in \mathbb{Z}^+$. Define the *height* of β to be $\operatorname{ht} \beta := \sum_{\alpha \in \Delta} c_\alpha$; so $\operatorname{ht} \beta = 1$ if and only if $\beta \in \Delta$.

Here are a few useful facts which emerge from the study of root systems with a given simple system Δ:

(1) The set $\Delta^\vee$ forms a simple system in $\Phi^\vee$.

(2) For any $\alpha, \beta \in \Phi$, the roots of the form $\alpha + k\beta$ form an unbroken *root string* $\alpha - r\beta, \ldots, \alpha - \beta, \alpha, \alpha + \beta, \ldots, \alpha + s\beta$, which involves at most four roots.

(3) If β is positive but not simple, there exists a simple root α for which $\langle \beta, \alpha^\vee \rangle > 0$; thus $s_\alpha \beta \in \Phi^+$ and $\operatorname{ht} s_\alpha \beta < \operatorname{ht} \beta$.

0.3. Weyl Groups

The natural symmetry group attached to a root system Φ is its *Weyl group* W, the (finite!) subgroup of $\mathrm{GL}(E)$ generated by all reflections s_α with $\alpha \in \Phi$ (or just the *simple reflections* s_α with $\alpha \in \Delta$ when Δ is a fixed simple system). Evidently the root lattice Λ_r is stable under the action of W.

Abstractly, W is a finite *Coxeter group*, having generators s_α $(\alpha \in \Delta)$ and defining relations of the form $(s_\alpha s_\beta)^{m(\alpha,\beta)} = 1$. Moreover, W satisfies the crystallographic restriction $m(\alpha, \beta) \in \{2, 3, 4, 6\}$ when $\alpha \neq \beta$. (In the Lie algebra setting there is a natural identification of W with a finite group of automorphisms of $\mathfrak{h}$ in $\mathfrak{g}$; this induces an action on $\mathfrak{h}^*$.) The Weyl group of $\Phi^\vee$ is naturally isomorphic to W. Moreover, $(w\alpha)^\vee = w\alpha^\vee$ when $\alpha \in \Phi$ and $w \in W$.

The subgroup of W fixing a point $\lambda \in E$ is itself a reflection group, being generated by those s_α for which $\langle \lambda, \alpha^\vee \rangle = 0$ (see for example [**129**, 1.12]).

Write $\ell(w) = n$ if $w = s_1 \cdots s_n$ with s_i simple reflections and n as small as possible; such an expression is called *reduced*. Standard facts about the length function on W include:

(1) The number of $\alpha \in \Phi^+$ for which $w\alpha < 0$ is precisely $\ell(w)$. In particular, when $\alpha \in \Delta$ (equivalent to $\ell(s_\alpha) = 1$), we have $s_\alpha \beta > 0$ for all $\beta \neq \alpha$ in Φ^+. Moreover, w is uniquely determined by the set of $\alpha > 0$ for which $w\alpha < 0$.

(2) If $w \in W$, then $\ell(w) = \ell(w^{-1})$. Thus $\ell(w) = |\Phi^+ \cap w(\Phi^-)|$.

(3) There is a unique element $w_\circ \in W$ of maximum length $|\Phi^+|$, sending Φ^+ to Φ^-. Moreover, $\ell(w_\circ w) = \ell(w_\circ) - \ell(w)$ for all $w \in W$.

(4) If $\alpha > 0$ and $w \in W$ satisfy $\ell(w s_\alpha) > \ell(w)$, then $w\alpha > 0$, while $\ell(w s_\alpha) < \ell(w)$ implies $w\alpha < 0$. It follows that $\ell(s_\alpha w) > \ell(w) \Leftrightarrow w^{-1}\alpha > 0$.

Given a subset $\mathrm{I} \subset \Delta$, the subgroup W_I it generates is called a "parabolic" subgroup of W. It is a Coxeter group in its own right and its length function ℓ_I agrees with the restriction of ℓ.

For these and other facts about Weyl groups (or Coxeter groups in general), see for example Bourbaki [**45**, VI, §1], Humphreys [**129**, Chap. 1].

0.4. Chevalley–Bruhat Ordering of W

There is a subtle (and very useful) way to partially order W. The discussion here applies equally well to an arbitrary Coxeter group, but in the case of Weyl groups the ordering arises first in the work of Chevalley and others on inclusions among closures of Bruhat cells for a semisimple group having Lie algebra $\mathfrak{g}$. The ordering is most often referred to just as the *Bruhat ordering*; see Björner–Brenti [**29**, Chap. 2], Humphreys [**129**, 5.9–5.11].

If S denotes the set of simple reflections s_α (with $\alpha \in \Delta$), the set T of all reflections in W is defined by $T := \bigcup_{w \in W} wSw^{-1}$. This is known to consist of the reflections s_α with $\alpha \in \Phi^+$. For $w, w' \in W$ and $t \in T$, write $w' \xrightarrow{t} w$ if $w = tw'$ and $\ell(w') < \ell(w)$. In turn, write $w' \to w$ if $w' \xrightarrow{t} w$ for some $t \in T$. Extend this relation to a partial ordering of W by defining $w' < w$ to mean that $w' = w_0 \to w_1 \to \cdots \to w_n = w$ for some $w_1, \ldots, w_{n-1}$.

The identity element of W is the unique minimal element for this ordering. It is clear from the definition that $w' < w \Rightarrow \ell(w') < \ell(w)$. (*Caution:* In the literature the ordering is sometimes defined with 1 as maximal rather than minimal element.)

In rank 2 cases, W is a dihedral group and the ordering is easy to describe: $w' < w$ if and only if $\ell(w') < \ell(w)$. But things get considerably more complicated in higher ranks. For a picture of the Bruhat ordering in the symmetric group S_4 (the Weyl group of type A_3), see [**29**, Fig. 2.4].

Although a parallel definition could be given using right rather than left multiplication by elements of T, the two definitions actually lead to the same ordering. Other important features can be summarized as follows.

Proposition. *The Bruhat ordering of a Coxeter group (W, S) satisfies:*

(a) *$w' \le w$ if and only if w' occurs as a subexpression in one (hence any) reduced expression $s_1 \cdots s_n$ (with $s_i \in S$) for w. Here a subexpression is a product $s_{i_1} \cdots s_{i_k}$ with $1 \le i_1 < \ldots < i_k \le n$.*

(b) *Adjacent elements in the Bruhat ordering differ in length by 1.*

(c) *If $w' < w$ and $s \in S$, then $w's \le w$ or $w's \le ws$ (or both).*

(d) *If $\ell(w_1) + 2 = \ell(w_2)$, the number of elements $w \in W$ satisfying $w_1 < w < w_2$ is 0 or 2.*

(e) *If $\mathrm{I} \subset \Delta$, the Bruhat ordering of the Coxeter group W_{I} is induced by the Bruhat ordering of W.*

Part (b) implies that the same ordering would result if we had defined $w' \xrightarrow{t} w$ more restrictively by requiring that $\ell(w) = \ell(w') + 1$. (This version will be useful in Chapter 6.)

Part (d) expresses a feature of the Möbius function for an arbitrary Coxeter group (which can be derived in various ways); this plays an essential role in 6.7.

0.5. Universal Enveloping Algebras

An essential tool in the construction and study of representations is the *universal enveloping algebra* $U(\mathfrak{a})$ of a Lie algebra $\mathfrak{a}$. This is an associative algebra with 1, infinite dimensional if $\mathfrak{a} \neq 0$ and noncommutative if $\mathfrak{a}$ is not abelian. The algebra $U(\mathfrak{a})$ is left and right noetherian and has no zero-divisors. Any Lie algebra homomorphism $\mathfrak{a}_1 \to \mathfrak{a}_2$ induces an associative algebra homomorphism between the enveloping algebras. The adjoint action of $\mathfrak{a}$ on itself induces an action of $\mathfrak{a}$ on $U(\mathfrak{a})$ by derivations: if $x \in \mathfrak{a}$, then $\operatorname{ad} x$ sends $u \mapsto xu - ux$ for all $u \in U(\mathfrak{a})$.

Given an ordered basis $(x_1, \ldots, x_n)$ of $\mathfrak{a}$, the monomials $x_1^{t_1} \cdots x_n^{t_n}$ with $t_i \in \mathbb{Z}^+$ form a basis of $U(\mathfrak{a})$. This is called a *PBW (Poincaré–Birkhoff–Witt) basis*. In particular, $\mathfrak{a}$ may be identified with a subspace of $U(\mathfrak{a})$. By extending a basis of a Lie subalgebra of $\mathfrak{a}$ to a basis of $\mathfrak{a}$, one gets a natural embedding of its enveloping algebra into $U(\mathfrak{a})$.

In the case of a semisimple Lie algebra $\mathfrak{g} = \mathfrak{n}^- \oplus \mathfrak{h} \oplus \mathfrak{n}$, write a standard basis in the order $y_1, \ldots, y_m$ (with $y_i \in \mathfrak{n}^-$), $h_1, \ldots, h_\ell$ (with $h_i = h_{\alpha_i}$ for simple roots α_i), $x_1, \ldots, x_m$. We say that a corresponding basis element

$$y_1^{r_1} \cdots y_m^{r_m} h_1^{s_1} \cdots h_\ell^{s_\ell} x_1^{t_1} \cdots x_m^{t_m}$$

is written in a *standard PBW ordering*. (Often it is convenient to let $y_1, \ldots, y_\ell$ correspond to negatives of simple roots.) Then $U(\mathfrak{g})$ is the direct sum of subspaces $U(\mathfrak{g})_\nu$ with $\nu \in \Lambda_r$: here $U(\mathfrak{g})_\nu$ is the span of those monomials for which $\nu = \sum_i (t_i - r_i)\alpha_i$.

Denote the *center* of $U(\mathfrak{g})$ by $Z(\mathfrak{g})$. This subalgebra plays an important role in representation theory, sometimes acting by scalars even on infinite dimensional modules to which Schur's Lemma does not apply. In Chapter 1 we will work out in some detail the structure of $Z(\mathfrak{g})$: it turns out to be just a polynomial algebra in ℓ indeterminates, though this is far from obvious.

In the classical theory one constructs directly a special element of $Z(\mathfrak{g})$ (unique up to scalar multiples) which is called a *Casimir element*. It plays a key role in algebraic proofs of Weyl's complete reducibility theorem (recalled below) and related study of finite dimensional representations. The definition uses the nondegeneracy of the Killing form on $\mathfrak{g}$. For example, when

$\mathfrak{g} = \mathfrak{sl}(2, \mathbb{C})$, with standard basis (h, x, y), a Casimir element is $h^2 + 2xy + 2yx$. Rewritten in a standard PBW ordering, it becomes $h^2 + 2h + 4yx$.

There is a standard anti-involution $\tau : \mathfrak{g} \rightarrow \mathfrak{g}$ which interchanges x_α with y_α for all $\alpha \in \Phi^+$ and fixes all $h \in \mathfrak{h}$; it extends canonically to an anti-automorphism of $U(\mathfrak{g})$ (see for example Jantzen [**147**, 1.2]). We may call τ the *transpose map*, since in the case when $\mathfrak{g} = \mathfrak{sl}(n, \mathbb{C})$ it is just the usual transpose map for matrices. The source of τ is an involution equal to $-\tau$, which is easily constructed using the Serre relations for $\mathfrak{g}$. (In some of the literature, such as [**144, 147**] and [**231**], the letter σ is used instead of τ.)

Later it will be shown that τ fixes $Z(\mathfrak{g})$ pointwise, using properties of the Harish-Chandra homomorphism: see Exercise 1.10. (There does not seem to be a more elementary proof.)

0.6. Integral Weights

The representations we study will involve *weights* relative to the action of $\mathfrak{h}$, as recalled in the following section. Here we assemble some basic facts which can be developed abstractly within the theory of root systems (see for example [**125**, §13]).

There is a natural dual lattice in E to the root lattice, defined by

$$\Lambda := \{\lambda \in E \mid \langle \lambda, \alpha^\vee \rangle \in \mathbb{Z} \text{ for all } \alpha \in \Phi\}.$$

Here it is enough to let α run over Δ. We call Λ the *integral weight lattice* associated to Φ. It lies in the $\mathbb{Q}$-span E_0 of the roots in $\mathfrak{h}^*$ and includes the root lattice Λ_r as a subgroup of finite index. (The index is given by the determinant of the matrix of Cartan invariants relative to Δ.) In its action on E, the Weyl group W keeps Λ stable.

When the simple system Δ is fixed, there is a natural partial ordering on Λ defined by $\mu \leq \lambda$ if and only if $\lambda - \mu \in \Gamma$, where $\Gamma \subset \Lambda$ is defined to be the set of all $\mathbb{Z}^+$-linear combinations of simple roots.

Write $\Delta = \{\alpha_1, \ldots, \alpha_\ell\}$. Then the group Λ is free abelian of rank ℓ, with a basis consisting of *fundamental weights* $\varpi_1, \ldots, \varpi_\ell$. These satisfy $\langle \varpi_i, \alpha_j^\vee \rangle = \delta_{ij}$. The subset $\Lambda^+ := \mathbb{Z}^+ \varpi_1 + \cdots + \mathbb{Z}^+ \varpi_\ell$ is called the set of *dominant integral weights*. From the fact that $\langle \beta, \alpha^\vee \rangle = \beta(h_\alpha)$ when $\beta \in \Phi$, one shows easily that $\langle \lambda, \alpha^\vee \rangle = \lambda(h_\alpha)$ for all $\lambda \in \Lambda$.

The special weight $\rho := \varpi_1 + \cdots + \varpi_\ell \in \Lambda^+$ can also be characterized as half the sum of positive roots. It satisfies $\langle \rho, \alpha^\vee \rangle = 1$, or $s_\alpha \rho = \rho - \alpha$, for all $\alpha \in \Delta$.

Here are two useful facts ([**125**, 13.2]):

(1) The number of dominant weights $\leq \lambda$ for a given $\lambda \in \Lambda^+$ is finite.

(2) Given $\lambda \in \Lambda^+$, all $w\lambda \leq \lambda$ for $w \in W$.

For a simple system Δ, there is a natural fundamental domain $\overline{C} \subset E$ for the action of W. Relative to the inner product on E,

$$C := \{\lambda \in E \mid (\lambda, \alpha) > 0 \text{ for all } \alpha \in \Delta\},$$

while its closure is obtained by changing $>$ to $\geq$. We call C a *Weyl chamber* in E. The Weyl chambers are in natural bijection with simple systems in Φ and are permuted simply transitively by W. They can be characterized as the connected components in the complement of the union of all hyperplanes orthogonal to roots.

For use in 7.9 we need an elementary lemma on the partial ordering of weights relative to a Weyl chamber (taken from Bourbaki [**45**, VI, §1, Prop. 18]).

Lemma. *Let $\lambda \in \Lambda$ and fix a chamber $C \subset E$ for the action of W corresponding to a simple system Δ. The following conditions are equivalent:*

> (a) $\lambda \in \overline{C}$.
>
> (b) $\lambda \geq s_\alpha \lambda$ *for all* $\alpha \in \Delta$.
>
> (c) $\lambda \geq w\lambda$ *for all* $w \in W$.

Proof. Evidently (a) is equivalent to (b), while (c) implies (b). To show that (a) implies (c), assume $\lambda \in \overline{C}$ and use induction on $\ell(w)$ to derive (c). The case $\ell(w) = 0$ is automatic. If $\ell(w) > 0$, write $w = w's_\alpha$ for some $\alpha \in \Delta$ so that $\ell(w') < \ell(w)$. This implies in particular that $w\alpha < 0$ and thus $w'\alpha > 0$ (0.2). Now consider

$$\lambda - w\lambda = (\lambda - w'\lambda) + w'(\lambda - s_\alpha\lambda).$$

The first summand on the right is ≥ 0 by induction, whereas the second equals $-\langle \lambda, \alpha^\vee \rangle w'\alpha$. The assumption that $\lambda \in \overline{C}$ forces $\langle \lambda, \alpha^\vee \rangle \geq 0$. Since $w'\alpha > 0$, the summand in question is ≥ 0. This shows that $\lambda \geq w\lambda$ as required. $\square$

0.7. Representations

Our main concern here is with representations of $\mathfrak{g}$ (or $U(\mathfrak{g})$), not necessarily finite dimensional. We mostly use the language of modules rather than representations: simple, semisimple, indecomposable, projective, etc. So the object of study is the category $\mathrm{Mod}\, U(\mathfrak{g})$ of all (left) $U(\mathfrak{g})$-modules, or rather a well-behaved subcategory.

In the finite dimensional case, the theory of integral weights outlined above is sufficient. In the infinite dimensional setting, we have to broaden the idea of "weight". If M is arbitrary, it still makes sense to define its *weight spaces* relative to the action of $\mathfrak{h}$. For each $\lambda \in \mathfrak{h}^*$, let

$$M_\lambda := \{v \in M \mid h \cdot v = \lambda(h)v \text{ for all } h \in \mathfrak{h}\}.$$

If $M_\lambda \neq 0$, we say that λ is a *weight* of M. The *multiplicity* of λ in M is then defined to be $\dim M_\lambda$ (possibly ∞). Define $\Pi(M) := \{\lambda \in \mathfrak{h}^* \mid M_\lambda \neq 0\}$, the set of all weights occurring in M. A partial ordering $\mu \leq \lambda$ of $\mathfrak{h}^*$ is defined just as in the integral case (0.6).

It is easily checked that weight vectors for distinct weights in M are linearly independent. Thus the sum $\sum_{\lambda \in \mathfrak{h}^*} M_\lambda$ is direct, though it may be 0 and moreover M need not be the direct sum of these subspaces. Our main focus will be on the case when M actually is the direct sum of its weight spaces, i.e., $\mathfrak{h}$ acts semisimply on M. Then we call M a *weight module*.

When Δ is fixed, the corresponding fundamental dominant weights in Λ form a convenient basis of $\mathfrak{h}^*$. If $\lambda = \sum_{i=1}^\ell c_i \varpi_i$, then $\lambda \in \Lambda$ (resp. Λ^+) precisely when all $c_i \in \mathbb{Z}$ (resp. $\mathbb{Z}^+$).

0.8. Finite Dimensional Modules

The theory of finite dimensional modules for $\mathfrak{g}$ has been well studied. In 1.6 and Chapter 2 many of the main results will be recovered in the more general setting of category $\mathcal{O}$. However, one key fact due to Weyl has no analogue in the infinite dimensional theory and will only be quoted here. It is usually referred to as *Weyl's Complete Reducibility Theorem* :

> *Every finite dimensional $U(\mathfrak{g})$-module is isomorphic to a direct sum of simple modules, the multiplicities of the latter being uniquely determined.*

In some treatments of the basic structure theory (such as [**125**]), Weyl's Theorem and its consequences play a key role from an early stage.

When $\dim M < \infty$, M is always a weight module. This follows from Weyl's Theorem: more precisely, elements of $\mathfrak{h}$ act on M via semisimple matrices while elements of $\mathfrak{n}$ or $\mathfrak{n}^-$ act via nilpotent matrices. This generalizes the usual notion of Jordan decomposition, which arises intrinsically in $\mathfrak{g}$, independent of any specific linear realization. All weights of M are in fact *integral.* The set $\Pi(M)$ of weights of M is W-invariant, with $\dim M_\lambda = \dim M_{w\lambda}$.

0.9. Simple Modules for $\mathfrak{sl}(2,\mathbb{C})$

While we do not presuppose any detailed knowledge of finite dimensional simple modules for an arbitrary semisimple Lie algebra $\mathfrak{g}$, we do need to refer at times to the results for $\mathfrak{g} = \mathfrak{sl}(2,\mathbb{C})$. These are easy to derive from scratch and play a key role in most developments of the structure theory outlined earlier.

Here is a quick summary (see for example [**125**, §7]). Thanks to Weyl's theorem on complete reducibility, the problem is to classify and construct (up to isomorphism) all simple finite dimensional modules. Fix a standard basis (h, x, y) for $\mathfrak{g}$, with $[hx] = 2x, [hy] = -2y, [xy] = h$. Since $\dim \mathfrak{h} = 1$, weights $\lambda \in \mathfrak{h}^*$ may be identified with complex numbers. In turn, the integral weight lattice Λ is identified with $\mathbb{Z}$ and Λ_r with $2\mathbb{Z}$.

The simple modules are in natural bijection with dominant integral weights $\lambda \in \Lambda^+$ and may be denoted $L(\lambda)$. It is not difficult to construct explicit matrix realizations for each of these. Here $L(\lambda)$ has 1-dimensional weight spaces, with weights $\lambda, \lambda - 2, \ldots, -(\lambda - 2), -\lambda$. In particular, $\dim L(\lambda) = \lambda + 1$. For example, $L(0)$ is the trivial module, $L(1)$ is the natural 2-dimensional realization of $\mathfrak{g}$, and $L(3)$ is the adjoint module. Basis vectors v_i $(0 \leq i \leq \lambda)$ for $L(\lambda)$ can be chosen so that (setting $v_{-1} = v_{\lambda+1} = 0$):

$$
\begin{aligned}
h \cdot v_i &= (\lambda - 2i)v_i, \\
x \cdot v_i &= (\lambda - i + 1)v_{i-1}, \\
y \cdot v_i &= (i + 1)v_{i+1}.
\end{aligned}
$$

Here v_0 is (up to scalars) the unique weight vector in $L(\lambda)$ killed by x. In this case W has order 2 and permutes weights of $L(\lambda)$ by sending $\lambda - 2i$ to its negative.

Part I

Highest Weight Modules

Chapter 1

Category $\mathcal{O}$: Basics

We begin by laying the foundations for study of the BGG category $\mathcal{O}$, starting with the axioms and their immediate consequences in 1.1. This category encompasses all finite dimensional $U(\mathfrak{g})$-modules, the traditional starting point for representation theory of semisimple Lie algebras. In 1.6 the finite dimensional simple modules are identified as those having dominant integral highest weights. Then in Chapter 2 we shall recover the classical theorems of Weyl and others in the category $\mathcal{O}$ setting. But our main concern is the study of $\mathcal{O}$ itself, which has wider implications in representation theory.

The most accessible infinite dimensional modules in $\mathcal{O}$ are the Verma modules (1.3), which arise first as an auxiliary tool for the construction of simple modules. Later their study deepens considerably, as we shall see in Chapters 4–5.

Some basic tools introduced in this chapter are central characters and related subcategories $\mathcal{O}_\chi$ (1.7–1.13), along with formal characters of weight modules (1.14–1.16).

One important aspect of category $\mathcal{O}$ is deferred until Chapter 9, largely for pedagogical reasons: the study of subcategories $\mathcal{O}^{\mathfrak{p}}$ attached to arbitrary parabolic subalgebras $\mathfrak{p}$ (not just the Borel subalgebra $\mathfrak{b}$). On one hand, $\mathcal{O}^{\mathfrak{p}}$ requires a cumbersome extra layer of notation; on the other hand, the results there are sometimes easier to derive once the corresponding results for $\mathcal{O}$ are in hand.

1.1. Axioms and Consequences

The **BGG category** $\mathcal{O}$ is defined to be the full subcategory of $\operatorname{Mod} U(\mathfrak{g})$ whose objects are the modules satisfying the following three conditions.

($\mathcal{O}$1) M is a finitely generated $U(\mathfrak{g})$-module.

($\mathcal{O}$2) M is $\mathfrak{h}$-semisimple, that is, M is a weight module: $M = \bigoplus_{\lambda \in \mathfrak{h}^*} M_\lambda$.

($\mathcal{O}$3) M is locally $\mathfrak{n}$-finite: for each $v \in M$, the subspace $U(\mathfrak{n}) \cdot v$ of M is finite dimensional.

As a matter of notation, we usually write $\mathrm{Hom}_{\mathcal{O}}(M, N)$ rather than $\mathrm{Hom}_{U(\mathfrak{g})}(M, N)$ or $\mathrm{Hom}_{\mathfrak{g}}(M, N)$ to emphasize that M and N lie in $\mathcal{O}$.

From 0.7 it is clear that all finite dimensional modules lie in $\mathcal{O}$. Before constructing further examples in 1.3, we explore the implications of the axioms. To begin, we deduce from the axioms two features of the weight structure of an arbitrary M in $\mathcal{O}$:

($\mathcal{O}$4) All weight spaces of M are finite dimensional.

($\mathcal{O}$5) In the notation of 0.7, the set $\Pi(M)$ of all weights of M is contained in the union of finitely many sets of the form $\lambda - \Gamma$, where $\lambda \in \mathfrak{h}^*$ and Γ is the semigroup in Λ_r generated by Φ^+.

First, it is obvious from ($\mathcal{O}$2) that a finite generating set in ($\mathcal{O}$1) can always be taken to consist of weight vectors. To verify ($\mathcal{O}$4) and ($\mathcal{O}$5) it will suffice to let M be generated by a single weight vector v of weight λ. Thanks to the PBW Theorem (0.5), we can write $U(\mathfrak{g}) = U(\mathfrak{n}^-)U(\mathfrak{h})U(\mathfrak{n})$. Applying a factor from $U(\mathfrak{n})$ to v we get a finite dimensional subspace V of M (thanks to ($\mathcal{O}$3)) spanned by weight vectors having weights of type $\lambda +$ sum of positive roots. Now V is stable under $U(\mathfrak{h})$, while the action of $U(\mathfrak{n}^-)$ on V produces only weights lower than these. Moreover, only a finite number of standard basis monomials $y_1^{i_1} \ldots y_m^{i_m}$ in $U(\mathfrak{n}^-)$ can yield the same weight when applied to a weight vector in V.

It is easy to derive a few further consequences of the axioms.

Theorem. *Category $\mathcal{O}$ satisfies:*

(a) *$\mathcal{O}$ is a noetherian category, i.e., each $M \in \mathcal{O}$ is a noetherian $U(\mathfrak{g})$-module.*

(b) *$\mathcal{O}$ is closed under submodules, quotients, and finite direct sums.*

(c) *$\mathcal{O}$ is an abelian category.*

(d) *If $M \in \mathcal{O}$ and L is finite dimensional, then $L \otimes M$ also lies in $\mathcal{O}$. Thus $M \mapsto L \otimes M$ defines an exact functor $\mathcal{O} \to \mathcal{O}$.*

(e) *If $M \in \mathcal{O}$, then M is $Z(\mathfrak{g})$-finite: for each $v \in M$, the span of $\{z \cdot v \mid z \in Z(\mathfrak{g})\}$ is finite dimensional.*

(f) *If $M \in \mathcal{O}$, then M is finitely generated as a $U(\mathfrak{n}^-)$-module.*

Proof. (a) Since $U(\mathfrak{g})$ is a noetherian ring (0.5), this follows from ($\mathcal{O}$1).

(b) Closure under taking quotients or finite direct sums is immediate. Since $U(\mathfrak{g})$ is noetherian, any submodule of a finitely generated $U(\mathfrak{g})$-module is also finitely generated, whereas $(\mathcal{O}2)$ and $(\mathcal{O}3)$ are automatic for a submodule of a module in $\mathcal{O}$.

(c) Since $\operatorname{Mod} U(\mathfrak{g})$ itself is an abelian category, we only need to check that $\mathcal{O}$ is closed under finite direct sums and under taking kernels and cokernels of homomorphisms. This follows from (b).

(d) It is clear that $L \otimes M$ satisfies axioms $(\mathcal{O}2)$ and $(\mathcal{O}3)$. To check finite generation, let $v_1, \ldots, v_n$ be a basis of L and let $w_1, \ldots, w_p$ generate the module M. We claim that the $v_i \otimes w_j$ generate $L \otimes M$. If N is the submodule they generate, then certainly all $v \otimes w_j$ with $v \in L$ lie in N. In turn, for each $x \in \mathfrak{g}$ we have $x \cdot (v \otimes w_j) = x \cdot v \otimes w_j + v \otimes x \cdot w_j \in N$, and then also $v \otimes x \cdot w_j \in N$. Iteration shows that $v \otimes u \cdot w_j \in N$ for all PBW monomials $u \in U(\mathfrak{g})$. Thus $L \otimes M \subset N$.

(e) Since each $v \in M$ is a sum of weight vectors, we may assume that $v \in M_\lambda$ for some $\lambda \in \mathfrak{h}^*$. The fact that $z \in Z(\mathfrak{g})$ commutes with the action of $\mathfrak{h}$ then implies that $z \cdot v \in M_\lambda$. Weight spaces of M being finite dimensional by $(\mathcal{O}4)$, the span of $\{z \cdot v \mid z \in Z(\mathfrak{g})\}$ is therefore finite dimensional.

(f) The axioms imply that M is generated by a finite dimensional $U(\mathfrak{b})$-submodule N. Thanks to the PBW Theorem, a basis of N then generates M as a $U(\mathfrak{n}^-)$-module. $\qquad\square$

For some applications it is enough to work in the subcategory $\mathcal{O}_{\mathrm{int}}$ whose objects all have *integral* weights: this encompasses for example all finite dimensional modules (1.6). In any case, proofs are often easier to carry out initially in $\mathcal{O}_{\mathrm{int}}$. Going in the opposite direction, we sometimes consider the larger category $\mathcal{C}$ of $U(\mathfrak{g})$-modules whose objects are the weight modules with finite dimensional weight spaces.

Exercise. (a) If $M \in \mathcal{O}$ and $[\lambda] := \lambda + \Lambda_r$ is any coset of $\mathfrak{h}^*$ modulo Λ_r, let $M^{[\lambda]}$ be the sum of all weight spaces M_μ for which $\mu \in [\lambda]$. Prove that $M^{[\lambda]}$ is a $U(\mathfrak{g})$-submodule of M and that M is the direct sum of (finitely many!) such submodules.

(b) Deduce that all weights of an *indecomposable* module $M \in \mathcal{O}$ lie in a single coset of $\mathfrak{h}^*$ modulo Λ_r.

1.2. Highest Weight Modules

Still lacking is a concrete construction of modules in $\mathcal{O}$, apart from familiar finite dimensional examples such as the adjoint module $\mathfrak{g}$. In this direction we introduce some helpful terminology. First define a nonzero vector v^+ in a $U(\mathfrak{g})$-module M to be a **maximal vector** of weight $\lambda \in \mathfrak{h}^*$ if $v^+ \in M_\lambda$

and $\mathfrak{n} \cdot v^+ = 0$. (Such a vector is also referred to in the literature as a **primitive vector** of weight λ.) Thanks to axioms $(\mathcal{O}2)$ and $(\mathcal{O}3)$, *every nonzero module in $\mathcal{O}$ has at least one maximal vector.*

Next define a $U(\mathfrak{g})$-module M to be a **highest weight module** of weight λ if there is a maximal vector $v^+ \in M_\lambda$ such that $M = U(\mathfrak{g}) \cdot v^+$. By the PBW Theorem, such a module satisfies $M = U(\mathfrak{n}^-) \cdot v^+$. It is easy to spell out the basic properties of an arbitrary highest weight module by specializing the discussion in the previous section:

Theorem. *Let M be a highest weight module of weight $\lambda \in \mathfrak{h}^*$, generated by a maximal vector v^+. Fix an ordering of the positive roots as $\alpha_1, \ldots, \alpha_m$ and choose corresponding root vectors y_i in $\mathfrak{g}_{-\alpha_i}$. Then:*

(a) *M is spanned by the vectors $y_1^{i_1} \ldots y_m^{i_m} \cdot v^+$ with $i_j \in \mathbb{Z}^+$, having respective weights $\lambda - \sum i_j \alpha_j$. Thus M is a semisimple $\mathfrak{h}$-module.*

(b) *All weights μ of M satisfy $\mu \le \lambda$: $\mu = \lambda - (\text{sum of positive roots})$.*

(c) *For all weights μ of M, we have $\dim M_\mu < \infty$, while $\dim M_\lambda = 1$. Thus M is a weight module, locally finite as $\mathfrak{n}$-module, and $M \in \mathcal{O}$.*

(d) *Each nonzero quotient of M is again a highest weight module of weight λ.*

(e) *Each submodule of M is a weight module. A submodule generated by a maximal vector of weight $\mu < \lambda$ is proper; in particular, if M is simple, its maximal vectors are all multiples of v^+.*

(f) *M has a unique maximal submodule and unique simple quotient. In particular, M is indecomposable.*

(g) *All simple highest weight modules of weight λ are isomorphic. If M is one of these, $\dim \operatorname{End}_{\mathcal{O}} M = 1$.*

Proof. Part (a) follows from the standard commutation relations involving elements of $\mathfrak{h}$ and $\mathfrak{n}^-$, together with the fact that weight vectors having distinct weights are linearly independent. Then (b) is immediate.

For (c), it is clear that only finitely many choices of $i_1, \ldots, i_m$ yield the same weight, by expressing sums of positive roots in terms of simple roots. Local finiteness of the action of $\mathfrak{n}$ follows from the fact that $\mathfrak{g}_\alpha$ maps M_μ into $M_{\mu+\alpha}$, coupled with (b).

Part (d) is clear, while part (e) follows from the fact that M lies in $\mathcal{O}$ (forcing its submodules to lie in $\mathcal{O}$) along with part (b).

Thanks to (e), each *proper* submodule of M is a weight module. It cannot have λ as a weight, since the 1-dimensional space M_λ generates M. Therefore the sum of all proper submodules is still proper, whence (f).

For (g), suppose M_1 and M_2 are simple highest weight modules of weight λ, with respective maximal vectors v_1^+, v_2^+. Set $M_0 := M_1 \oplus M_2$ and $v^+ := (v_1^+, v_2^+)$. Evidently v^+ is a maximal vector in M_0, so the submodule N it generates is a highest weight module of weight λ. In turn, the two projections $N \to M_1$ and $N \to M_2$ are surjective. As simple quotients of a highest weight module, M_1 and M_2 must then be isomorphic, thanks to (f). [In fact, they are both isomorphic to N under the projections, since N is semisimple.]

Finally, since M is simple, any nonzero endomorphism φ must be an isomorphism and also take v^+ to a multiple cv^+. Since v^+ generates M, it follows that φ is just multiplication by c. (This property is analogous to Schur's Lemma, even though M need not be finite dimensional.) $\qquad\square$

In a loose sense, highest weight modules are the building blocks for all objects in $\mathcal{O}$. Since we cannot expect modules here to be semisimple, we look instead for a finite **filtration**: a chain of submodules $0 \subset M_1 \subset M_2 \subset \cdots \subset M_n = M$ for which the quotients M_{i+1}/M_i have a known structure.

Corollary. *Let M be any nonzero module in $\mathcal{O}$. Then M has a finite filtration $0 \subset M_1 \subset M_2 \subset \cdots \subset M_n = M$ with nonzero quotients each of which is a highest weight module.*

Proof. Since M can be generated by finitely many weight vectors and is moreover locally $\mathfrak{n}$-finite, the $\mathfrak{n}$-submodule V generated by such a generating family of weight vectors is finite dimensional. If $\dim V = 1$, it is clear that M itself is a highest weight module. Otherwise proceed by induction on $\dim V$.

Start with a nonzero weight vector $v \in V$ of weight λ which is maximal among all weights of V and is therefore a maximal vector in M. It generates a submodule M_1, while the quotient $\overline{M} := M/M_1$ again lies in $\mathcal{O}$ and is generated by the image $\overline{V}$ of V. Since $\dim \overline{V} < \dim V$, the induction hypothesis can be applied to $\overline{M}$, yielding a chain of highest weight submodules whose pre-images in M are the desired $M_2, \ldots, M_n$. $\qquad\square$

1.3. Verma Modules and Simple Modules

We can construct a large family of highest weight modules by exploiting the technique of **induction**. The idea, which arises in similar ways in other types of representation theory, is to start with an easily constructed family of modules for a subalgebra and then "induce" to $\mathfrak{g}$.

Here we start with the Borel subalgebra $\mathfrak{b}$ corresponding to a fixed choice of positive roots, which in turn has an abelian quotient algebra $\mathfrak{b}/\mathfrak{n}$ isomorphic to $\mathfrak{h}$. Any $\lambda \in \mathfrak{h}^*$ then defines a 1-dimensional $\mathfrak{b}$-module with trivial $\mathfrak{n}$-action, denoted $\mathbb{C}_\lambda$. Now set $M(\lambda) := U(\mathfrak{g}) \otimes_{U(\mathfrak{b})} \mathbb{C}_\lambda$, which has a natural structure of left $U(\mathfrak{g})$-module. This is called a **Verma module** and may also be written as $\mathrm{Ind}_{\mathfrak{b}}^{\mathfrak{g}}\,\mathbb{C}_\lambda$ to emphasize the functorial nature of induction. Thanks to the PBW Theorem, $U(\mathfrak{g}) \cong U(\mathfrak{n}^-) \otimes U(\mathfrak{b})$, which allows us to write $M(\lambda) \cong U(\mathfrak{n}^-) \otimes \mathbb{C}_\lambda$ as a left $U(\mathfrak{n}^-)$-module. Therefore $M(\lambda)$ is a free $U(\mathfrak{n}^-)$-module of rank one. In particular, the vector $v^+ := 1 \otimes 1$ in the definition of $M(\lambda)$ is nonzero and is acted on freely by $U(\mathfrak{n}^-)$, while $\mathfrak{n} \cdot v^+ = 0$ and $h \cdot v^+ = \lambda(h)v^+$ for all $h \in \mathfrak{h}$. Thus v^+ is a maximal vector and generates the $U(\mathfrak{g})$-module $M(\lambda)$. Moreover, the set of weights of $M(\lambda)$ is visibly $\lambda - \Gamma$. It follows that $M(\lambda)$ lies in $\mathcal{O}$.

Remark. More generally, one can start with an arbitrary finite dimensional $U(\mathfrak{b})$-module N on which $\mathfrak{h}$ acts semisimply and get an induced module $U(\mathfrak{g}) \otimes_{U(\mathfrak{b})} N$ in $\mathcal{O}$. This defines an *exact* functor from such $U(\mathfrak{b})$-modules to $U(\mathfrak{g})$-modules, since (as above) an induced module is free as a $U(\mathfrak{n}^-)$-module.

One can alternatively describe $M(\lambda)$ by generators and relations. From the PBW Theorem, we see that the left ideal I of $U(\mathfrak{g})$ which annihilates v^+ is generated by $\mathfrak{n}$ together with all $h - \lambda(h) \cdot 1$ with $h \in \mathfrak{h}$. Thus $M(\lambda) \cong U(\mathfrak{g})/I$. Obviously I also annihilates a maximal vector of weight λ generating an arbitrary highest weight module M. So $M(\lambda)$ maps naturally onto M and therefore plays the role of **universal highest weight module** of weight λ.

Using Theorem 1.2(f), we can write unambiguously $L(\lambda)$ (resp. $N(\lambda)$) for the unique simple quotient (resp. unique maximal submodule) of $M(\lambda)$. Since every nonzero module in $\mathcal{O}$ has at least one maximal vector, we conclude from this discussion and Theorem 1.2(g):

Theorem. *Every simple module in $\mathcal{O}$ is isomorphic to a module $L(\lambda)$ with $\lambda \in \mathfrak{h}^*$ and is therefore determined uniquely up to isomorphism by its highest weight. Moreover,* $\dim \mathrm{Hom}_{\mathcal{O}}\big(L(\mu), L(\lambda)\big) = \delta_{\lambda\mu}.$ $\qquad\square$

Exercise. Show that $M(\lambda)$ has the following property: For any M in $\mathcal{O}$,

$$\mathrm{Hom}_{U(\mathfrak{g})}\big(M(\lambda), M\big) = \mathrm{Hom}_{U(\mathfrak{g})}\big(\mathrm{Ind}_{\mathfrak{b}}^{\mathfrak{g}}\,\mathbb{C}_\lambda, M\big) \cong \mathrm{Hom}_{U(\mathfrak{b})}\big(\mathbb{C}_\lambda, \mathrm{Res}_{\mathfrak{b}}^{\mathfrak{g}}\,M\big),$$

where $\mathrm{Res}_{\mathfrak{b}}^{\mathfrak{g}}$ is the restriction functor. [Use the universal mapping property of tensor products.] In the context of induced modules for group algebras of finite groups, this adjointness property is known as *Frobenius reciprocity*.

1.4. Maximal Vectors in Verma Modules

In order to explore the submodule structure of a Verma module $M(\lambda)$, it is natural to begin by looking for maximal vectors of weight $\mu < \lambda$. This theme will be developed further in Chapter 4, but for our immediate purposes it is enough to carry out just the most elementary construction. Fix a standard basis (for example, a Chevalley basis) for $\mathfrak{g}$ as in 0.1, consisting of $h_1, \ldots, h_\ell$ together with root vectors $x_\alpha \in \mathfrak{g}_\alpha$ and $y_\alpha \in \mathfrak{g}_{-\alpha}$ for $\alpha \in \Phi^+$ (abbreviated to x_i, y_i for a numbering of simple roots $\alpha_1, \ldots, \alpha_\ell$).

Proposition. *Given $\lambda \in \mathfrak{h}^*$ and a fixed simple root α, suppose $n := \langle \lambda, \alpha^\vee \rangle$ lies in $\mathbb{Z}^+$. If v^+ is a maximal vector of weight λ in $M(\lambda)$, then $y_\alpha^{n+1} \cdot v^+$ is a maximal vector of weight $\mu := \lambda - (n+1)\alpha < \lambda$. Thus there exists a nonzero homomorphism $M(\mu) \to M(\lambda)$ whose image lies in the maximal submodule $N(\lambda)$.*

Here it is essential that α be *simple*; in this case it is easily seen that all weight spaces $M(\lambda)_{\lambda - k\alpha}$ with $k \in \mathbb{Z}^+$ are 1-dimensional and spanned by a vector $y_\alpha^k \cdot v^+$. Otherwise a less direct approach to the existence of such a maximal vector is required (to be discussed in Chapter 4). The problem is that for a nonsimple positive root α with $\langle \lambda, \alpha^\vee \rangle \in \mathbb{Z}^{>0}$, a weight space $M(\lambda)_{\lambda - k\alpha}$ is typically of dimension > 1 and spanned by the images of v^+ under a number of monomials in $U(\mathfrak{n})^-$ involving powers $y_\beta^{a(\beta)}$ with $\sum a(\beta) = k\alpha$. To write down explicitly a maximal vector in such a weight space is quite challenging.

The proof of the proposition depends on some standard commutation formulas in $U(\mathfrak{g})$ (see for example [**125**, 21.2]):

Lemma. *Let x_i, y_i be standard basis vectors as above, corresponding to simple roots $\alpha_1, \ldots, \alpha_\ell$. Then for all $k \geq 0$ and $1 \leq i, j \leq \ell$, we have:*

(a) $[x_j, y_i^{k+1}] = 0$ *whenever* $j \neq i$.

(b) $[h_j, y_i^{k+1}] = -(k+1)\alpha_i(h_j)y_i^{k+1}$.

(c) $[x_i, y_i^{k+1}] = -(k+1)y_i^k(k \cdot 1 - h_i)$.

Proof. We outline the steps, which are elementary, leaving the details as an exercise for the reader. When $k = 0$, (a) follows from the fact that $\alpha_j - \alpha_i$ is not a root if $j \neq i$, while (b) just expresses the fact that $[h_j, y_i] = -\alpha_i(h_j)y_i$ and (c) is obvious. Now proceed by induction. For example, in (c) rewrite the left side in $U(\mathfrak{g})$:

$$[x_i, y_i^{k+1}] = x_i y_i^{k+1} - y_i^{k+1} x_i = [x_i, y_i]y_i^k + y_i[x_i, y_i^k] = h_i y_i^k + y_i[x_i, y_i^k].$$

Then apply the induction hypothesis, together with (b) (replacing $k+1$ there by k). $\qquad\square$

Proof of Proposition. Recalling the isomorphism of $U(\mathfrak{n}^-)$-modules between $U(\mathfrak{n}^-)$ and $M(\lambda)$, part (c) of the lemma translates into the statement of the proposition in view of part (b). $\qquad\square$

Corollary. *With λ as in the proposition, let v^+ be instead a maximal vector of weight λ in $L(\lambda)$. Then $y_\alpha^{n+1} \cdot v^+ = 0$.*

Proof. Since $\lambda - (n+1)\alpha < \lambda$, no maximal vector of this weight can exist in the simple module $L(\lambda)$ (by Theorem 1.2(e)). So the maximal vector of this weight in $M(\lambda)$ constructed in the proposition must lie in the maximal submodule and map to 0 in the quotient $L(\lambda)$. $\qquad\qquad\square$

1.5. Example: $\mathfrak{sl}(2, \mathbb{C})$

The description of Verma modules and simple modules in the case $\mathfrak{g} = \mathfrak{sl}(2, \mathbb{C})$ serves as a (much simplified) prototype of the general theory to be developed in the following sections. The reader is encouraged to fill in the following outline now, although some steps will become considerably easier later on. (The finite dimensional case has already been reviewed in 0.9.)

Fix a standard basis (h, x, y) for $\mathfrak{g}$, with $[hx] = 2x$, $[hy] = -2y$, $[xy] = h$. Since $\dim \mathfrak{h} = 1$, weights $\lambda \in \mathfrak{h}^*$ may be identified with complex numbers. In turn, the integral weight lattice Λ is identified with $\mathbb{Z}$ and Λ_r with $2\mathbb{Z}$.

- $M(\lambda)$ has weights $\lambda, \lambda{-}2, \lambda{-}4, \ldots$, each with multiplicity one. Basis vectors v_i $(i \geq 0)$ for $M(\lambda)$ can be chosen so that (setting $v_{-1} = 0$):

$$
\begin{aligned}
h \cdot v_i &= (\lambda - 2i)v_i, \\
x \cdot v_i &= (\lambda - i + 1)v_{i-1}, \\
y \cdot v_i &= (i + 1)v_{i+1}.
\end{aligned}
$$

  (This is not the only way to choose a basis, but is convenient when working with integral weights and then reducing modulo a prime.)
- $\dim L(\lambda) < \infty$ if and only if $\lambda \in \mathbb{Z}^+$. In this case, the maximal submodule of $M(\lambda)$ is isomorphic to $L(-\lambda - 2)$.
- $M(\lambda)$ is simple if and only if λ is not in $\mathbb{Z}^+$.

Exercise. When $\mathfrak{g} = \mathfrak{sl}(2, \mathbb{C})$, show that $M(\lambda) \otimes M(\mu)$ cannot lie in $\mathcal{O}$.

1.6. Finite Dimensional Modules

Relying on the representation theory of $\mathfrak{sl}(2, \mathbb{C})$, we can now determine precisely which of the simple modules $L(\lambda)$ with $\lambda \in \mathfrak{h}^*$ are finite dimensional. For each $\alpha \in \Phi^+$, let $\mathfrak{s}_\alpha$ be the copy of $\mathfrak{sl}(2, \mathbb{C})$ in $\mathfrak{g}$ spanned by $h_\alpha, x_\alpha, y_\alpha$. Given an enumeration of simple roots as $\alpha_1, \ldots, \alpha_\ell$, abbreviate $\mathfrak{s}_{\alpha_i}$ by $\mathfrak{s}_i$, h_{α_i} by h_i, and s_{α_i} by s_i.

For use in the proof below, observe that if $M \in \mathcal{O}$ and $N \subset M$ is a finite dimensional $\mathfrak{s}_i$-submodule of M generated by a weight vector, then $\mathfrak{h}$ automatically stabilizes N. Indeed, if $v \in N$ has weight μ, then the commutation relations in $\mathfrak{g}$ force $h \cdot (x_i \cdot v) = x_i \cdot (h \cdot v) + \alpha_i(h)x_i \cdot v$. Since $h \cdot v = \mu(h)v \in N$, the left side also lies in N. (Similarly for $y_i \cdot v$.)

Theorem. *The simple module $L(\lambda)$ in $\mathcal{O}$ is finite dimensional if and only if $\lambda \in \Lambda^+$. This is the case if and only if $\dim L(\lambda)_\mu = \dim L(\lambda)_{w\mu}$ for all $\mu \in \mathfrak{h}^*$ and all $w \in W$.*

Proof. Proceed in steps.

(1) In one direction, assume that $\dim L(\lambda) < \infty$. For each fixed i, the structure of the $\mathfrak{s}_i$-module $L(\lambda)$ is then transparent. In particular, $\lambda(h_i) \in \mathbb{Z}^+$, forcing $\lambda \in \Lambda^+$ (0.7).

(2) In the other direction, assume that $\lambda \in \Lambda^+$. To show that $\dim L(\lambda) < \infty$ requires more work. Again we look at the $\mathfrak{s}_i$-module structure of $L(\lambda)$ for each $1 \leq i \leq \ell$. Since $n := \langle \lambda, \alpha_i^\vee \rangle \in \mathbb{Z}^+$, Corollary 1.4 implies that the $\mathfrak{s}_i$-submodule generated by v^+ is finite dimensional (and is in fact the unique simple module of dimension $n + 1$).

(3) In turn, we claim that for each fixed i, $L(\lambda)$ is the sum of all its finite dimensional $\mathfrak{s}_i$-submodules; call this sum M. By step (2), $M \neq 0$. On the other hand, take any *finite dimensional* $\mathfrak{s}_i$-submodule N of $L(\lambda)$; so $N \subset M$. Then $\mathfrak{g} \otimes N$ is a finite dimensional $\mathfrak{s}_i$-module, where we use the adjoint action of $\mathfrak{s}_i$ on $\mathfrak{g}$. It is routine to check that the map $\mathfrak{g} \otimes N \to L(\lambda)$ sending $x \otimes v$ to $x \cdot v$ is an $\mathfrak{s}_i$-module homomorphism, so its image lies in M. Thus M is a nonzero $\mathfrak{g}$-submodule of the simple module $L(\lambda)$, forcing $L(\lambda) = M$.

(4) Since each $v \in L(\lambda)$ lies in a finite dimensional $\mathfrak{s}_i$-submodule, by (3), it follows that each x_i or y_i acts on $L(\lambda)$ as a *locally nilpotent* operator. If we denote the representation afforded by $L(\lambda)$ as $\varphi : U(\mathfrak{g}) \to \mathrm{End}\, L(\lambda)$, it now makes sense to patch together the unipotent operators $\exp \varphi(x_i) \exp \varphi(-y_i) \exp \varphi(x_i)$ which act on finite dimensional $\mathfrak{s}_i$-submodules. This yields a well-defined automorphism r_i of $L(\lambda)$.

(5) If μ is any weight of $L(\lambda)$, we claim that $r_i(L(\lambda)_\mu) = L(\lambda)_{s_i\mu}$ for all i. Indeed, the weight space $L(\lambda)_\mu$ lies in a finite dimensional $\mathfrak{s}_i$-submodule by step (3). This in turn is a weight module, thanks to the observation just before the theorem. Now the claim follows from the $\mathfrak{sl}(2, \mathbb{C})$ theory in 0.9.

(6) Because W is generated by the simple reflections s_i, step (5) implies that all weight spaces $L(\lambda)_{w\mu}$ have the same dimension. Since λ is dominant integral, by assumption, all weights of $L(\lambda)$ are therefore W-conjugates of dominant integral weights $\mu \leq \lambda$. But there exist only finitely many of these (0.6). It follows that $L(\lambda)$ has only finitely many weights, thus is finite dimensional.

(7) From the proof it also follows that $\dim L(\lambda)_\mu = \dim L(\lambda)_{w\mu}$ for all $w \in W$ whenever $\dim L(\lambda) < \infty$. $\qquad\square$

Variants of this proof are found in numerous texts, including Bourbaki [**46**, Chap. VIII, §7], Carter [**60**, Chap. 10], Humphreys [**125**, 21.2].

Exercise. Let M be a highest weight module of weight $\lambda \in \Lambda^+$. If all x_i and y_i $(1 \le i \le \ell)$ act locally nilpotently on M (which is automatic for the x_i), rework steps in the proof above to show that $\dim M < \infty$. Conclude that $M \cong L(\lambda)$. (This situation comes up in 2.6 below.)

Example. It is easy to recover from step (2) of the proof the familiar fact that $L(0)$ is the trivial one-dimensional $\mathfrak{g}$-module: the subalgebras $\mathfrak{s}_i$ (which generate $\mathfrak{g}$) act trivially on v^+.

Coupled with Weyl's Complete Reducibility Theorem (0.7), the theorem gives a complete parametrization of the finite dimensional modules in $\mathcal{O}$. Of course, there are many concrete features yet to be investigated: notably, formal characters and dimensions. These will be worked out in Chapter 2 after we assemble more information about $\mathcal{O}$. Here we just note a few standard facts for later use in the case $\lambda \in \Lambda^+$:

- Let $\mu := w\lambda$ with $w \in W$. For any $\alpha \in \Phi$, not both $\mu - \alpha$ and $\mu + \alpha$ can occur as weights of $L(\lambda)$. [Recall that W permutes the weights of $L(\lambda)$. If $\beta := w^{-1}\alpha$, not both $\lambda + \beta$ and $\lambda - \beta$ can be $< \lambda$.]

- If μ and $\mu + k\alpha$ (with $k \in \mathbb{Z}, \alpha \in \Phi$) are weights of $L(\lambda)$, then so are all intermediate weights $\mu + i\alpha$. [View $L(\lambda)$ as an $\mathfrak{sl}(2, \mathbb{C})$-module and use the theory in 0.9.]

- The dual space $L(\lambda)^*$, with the standard action $(x \cdot f)(v) = -f(x \cdot v)$ for $x \in \mathfrak{g}, v \in L(\lambda), f \in L(\lambda)^*$, is isomorphic to $L(-w_\circ\lambda)$ (where $w_\circ \in W$ is the longest element). [Observe that $L(\lambda)^*$ is again simple; its weights relative to $\mathfrak{h}$ are the negatives of those for $L(\lambda)$. On the other hand, $w_\circ\lambda$ is the lowest weight of $L(\lambda)$.]

1.7. Action of the Center

Our discussion of Verma modules has shown the existence of a unique maximal submodule and simple quotient, but beyond this (and Proposition 1.4) we have no insight into the submodule structure. Without using deeper information about the universal enveloping algebra we cannot say much more. Indeed, our arguments so far have relied mainly on the triangular decomposition $\mathfrak{g} = \mathfrak{n}^- \oplus \mathfrak{h} \oplus \mathfrak{n}$ and the resulting PBW Theorem. But this kind of set-up is found much more generally in infinite dimensional Kac–Moody algebras (see Carter [**60**], Kac [**165**], Moody–Pianzola [**223**]); here there is a good analogue of the category $\mathcal{O}$, but for example "Verma modules" typically fail to have finite Jordan–Hölder length.

To prove a finite length theorem in our situation, we have to look closer at the action of the center $Z(\mathfrak{g})$ of $U(\mathfrak{g})$. By contrast, the study of finite dimensional modules (including the proof of Weyl's Complete Reducibility Theorem), requires only the action of the *Casimir element* of $Z(\mathfrak{g})$ (0.5). Recall from Theorem 1.1(e) that every $M \in \mathcal{O}$ is locally finite as a $Z(\mathfrak{g})$-module. The situation simplifies when M is a highest weight module, generated by a maximal vector v^+ of weight λ. If $z \in Z(\mathfrak{g})$ and $h \in \mathfrak{h}$, we have

$$h \cdot (z \cdot v^+) = z \cdot (h \cdot v^+) = z \cdot (\lambda(h)v^+) = \lambda(h)z \cdot v^+.$$

Since $\dim M_\lambda = 1$, this forces $z \cdot v^+ = \chi_\lambda(z)v^+$ for some scalar $\chi_\lambda(z) \in \mathbb{C}$. In turn, z acts on an arbitrary element $u \cdot v^+$ of M (with $u \in U(\mathfrak{n}^-)$) by the same scalar, since $zu = uz$.

For fixed λ, the function $z \mapsto \chi_\lambda(z)$ defines an algebra homomorphism $\chi_\lambda : Z(\mathfrak{g}) \to \mathbb{C}$, whose kernel is a maximal ideal of $Z(\mathfrak{g})$. We call it the **central character** associated with λ. More generally, we call any algebra homomorphism $Z(\mathfrak{g}) \to \mathbb{C}$ a central character. The set of all these is in natural bijection with the set $\mathrm{Max}\, Z(\mathfrak{g})$ of maximal ideals.

To describe χ_λ more concretely, write $z \in Z(\mathfrak{g})$ as a linear combination of PBW monomials based on the decomposition $\mathfrak{g} = \mathfrak{n}^- \oplus \mathfrak{h} \oplus \mathfrak{n}$. Any monomial having a nonzero factor from $\mathfrak{n}$ will kill v^+. In turn, factors from $\mathfrak{h}$ just multiply v^+ by scalars, while $\mathfrak{n}^-$ then takes v^+ to vectors of lower weight. The upshot is that $z \cdot v^+$ depends just on the PBW monomials with factors in $\mathfrak{h}$. Denoting by $\mathrm{pr} : U(\mathfrak{g}) \to U(\mathfrak{h})$ the projection onto the subspace $U(\mathfrak{h})$ given by setting all other monomials equal to 0, we see that

$$\chi_\lambda(z) = \lambda(\mathrm{pr}(z)) \text{ for all } z \in Z(\mathfrak{g}).$$

Here the linear function λ is extended canonically to an algebra homomorphism $\lambda : U(\mathfrak{h}) \to \mathbb{C}$.

Using the obvious fact that $\bigcap_{\lambda \in \mathfrak{h}^*} \mathrm{Ker}\, \lambda = 0$, it follows that *the restriction of* pr *to* $Z(\mathfrak{g})$ *is an algebra homomorphism*. Denote this restriction by ξ; it is called the **Harish-Chandra homomorphism**.

Remark. To understand more intrinsically why ξ is a homomorphism (as in the original study of the center by Harish-Chandra), without resort to representation theory, look at the centralizer $U(\mathfrak{g})_0$ of $\mathfrak{h}$ in the algebra $U(\mathfrak{g})$. This subalgebra obviously includes $Z(\mathfrak{g})$ as well as $U(\mathfrak{h})$. As the notation suggests, it is actually the 0-graded subspace of $U(\mathfrak{g})$ relative to the natural grading induced by the extension to $U(\mathfrak{g})$ of the adjoint representation of $\mathfrak{g}$ (0.5). In turn, $L := U(\mathfrak{g})\mathfrak{n} \cap U(\mathfrak{g})_0$ is seen to coincide with $\mathfrak{n}^- U(\mathfrak{g}) \cap U(\mathfrak{g})_0$. This is a *two-sided* ideal of $U(\mathfrak{g})$, complementary to the subspace $U(\mathfrak{h})$ of

$U(\mathfrak{g})_0$, and the resulting projection homomorphism $U(\mathfrak{g})_0 \to U(\mathfrak{h})$ restricts to $\xi : Z(\mathfrak{g}) \to U(\mathfrak{h})$.

Two natural questions arise at this point: (1) Is the Harish-Chandra homomorphism injective? (2) What is the image of ξ in $U(\mathfrak{h})$?

1.8. Central Characters and Linked Weights

Since $\chi_\lambda = \chi_\mu$ whenever $L(\mu)$ occurs as a subquotient of $M(\lambda)$, it is important to determine precisely when this equality of central characters occurs. It is also natural to ask whether there are any other central characters besides the χ_λ. To deal with both of these issues we have to formalize better the relationship here between λ and μ, then bring the Harish-Chandra homomorphism into the picture.

A valuable clue is provided by Proposition 1.4: If $n := \langle \lambda, \alpha^\vee \rangle \in \mathbb{Z}^+$ for some $\alpha \in \Delta$, then $M(\lambda)$ has a maximal vector of weight $\lambda - (n+1)\alpha < \lambda$. In this case, we have $\chi_\lambda = \chi_\mu$ where $\mu := \lambda - (n+1)\alpha$. To express this more suggestively, recall from 0.6 that $s_\alpha \rho = \rho - \alpha$ whenever α is simple. Thus $s_\alpha(\lambda + \rho) = \lambda - \langle \lambda, \alpha^\vee \rangle \alpha + \rho - \alpha = \mu + \rho$, or $s_\alpha(\lambda + \rho) - \rho = \mu$.

Definition. For $w \in W$ and $\lambda \in \mathfrak{h}^*$, define a shifted action of W (called the **dot action**) by $w \cdot \lambda = w(\lambda + \rho) - \rho$. If $\lambda, \mu \in \mathfrak{h}^*$, we say that λ and μ are **linked** (or W-linked) if for some $w \in W$, we have $\mu = w \cdot \lambda$. Linkage is clearly an equivalence relation on $\mathfrak{h}^*$. The orbit $\{w \cdot \lambda \mid w \in W\}$ of λ under the dot action is called the **linkage class** (or W-linkage class) of λ.

For example, the fixed point $-\rho$ under the dot action lies in a linkage class by itself (and no other weight does). The usual notion of *regular weight* for $\lambda \in \mathfrak{h}^*$, requiring that the isotropy group of λ in W be trivial, or $|W\lambda| = |W|$, is replaced in the setting of linkage classes by a shifted notion of **regular weight** (which might also be called **dot-regular**): the weight $\lambda \in \mathfrak{h}^*$ is regular if $|W \cdot \lambda| = |W|$, or in other words, $\langle \lambda + \rho, \alpha^\vee \rangle \neq 0$ for all $\alpha \in \Phi$. Weights which are not regular may be called **singular**. In this sense, 0 is regular while $-\rho$ is singular.

In general, each $\lambda + \rho \in \Lambda$ is W-conjugate (by 0.6) to a unique element in the closure $\overline{C}$ of the Weyl chamber

$$C := \{\mu \in E \mid \langle \mu, \alpha^\vee \rangle > 0 \text{ for all } \alpha \in \Delta\}.$$

So the linkage class of λ has a unique element in $\overline{C} - \rho$; this weight is then the unique maximal element in its linkage class. (For arbitrary $\lambda \in \mathfrak{h}^*$, we shall fine-tune the parametrization in Chapter 3.)

Exercise. Unlike the usual action of W on $\mathfrak{h}^*$, the dot action is not additive. If $\lambda, \mu \in \mathfrak{h}^*$ and $w \in W$, verify that

$$w \cdot (\lambda + \mu) = w \cdot \lambda + \mu,$$
$$w \cdot \lambda - w \cdot \mu = w(\lambda - \mu).$$

This language allows us to reformulate Proposition 1.4:

> *If $\alpha \in \Delta$, $\lambda \in \mathfrak{h}^*$ and $\langle \lambda + \rho, \alpha^\vee \rangle = 0$, then $M(s_\alpha \cdot \lambda) = M(\lambda)$. If $\langle \lambda + \rho, \alpha^\vee \rangle \in \mathbb{Z}^{>0}$, then there exists a nonzero homomorphism $M(s_\alpha \cdot \lambda) \to M(\lambda)$ with image in $N(\lambda)$.*

Now we can strengthen somewhat our earlier conclusion:

Proposition. *If $\lambda \in \Lambda$ and μ is linked to λ, then $\chi_\lambda = \chi_\mu$.*

Proof. Start as before with a *simple* root α. Since $\lambda \in \Lambda$, $n := \langle \lambda, \alpha^\vee \rangle \in \mathbb{Z}$. In case $n \in \mathbb{Z}^+$, we already saw that $\chi_\lambda = \chi_\mu$ with $\mu = s_\alpha \cdot \lambda$. Suppose on the other hand $n \in \mathbb{Z}^{<0}$. If $n = -1$, then $s_\alpha \cdot \lambda = \lambda$ and there is nothing to prove. If $n < -1$, then setting $\mu := s_\alpha \cdot \lambda$ (which forces $s_\alpha \cdot \mu = \lambda$), we have $\langle \mu, \alpha^\vee \rangle = -n - 2 \geq 0$. So by the first case, $\chi_\lambda = \chi_\mu$.

Now W is generated by the simple reflections (0.3), while the linkage relation is transitive, so the proposition follows by induction on $\ell(w)$. $\qquad\square$

1.9. Harish-Chandra Homomorphism

In order to remove the restriction in Proposition 1.8 that λ be an *integral* weight, we can invoke a simple density argument in affine algebraic geometry. Here we view $\mathfrak{h}^*$ as the affine space $\mathbb{A}^\ell$ over $\mathbb{C}$, while $U(\mathfrak{h}) = S(\mathfrak{h})$ (identified in turn with $P(\mathfrak{h}^*)$) is the algebra of polynomial functions in ℓ variables acting naturally on $\mathbb{A}^\ell$. To make this concrete, take for example the fundamental weights $\varpi_1, \ldots, \varpi_\ell$ as a basis for $\mathfrak{h}^*$, along with generators $h_1, \ldots, h_\ell$ for $S(\mathfrak{h})$.

Whereas Λ is a discrete subspace of $\mathfrak{h}^*$ in the usual topology, it becomes *dense* in the Zariski topology; here the closed sets are the zero sets of finite sets of polynomials. Recall the argument: In $\mathbb{A}^\ell$ we can identify Λ with $\mathbb{Z}^\ell$. Now use induction on ℓ to see that a polynomial function f on $\mathbb{A}^\ell$ vanishing on $\mathbb{Z}^\ell$ must be zero. If $\ell = 1$, use the fact that a nonzero polynomial can have only finitely many roots. If $\ell > 1$, write f as a polynomial in the last variable. Substituting fixed integers for the first $\ell - 1$ variables produces a polynomial in one variable vanishing on $\mathbb{Z}$ (therefore zero). So the induction hypothesis for $\ell - 1$ can be applied, showing that $f = 0$. From this argument we conclude that $\mathbb{Z}^\ell$ is dense in $\mathbb{A}^\ell$.

We know that $\chi_\lambda = \chi_{w \cdot \lambda}$ for all $w \in W$ when $\lambda \in \Lambda$. Since $\chi_\lambda(z) = \lambda(\xi(z))$ for $z \in Z(\mathfrak{g})$, this translates into the statement that the polynomial functions $\xi(z)$ and $w \cdot \xi(z)$ agree on Λ. By density, this forces them to agree everywhere on $\mathfrak{h}^*$. In other words, *Proposition 1.8 is true for all $\lambda \in \mathfrak{h}^*$.*

Using this we can say more about the image of ξ. To take account of the ρ-shift involved in the dot action, first compose ξ with the algebra automorphism of $S(\mathfrak{h})$ induced by the substitution $p(\lambda) \mapsto p(\lambda - \rho)$. Call this composite homomorphism $\psi : Z(\mathfrak{g}) \to S(\mathfrak{h})$ the **twisted Harish-Chandra homomorphism**. Now we have

$$\chi_\lambda(z) = (\lambda + \rho)(\psi(z)) \text{ for all } z \in Z(\mathfrak{g}).$$

We have shown that the polynomial functions on $\mathfrak{h}^*$ in the image of ψ are constant on W-orbits (linkage classes), in other words lie in the algebra $S(\mathfrak{h})^W$ of W-invariants. To summarize:

Theorem.   (a) *Whenever $\lambda, \mu \in \mathfrak{h}^*$ are W-linked, $\chi_\lambda = \chi_\mu$.*

> (b) *The image of the twisted Harish-Chandra homomorphism $\psi : Z(\mathfrak{g}) \to U(\mathfrak{h}) = S(\mathfrak{h})$ lies in the subalgebra $S(\mathfrak{h})^W$ of W-invariant polynomials.* $\square$

Example. Let $\mathfrak{g} = \mathfrak{sl}(2, \mathbb{C})$ and identify $\mathfrak{h}^*$ as usual with $\mathbb{C}$ via $\lambda \mapsto \lambda(h)$. Recall from 0.5 that a Casimir element in $Z(\mathfrak{g})$ is given explicitly by $c = 2xy + h^2 + 2yx = h^2 + 2h + 4yx$. Recalling that $\rho(h) = 1$, we get $\psi(c) = (h-1)^2 + 2(h-1) = h^2 - 1$, which visibly lies in $S(\mathfrak{h})^W$. In turn, $\chi_\lambda(c) = (\lambda + \rho)(h^2 - 1) = (\lambda + 1)^2 - 1$. It follows that $\chi_\lambda = \chi_\mu$ if and only if λ and μ are linked: $\mu = \lambda$ or $\mu = -\lambda - 2$.

Exercise. Show that the homomorphism ψ is independent of the choice of a simple system in Φ. [Any simple system has the form $w\Delta$ for some $w \in W$.]

1.10. Harish-Chandra's Theorem

What we really need is the converse of Theorem 1.9(a), in order to get a workable necessary condition for $L(\lambda)$ and $L(\mu)$ to occur as composition factors of a Verma module. This is contained in the following basic theorem.

Theorem (Harish-Chandra). *Let $\psi : Z(\mathfrak{g}) \to S(\mathfrak{h})$ be the twisted Harish-Chandra homomorphism.*

> (a) *The homomorphism ψ is an isomorphism of $Z(\mathfrak{g})$ onto $S(\mathfrak{h})^W$.*

> (b) *For all $\lambda, \mu \in \mathfrak{h}^*$, we have $\chi_\lambda = \chi_\mu$ if and only if $\mu = w \cdot \lambda$ for some $w \in W$.*

> (c) *Every central character $\chi : Z(\mathfrak{g}) \to \mathbb{C}$ is of the form χ_λ for some $\lambda \in \mathfrak{h}^*$.*

(a) This is the key point. Its proof depends mainly on arguments involving the enveloping algebra, independent of category $\mathcal{O}$. Here we just outline the basic ideas, referring in the Notes below to several thorough discussions in the literature.

We have seen that ψ maps $Z(\mathfrak{g})$ into the algebra $S(\mathfrak{h})^W$. Instead of trying to prove injectivity and surjectivity directly, it is better to make a comparison with a similar but somewhat more transparent map. Consider the algebra $P(\mathfrak{g})$ of polynomial functions on the vector space $\mathfrak{g}$, which is naturally isomorphic to $S(\mathfrak{g}^*)$. The restriction map $\theta : P(\mathfrak{g}) \to P(\mathfrak{h})$ is an algebra homomorphism. Next introduce the **adjoint group** $G \subset \mathrm{Aut}\,\mathfrak{g}$ generated by the operators $\exp\mathrm{ad}\,x$ with $x \in \mathfrak{g}$ nilpotent. (Since $\mathrm{ad}\,x$ is nilpotent, the exponential power series is just a polynomial.) The group G is a Lie group (or a semisimple algebraic group if $\mathbb{C}$ is replaced by another algebraically closed field of characteristic 0). It acts naturally on $P(\mathfrak{g})$ as a group of automorphisms, with algebra of fixed points denoted $P(\mathfrak{g})^G$. Similarly, W acts on $P(\mathfrak{h})$.

In this setting Chevalley proved a fundamental **Restriction Theorem**:

$$\theta \ \text{maps} \ P(\mathfrak{g})^G \ \text{isomorphically onto} \ P(\mathfrak{h})^W.$$

Since the Killing form is nondegenerate on $\mathfrak{g}$ and restricts to a nondegenerate form on $\mathfrak{h}$, we can further identify $P(\mathfrak{g})$ with $S(\mathfrak{g})$ and $P(\mathfrak{h})$ with $S(\mathfrak{h})$. The proof of Chevalley's theorem involves some use of finite dimensional representation theory: $P(\mathfrak{g})^G$ is generated by traces of powers of operators representing elements of $\mathfrak{g}$ in such representations.

At this point the rewritten isomorphism $S(\mathfrak{g})^G \to S(\mathfrak{h})^W$ somewhat resembles the Harish-Chandra homomorphism ξ. In fact, $S(\mathfrak{g})^G$ can even be identified naturally as a *vector space* with $Z(\mathfrak{g})$, though not as an algebra. Even though the Chevalley map θ does not coincide with ξ (as illustrated when $\mathfrak{g} = \mathfrak{sl}(2,\mathbb{C})$ in [**125**, 23.3]), one gets enough information from the comparison to see that ξ is *bijective* and thus to complete the proof. The trick is to see that the various maps here are compatible with the natural filtrations of the algebras, then to observe that the induced *graded* versions of ξ and the Chevalley isomorphism agree.

(b) Suppose that for some $\lambda, \mu \in \mathfrak{h}^*$, the linkage classes of λ and μ are disjoint. Using Lagrange interpolation, find a polynomial function f on $\mathfrak{h}^*$ which takes value 1 on $W \cdot \lambda$ and 0 on $W \cdot \mu$. Then replace f by its "average"

$$\frac{1}{|W|} \sum_{w \in W} wf$$

to get a W-*invariant* g having the same values on these dot orbits. Using the assumption about ψ, take any pre-image $z \in Z(\mathfrak{g})$ of g under ψ. Then check that $\chi_\lambda(z) = \lambda(\psi(z)) = g(\lambda) = 1$, whereas $\chi_\mu(z) = \mu(\psi(z)) = g(\mu) = 0$. This forces $\chi_\lambda \neq \chi_\mu$.

(c) Here the argument relies on some standard commutative algebra. To the given χ corresponds (via ψ) a homomorphism $\varphi : S(\mathfrak{h})^W \to \mathbb{C}$. Since W is a finite group, it is easy to see that $S(\mathfrak{h})$ is an *integral* extension of the ring of invariants: any $f \in S(\mathfrak{h})$ is a root of the monic polynomial $\prod_{w \in W}(t - wf)$ (with t an indeterminate), whose coefficients clearly lie in $S(\mathfrak{h})^W$. So a standard consequence of the Going Up Theorem in commutative algebra can be invoked (our field being algebraically closed): φ extends to a homomorphism $\widehat{\varphi} : S(\mathfrak{h}) \to \mathbb{C}$.

This translates into the desired conclusion as follows. Since $S(\mathfrak{h})$ is the algebra of polynomial functions on $\mathfrak{h}^*$, any homomorphism $S(\mathfrak{h}) \to \mathbb{C}$ such as $\widehat{\varphi}$ is the evaluation map $f \mapsto f(\lambda)$ for some $\lambda \in \mathfrak{h}^*$. (The kernel of the homomorphism is the obvious maximal ideal vanishing at the point λ.) In turn, $\chi(z) = \lambda(\psi(z)) = \chi_\lambda(z)$ for all $z \in Z(\mathfrak{g})$, forcing $\chi = \chi_\lambda$. $\qquad\square$

Remark. Chevalley made more precise the structure of the algebra $S(\mathfrak{h})^W$ in the context of his study of invariants under arbitrary finite reflection groups: $S(\mathfrak{h})^W$ is isomorphic to a polynomial algebra in ℓ variables, which can be generated by homomogenous elements of the polynomial algebra $S(\mathfrak{h})$ of degrees $d_1, \ldots, d_\ell$ satisfying $d_1 \cdots d_\ell = |W|$. The smallest degree is 2. (These d_i have other interesting interpretations as well.) See for example Bourbaki [**45**, V, §5], Humphreys [**129**, Chap. 3].

Exercise. Prove that the transpose map τ (see 0.5) fixes $Z(\mathfrak{g})$ pointwise. [Check that τ commutes with the Harish-Chandra homomorphism ξ and use the fact that ξ is injective.]

1.11. Category $\mathcal{O}$ is Artinian

Now it is easy to see that category $\mathcal{O}$ is *artinian* as well as noetherian: each $M \in \mathcal{O}$ is both artinian and noetherian.

Theorem. *Each $M \in \mathcal{O}$ is artinian (as well as noetherian). Moreover, $\dim \mathrm{Hom}_\mathcal{O}(M, N) < \infty$ for all $M, N \in \mathcal{O}$.*

Proof. Thanks to Corollary 1.2, it is enough to prove the first statement when $M = M(\lambda)$ is a Verma module. Let $V := \sum_{w \in W} M(\lambda)_{w \cdot \lambda}$, so $\dim V < \infty$. Suppose $N \supset N'$ is a proper inclusion of submodules of $M(\lambda)$, so $Z(\mathfrak{g})$ acts on the quotient N/N' by the character χ_λ. The nonzero module N/N' has a maximal vector of some weight $\mu \leq \lambda$, so $\chi_\mu = \chi_\lambda$. But then by Theorem 1.10(b), $\mu = w \cdot \lambda$ for some $w \in W$, forcing $N \cap V \neq 0$ and $\dim(N \cap V) > \dim(N' \cap V)$. It follows that any properly descending chain of submodules in $M(\lambda)$ terminates after finitely many steps.

Now let $M, N \in \mathcal{O}$. Starting with the fact that $\dim \mathrm{End}_\mathcal{O} L(\lambda) = 1$ for all $\lambda \in \mathfrak{h}^*$ (Theorem 1.2(g)), induction on the length of M shows that $\dim \operatorname{Hom}_{\mathcal{O}}\big(M, L(\lambda)\big) < \infty$. In turn, a similar induction shows that $\dim \operatorname{Hom}_{\mathcal{O}}(M, N) < \infty$. $\qquad\square$

As a consequence of the theorem, each $M \in \mathcal{O}$ possesses a **composition series** with simple quotients isomorphic to various $L(\lambda)$. The multiplicity of $L(\lambda)$ is independent of the choice of composition series and is denoted by $[M : L(\lambda)]$. For example, $[M(\lambda) : L(\lambda)] = 1$ in view of the fact that $\dim M(\lambda)_\lambda = 1$. We call the common length of all such composition series the **length** of M (short for **Jordan–Hölder length**).

In particular, Harish-Chandra's Theorem implies that *each $M(\lambda)$ has a composition series with $[M(\lambda) : L(\mu)] \neq 0$ only if $\mu \leq \lambda$ and λ, μ are linked.*

Example. As we observed in 1.8, the weight $-\rho$ lies in a linkage class by itself. Therefore $M(-\rho)$ has no composition factor of lower weight, forcing $M(-\rho) = L(-\rho)$.

Since $\mathcal{O}$ is artinian and noetherian, a standard theorem (associated in various contexts with the names Azumaya, Krull, Remak, Schmidt) is valid here: Each $M \in \mathcal{O}$ can be written as a direct sum of indecomposable modules, with the summands being unique up to isomorphism and order. This will be made more precise in the following section.

Similarly, the chain conditions imply a straightforward description of the **Grothendieck group** $K(\mathcal{O})$, defined to be the quotient of the free abelian group on symbols $[M]$ (for $M \in \mathcal{O}$) by the subgroup generated by all $[B] - [A] - [C]$ for which $0 \to A \to B \to C \to 0$ is a short exact sequence in $\mathcal{O}$. Whenever $M, N \in \mathcal{O}$, it follows from the construction that $[M] = [N]$ if and only if M and N have the same composition factor multiplicities. Note too that any exact functor $\mathcal{O} \to \mathcal{O}$ induces an endomorphism of $K(\mathcal{O})$: for example, tensoring with a fixed finite dimensional module (Theorem 1.1(d)).

In our case, $K(\mathcal{O})$ has a $\mathbb{Z}$-basis consisting of the symbols $[L(\lambda)]$, where $\lambda \in \mathfrak{h}^*$. Indeed, for any $M \in \mathcal{O}$, the corresponding element $[M]$ of $K(\mathcal{O})$ can be written uniquely as a (finite!) $\mathbb{Z}^+$-linear combination of this basis; here the coefficient of $[L(\lambda)]$ is the composition factor multiplicity $[M : L(\lambda)]$.

Although we often have to settle for formal information about modules in $\mathcal{O}$, the fact that modules have finite length encourages us to look for some natural submodules. For example, the **socle** of a module M is the sum (automatically direct) of all simple submodules of M, denoted $\operatorname{Soc} M$; it is the largest semisimple submodule of M. The **radical** of M is the intersection of all maximal submodules in M, denoted $\operatorname{Rad} M$. It may also be characterized as the smallest submodule for which $M/\operatorname{Rad} M$ is semisimple; this quotient is called the **head**, denoted $\operatorname{Hd} M$. For example, $\operatorname{Rad} M(\lambda) = N(\lambda)$ and $\operatorname{Hd} M(\lambda) \cong L(\lambda)$.

1.12. Subcategories $\mathcal{O}_\chi$

The action of $Z(\mathfrak{g})$ on an arbitrary $M \in \mathcal{O}$ is usually complicated. But in view of Corollary 1.2 we can expect this action to involve only a finite number of central characters. To make this precise, define a subspace of M for each fixed χ by

$$M^\chi := \{v \in M \mid (z - \chi(z))^n \cdot v = 0 \text{ for some } n > 0 \text{ depending on } z\}.$$

In words, each element of $Z(\mathfrak{g})$ acts on v as the scalar $\chi(z)$ plus a nilpotent operator. It is clear that M^χ is a $U(\mathfrak{g})$-submodule of M, while the subspaces M^χ are independent.

Now $Z(\mathfrak{g})$ stabilizes each weight space M_μ, since $Z(\mathfrak{g})$ and $U(\mathfrak{h})$ commute. Since we are working over an algebraically closed field, a standard result from linear algebra on families of commuting operators implies that $M_\mu = \bigoplus_\chi (M_\mu \cap M^\chi)$. Because M is generated by finitely many weight vectors, it must therefore be the direct sum of finitely many nonzero submodules M^χ. Thanks to Harish-Chandra's Theorem, each central character χ occurring in this way must be of the form χ_λ for some $\lambda \in \mathfrak{h}^*$.

Denote by $\mathcal{O}_\chi$ the (full) subcategory of $\mathcal{O}$ whose objects are the modules M for which $M = M^\chi$. From the discussion above we conclude:

Proposition. *Category $\mathcal{O}$ is the direct sum of the subcategories $\mathcal{O}_\chi$ as χ ranges over the central characters of the form χ_λ. Therefore each indecomposable module lies in a unique $\mathcal{O}_\chi$. In particular, each highest weight module of weight λ lies in $\mathcal{O}_{\chi_\lambda}$.* $\qquad\qquad\square$

The decomposition $\mathcal{O} = \bigoplus_\chi \mathcal{O}_\chi$ is a major step in the direction of reducing the study of category $\mathcal{O}$ to the study of more manageable subcategories. In view of the proposition, each indecomposable module lies in a single $\mathcal{O}_\chi$. Moreover, $\mathcal{O}_\chi$ involves only finitely many simple modules and the corresponding Verma modules. (In Chapter 3 projectives and injectives will be added to this list.)

Exercise. Fix a central character χ and let $\{V^{(\lambda)}\}$ be a collection of modules in $\mathcal{O}_\chi$ indexed by the weights λ for which $\chi = \chi_\lambda$ and satisfying: (1) $\dim V^{(\lambda)}_\lambda = 1$; (2) $\mu \leq \lambda$ for all weights μ of $V^{(\lambda)}$. Then the symbols $[V^{(\lambda)}]$ form a $\mathbb{Z}$-basis of the Grothendieck group $K(\mathcal{O}_\chi)$. For example, take $V^{(\lambda)} = M(\lambda)$ or $L(\lambda)$.

1.13. Blocks

In module categories which fail to be semisimple, the notion of **block** often helps to organize modules. Blocks play a major role, for example, in the modular representation theory of finite groups as developed by Brauer and others. In that theory the blocks are as efficient as possible in separating indecomposable modules which are unrelated homologically.

In general, one looks at the natural Ext functors on a module category which is both artinian and noetherian. If two simple modules M_1 and M_2 can be extended nontrivially, i.e., if there is a nonsplit short exact sequence $0 \to M_i \to M \to M_j \to 0$ with $\{i, j\} = \{1, 2\}$, then we place M_1 and M_2 in the same block. More generally, if simple modules M and N belong to a finite sequence $M = M_1, M_2, \ldots, M_n = N$ such that adjacent pairs belong to the same block, we place M and N in the same block. This partitions the simple modules into blocks.

Now if M is arbitrary, we say it belongs to a block if all its composition factors do. A formal argument (which we leave to the reader) shows that M decomposes uniquely as a direct sum of submodules, each belonging to a single block. In particular, each *indecomposable* module belongs to a single block. (See for example Jantzen [**152**, II.7.1]; his finite dimensionality assumption can be replaced here by the chain conditions on modules.)

Our decomposition of $\mathcal{O}$ using central characters obviously goes in this direction: in the partition of modules just described, each block of $\mathcal{O}$ will lie in some $\mathcal{O}_\chi$. When the weights are all *integral*, these are precisely the blocks:

Proposition. *If $\lambda \in \Lambda$, the subcategory $\mathcal{O}_{\chi_\lambda}$ is a block of $\mathcal{O}$.*

Proof. It just has to be shown that all $L(w \cdot \lambda)$ lie in the same block. If $\alpha \in \Delta$, set $\mu := s_\alpha \cdot \lambda$. Unless $\mu = \lambda$, we may assume $\mu < \lambda$ (otherwise reverse the roles of λ and μ); cf. the proof of Proposition 1.8. We know from Proposition 1.4 that there is a nonzero homomorphism $M(\mu) \to N(\lambda) \subset M(\lambda)$. Factoring out $N(\mu)$ (and its image N in $M(\lambda)$), this induces an embedding of $L(\mu)$ into $M(\lambda)/N$, which is a highest weight module with quotient $L(\lambda)$. Since highest weight modules are indecomposable, it follows that $L(\lambda)$ and $L(\mu)$ lie in the same block.

Now iterate this procedure, using a reduced expression for $w \in W$, to conclude that $L(\lambda)$ and $L(w \cdot \lambda)$ lie in the same block. $\square$

When $\chi = \chi_0$, $\mathcal{O}_\chi$ is called the **principal block** (sometimes denoted $\mathcal{O}_0$), by analogy with the case of modules for a finite group.

This neat description of blocks breaks down when *nonintegral* weights are involved, as the example $\mathfrak{g} = \mathfrak{sl}(2, \mathbb{C})$ shows: a typical Verma module $M(\lambda)$ with λ positive but not integral will be simple, even though $-\lambda - 2$ is linked to λ (1.5). The problem here is that the linkage class intersects more than one coset of Λ_r in $\mathfrak{h}^*$. The following easy exercise gives a refinement of the linkage classes in the nonintegral case, which will be studied further in

3.5 below. But a definitive description of blocks requires more information about the submodule structure of Verma modules (4.9).

Exercise. Suppose $\lambda \notin \Lambda$, so the linkage class $W \cdot \lambda$ is the disjoint union of its nonempty intersections with various cosets of Λ_r in $\mathfrak{h}^*$. Prove that each $M \in \mathcal{O}_{\chi_\lambda}$ has a corresponding direct sum decomposition $M = \bigoplus M_i$, in which all weights of M_i lie in a single coset. [Recall Exercise 1.1(b).]

Caution: In the category $\mathcal{O}$ literature the term "block" is sometimes defined more broadly to mean a subcategory $\mathcal{O}_\chi$.

1.14. Formal Characters of Finite Dimensional Modules

Beyond the parametrization of simple modules $L(\lambda)$ in $\mathcal{O}$ by highest weights, how much can one expect to say about their internal structure? In the traditional finite dimensional representation theory of finite groups, compact Lie groups, or associated Lie algebras, a natural goal is to write down explicitly the matrices giving the action of individual group or Lie algebra elements on a conveniently chosen basis. This is especially useful for applications in classical physics or chemistry, but can be carried out effectively mainly in low ranks or other special cases such as symmetric or exterior powers of a natural representation.

A more modest goal in finite dimensional representation theory is to determine the *characters* of irreducible representations: the traces of representing matrices (which are independent of the choice of basis). Here the theory is well developed. In well-behaved cases, the character of a representation determines it uniquely up to equivalence. For a semisimple Lie algebra $\mathfrak{g}$ over $\mathbb{C}$, the notion of character is formulated more abstractly: here it is enough to keep track of multiplicities of weights, which indirectly determine traces of matrices representing group elements in associated highest weight representations. (In a compact Lie group, each element is conjugate to a semisimple one.)

It is easy to introduce a suitable **formal character** in the finite dimensional setting. Start with the integral group ring $\mathbb{Z}\Lambda$ of the weight lattice Λ. Although the group structure of Λ is usually written additively, this would confuse the situation in the group ring where the elements are formal $\mathbb{Z}$-linear combinations of group elements. So we associate to each weight λ a symbol $e(\lambda)$, suggestive of an exponential. Given a (finite dimensional!) $U(\mathfrak{g})$-module M, which is automatically a weight module, its formal character can then be defined as

$$\operatorname{ch} M := \sum_{\lambda \in \Lambda} \dim M_\lambda \, e(\lambda).$$

Using the ring structure of $\mathbb{Z}\Lambda$, we have $\mathrm{ch}(M \otimes N) = \mathrm{ch}\, M \,\mathrm{ch}\, N$. Thanks to Weyl's Complete Reducibility Theorem, all such characters are known once we find $\mathrm{ch}\, L(\lambda)$. In Chapter 2 the classical results on these characters will be recovered in the more general setting of category $\mathcal{O}$.

1.15. Formal Characters of Modules in $\mathcal{O}$

In order to make effective use of Verma modules, we need to introduce the language of formal characters in the infinite dimensional setting. The idea is obvious: write down a formal sum as above, which no longer need be finite. To manipulate such formal sums rigorously, we reformulate the definition as follows. Specifying dimensions of weight spaces amounts to assigning a nonnegative integer to each weight of a module $M \in \mathcal{O}$. Thus we view $\mathrm{ch}\, M$ as a $\mathbb{Z}^+$-valued *function* ch_M on $\mathfrak{h}^*$. (More generally, any module in the larger category of weight modules having finite dimensional weight spaces can be assigned a formal character in this way.)

In place of the basis element $e(\lambda)$ of $\mathbb{Z}\Lambda$ we now have a characteristic function e_λ which takes value 1 at λ and value 0 at $\mu \neq \lambda$. In the finite dimensional case the functions corresponding to elements of $\mathbb{Z}\Lambda$ vanish outside a finite subset of Λ, allowing the multiplication in $\mathbb{Z}\Lambda$ to be reinterpreted as the **convolution** product:

$$(f * g)(\lambda) := \sum_{\mu+\nu=\lambda} f(\mu)\, g(\nu).$$

To deal with modules in $\mathcal{O}$, we simply carry over this definition to the additive group $\mathcal{X}$ of functions $f : \mathfrak{h}^* \to \mathbb{Z}$ whose support (outside which f is 0) lies in a finite union of sets of the form $\lambda - \Gamma$. Obviously all functions e_λ (with $\lambda \in \mathfrak{h}^*$) lie in $\mathcal{X}$. It is easy to check that $\mathcal{X}$ is a commutative ring under convolution, in which e_0 plays the role of multiplicative identity while $e_\lambda e_\mu = e_{\lambda+\mu}$ for $\lambda, \mu \in \mathfrak{h}^*$.

For typographical reasons we continue to write $\mathrm{ch}\, M$ (rather than ch_M) when M is an arbitrary weight module, viewing this as the function whose value at $\lambda \in \mathfrak{h}^*$ is $\dim M_\lambda$. Similarly, we may write $e(\lambda)$ in place of e_λ. When convenient we also express $\mathrm{ch}\, M$ as a (possibly infinite) formal linear combination of the $e(\lambda)$. For example,

$$\mathrm{ch}\, L(\lambda) = e(\lambda) + \sum_{\mu<\lambda} m_{\lambda\mu} e(\mu) \text{ for some } m_{\lambda\mu} \in \mathbb{Z}^+.$$

Denote by $\mathcal{X}_0$ the additive subgroup of $\mathcal{X}$ generated by all $\mathrm{ch}\, M$.

Proposition. *Formal characters of modules in $\mathcal{O}$ have the following properties.*

(a) *If $0 \to M' \to M \to M'' \to 0$ is a short exact sequence in $\mathcal{O}$, we have $\operatorname{ch} M = \operatorname{ch} M' + \operatorname{ch} M''$. Thus any $\operatorname{ch} M$ is determined uniquely by the formal characters of the composition factors of M, taken with multiplicity.*

(b) *$\mathcal{X}_0$ can be identified naturally with the Grothendieck group $K(\mathcal{O})$ by sending $\operatorname{ch} M \mapsto [M]$.*

(c) *If $M \in \mathcal{O}$ and $\dim L < \infty$, then $\operatorname{ch}(L \otimes M) = \operatorname{ch} L * \operatorname{ch} M$.*

Proof. (a) This follows from the fact that $\dim M_\mu = \dim M'_\mu + \dim M''_\mu$. Then (b) is an easy consequence.

(c) Recall from Theorem 1.1(d) that $L \otimes M$ also lies in $\mathcal{O}$, so its formal character is defined. The stated formula is clear from the distributive law when formal characters are thought of as formal sums, as in the finite dimensional setting. To make this precise in general using convolution, just observe that $(L \otimes M)_\lambda = \sum_{\mu+\nu=\lambda} L_\mu \otimes M_\nu$. $\qquad\square$

Remark. In the setting of $\mathbb{Z}\Lambda$, the natural action of W on Λ extends to an action on $\mathbb{Z}\Lambda$. Each character of a finite dimensional module such as $L(\lambda)$ with $\lambda \in \Lambda^+$ is then W-invariant (Theorem 1.6): $w \operatorname{ch} L(\lambda) = \operatorname{ch} L(\lambda)$. This is no longer true in general for $\mathcal{X}$, which creates an obstacle to using Verma modules as a tool in the proof of Weyl's Character Formula (Chapter 2).

1.16. Formal Characters of Verma Modules

What is the formal character $\operatorname{ch} M(\lambda)$ of a Verma module? This is easily computed. In fact, all such characters can be described uniformly in terms of a single function $p \in \mathcal{X}$: let $p(\gamma)$ equal the number of families $(c_\alpha)_{\alpha>0}$ with $c_\alpha \in \mathbb{Z}^+$ for which $\gamma = -\sum_{\alpha>0} c_\alpha \alpha$. Call this the **Kostant function** (though it is actually the negative of the "partition function" $\mathcal{P}$ originally introduced by Kostant in his study of weight multiplicities for finite dimensional modules). By definition, p vanishes outside Λ_r. Now p is just the formal character of $M(0)$: indeed, $\dim M(0)_\nu$ is the number of distinct PBW monomials $\prod_{\alpha>0} y_\alpha^{c_\alpha}$ in $U(\mathfrak{n}^-)$ (for a fixed ordering of Φ^+) which take a maximal vector v^+ into the weight space $M(0)_\nu$. Similar reasoning shows in general:

Proposition. *For any $\lambda \in \mathfrak{h}^*$, we have $\operatorname{ch} M(\lambda) = p * e(\lambda)$. In particular, $\operatorname{ch} M(0) = p$.* $\qquad\square$

While the formal characters of Verma modules are quite transparent, it is far from clear how to determine $\operatorname{ch} L(\lambda)$ for arbitrary $\lambda \in \mathfrak{h}^*$. This is a fundamental problem in the study of $\mathcal{O}$. To go beyond the classical finite dimensional case treated below in Chapter 2, we will need to undertake a deeper study of Verma modules in Chapters 4–5. But ultimately the computation of formal characters of all simple modules in $\mathcal{O}$ turns out to require a transition to a more sophisticated geometrically defined category (Chapter 8 below).

Working in $\mathcal{X}_0$ does make it possible to reformulate the problem in a useful way, taking advantage of Harish-Chandra's Theorem. First write

$$(1) \qquad \operatorname{ch} M(\lambda) = \sum_{\mu} a(\lambda, \mu) \operatorname{ch} L(\mu) \text{ with } a(\lambda, \mu) \in \mathbb{Z}^+ \text{ and } a(\lambda, \lambda) = 1.$$

Here μ ranges over weights $\leq \lambda$ and linked to λ by W, while $a(\lambda, \mu) = [M(\lambda) : L(\mu)]$. The partial ordering permits us to invert the resulting triangular system of equations:

$$(2) \qquad \operatorname{ch} L(\lambda) = \sum_{\mu} b(\lambda, \mu) \operatorname{ch} M(\mu) \text{ with } b(\lambda, \mu) \in \mathbb{Z} \text{ and } b(\lambda, \lambda) = 1.$$

The sum is again taken over weights $\mu \leq \lambda$ linked to λ. To make the role of W more explicit we can recast this as:

$$(3) \qquad \operatorname{ch} L(\lambda) = \sum_{w \cdot \lambda \leq \lambda} b(\lambda, w) \operatorname{ch} M(w \cdot \lambda) \text{ with } b(\lambda, w) \in \mathbb{Z}, \ b(\lambda, 1) = 1.$$

This last format turns out to be most useful in practice, though the determination of the coefficients is typically quite subtle. Even in the case $\lambda = 0$, where the left side is known to be equal to $e(0)$, it is nontrivial to fill in the right side.

Remark. In the simplest case, when $\mathfrak{g} = \mathfrak{sl}(2, \mathbb{C})$, it is easy to see that for $\lambda \geq 0$ we have $\operatorname{ch} L(\lambda) = \operatorname{ch} M(\lambda) - \operatorname{ch} M(s_\alpha \cdot \lambda)$, where $s_\alpha \cdot \lambda = -\lambda - 2$ (see 1.5 and Example 1.9). Thus $b(\lambda, w) = \pm 1$ is independent of the choice of dominant λ. On the other hand, for $\lambda < 0$ we have $b(\lambda, w) = 1$ for $w = 1$ and 0 otherwise; here again the result is independent of λ. This raises the question in general as to how $b(\lambda, w)$ depends on each of λ and w. For $\lambda \in \Lambda^+$ the answer will be worked out in the next chapter, but in other cases deeper ideas are involved.

Exercise. Show that the formal characters $\operatorname{ch} L(\lambda)$ with $\lambda \in \mathfrak{h}^*$ are linearly independent in $\mathcal{X}$ and form a $\mathbb{Z}$-basis of the additive subgroup $\mathcal{X}_0$ of $\mathcal{X}$ generated by all $\operatorname{ch} M$, while the formal characters of Verma modules provide an alternative $\mathbb{Z}$-basis. Similar statements hold in the Grothendieck group $K(\mathcal{O})$ for the corresponding elements $[L(\lambda)]$ and $M(\lambda)]$.

Notes

The axiomatic treatment of category $\mathcal{O}$ originates in the papers [**25, 26, 27**] of Bernstein–Gelfand–Gelfand (abbreviated as BGG in what follows). The letter chosen to label the category is the first letter of a Russian word meaning "basic".

Concerning the notion of "induced module" (1.3) see Dixmier [**84**, 5.1].

Step (3) in the proof of Theorem 1.6 is adapted from Dixmier [**84**, Prop. 1.7.9], as suggested by Jantzen.

The structure of $Z(\mathfrak{g})$, together with applications to representation theory (1.7–1.10), is explained in numerous sources, starting with the fundamental paper of Harish-Chandra [**121**]. See the texts (in roughly chronological order): Humphreys [**125**, §23], Dixmier [**84**, 7.4–7.5], Varadarajan [**250**, 4.10], Bourbaki [**46**, VIII, §8], Carter [**60**, Chap. 11].

(1.7) We have emphasized highest weight modules and their central characters. In fact, every *simple* $U(\mathfrak{g})$-module M satisfies $\dim \operatorname{End}_{U(\mathfrak{g})} M = 1$ and therefore has a central character. This is a general fact (Quillen's Lemma) for a finite dimensional Lie algebra over an algebraically closed field; see Dixmier [**84**, 2.6.8].

Chapter 2

Characters of Finite Dimensional Modules

As promised in Chapter 1, we pause here to look more closely at the subcategory of $\mathcal{O}$ consisting of finite dimensional modules. These are direct sums of various $L(\lambda)$ with $\lambda \in \Lambda^+$, so the main question at first is to work out the formal characters and dimensions of these simple modules. To derive the classical formulas of Weyl and Kostant, we follow the approach of BGG [**25**] in studying $L(\lambda)$ formally as a $\mathbb{Z}$-linear combination of Verma modules.

2.1. Summary of Prerequisites

In addition to the general background recalled in Chapter 0, our treatment of the finite dimensional theory requires most of the machinery developed in Chapter 1—but only in the case of *integral* weights.

The construction of simple modules as quotients of Verma modules is essential, together with the language of formal characters.

From 1.6 one has to know that finite dimensional simple modules have highest weights in Λ^+. This is the easier half of the classification theorem, but of course the other half is needed to ensure the existence of enough such modules to make the theory nontrivial.

One substantial shortcut could be taken here to bypass the use of Harish-Chandra's Theorem on central characters. As explained in detail in the Appendix to §23 in [**125**] (second and later printings), the study by V. Kac of "integrable" modules for Kac–Moody Lie algebras relies just on the Casimir elements associated with representations to separate highest weights. There the full center of the enveloping algebra is both unwieldy and unnecessary for this special class of modules. While the method of Kac can be used equally well in the finite dimensional situation and cuts down the prerequisites, it would be artificial in the wider context of category $\mathcal{O}$ to delay the discussion of central characters.

2.2. Formal Characters Revisited

Beyond rank 1, it becomes impractical to write down simultaneously all weight multiplicities of an arbitrary finite dimensional $L(\lambda)$. Weyl's classical character formula captures the essential information indirectly but in closed form by expressing $\operatorname{ch} L(\lambda)$ as a quotient in which the denominator depends only on root system data. While this approach leaves the individual weight multiplicities concealed, it has conceptual elegance and also leads to an easily computable formula for $\dim L(\lambda)$.

The approach by BGG to Weyl's formula requires some "thinking outside the box", the box in this case being the category of finite dimensional $U(\mathfrak{g})$-modules. The idea is already implicit in the formulation given in 1.16(3):

$$\operatorname{ch} L(\lambda) = \sum_{w \in W, w \cdot \lambda \leq \lambda} b(\lambda, w) \operatorname{ch} M(w \cdot \lambda), \quad \text{with } b(\lambda, w) \in \mathbb{Z} \text{ and } b(\lambda, 1) = 1.$$

In the finite dimensional case, the sum here is taken over all of W: since $\lambda \in \Lambda^+$ (and $\rho \in \Lambda^+$), we have $w \cdot \lambda \leq \lambda$ for all $w \in W$ (0.6).

In the simplest case $\mathfrak{g} = \mathfrak{sl}(2, \mathbb{C})$, with $\Lambda^+ \cong \mathbb{Z}^+$, the situation is quite transparent (0.9): Here $|W| = 2$ and we find that

$$\operatorname{ch} L(\lambda) = \operatorname{ch} M(\lambda) - \operatorname{ch} M(-\lambda - 2), \quad \text{with } s_\alpha \cdot \lambda = -\lambda - 2.$$

But in general it turns out to be a deep problem to determine all coefficients occurring for arbitrary λ (even when $\lambda \in \Lambda$).

We know from Theorem 1.6 that the character of a finite dimensional module is actually *W-invariant*, but in general this fails for modules in $\mathcal{O}$. In particular, we cannot apply W to the character of a Verma module without leaving $\mathcal{O}$; so there is no easy way to exploit the W-invariance in the formula above. Instead a less direct way must be developed.

2.3. The Functions p and q

To investigate the formal character of a finite dimensional $L(\lambda)$, we work with two special functions in the ring $\mathcal{X}$ of formal characters.

Recall from 1.16 the Kostant function, defined by $p(\nu) =$ number of tuples $(c_\alpha)_{\alpha>0}$ with $c_\alpha \in \mathbb{Z}^+$ and $-\sum_{\alpha>0} c_\alpha \alpha = \nu$; so $p(\nu) = 0$ outside Λ_r.

There we proved:

$$p = \operatorname{ch} M(0) \text{ and } p * e(\lambda) = \operatorname{ch} M(\lambda).$$

Below we shall apply the convolution operation $*$ in $\mathcal{X}$ to longer products, writing just $\prod$ for convenience.

To express p in terms of the individual positive roots, we introduce further elements $f_\alpha \in \mathcal{X}$, $\alpha > 0$. By definition,

$$f_\alpha(\lambda) := \begin{cases} 1 \text{ if } \lambda = -k\alpha \text{ for some } k \in \mathbb{Z}^+, \\ 0 \text{ otherwise.} \end{cases}$$

Symbolically, $f_\alpha = e(0) + e(-\alpha) + e(-2\alpha) + \cdots$. It is clear that $f_\alpha \in \mathcal{X}$.

Lemma (A). *With the above notation,*

(a) $p = \prod_{\alpha>0} f_\alpha$.
(b) $\big(e(0) - e(-\alpha)\big) * f_\alpha = e(0)$.

Proof. (a) follows at once from the definition of convolution.

(b) Symbolically, $\big(e(0) - e(-\alpha)\big) * \big(e(0) + e(-\alpha) + e(-2\alpha) + \cdots\big) = e(0)$, which is clear after cancelling other terms in pairs. (A formal proof is easy.) $\qquad\square$

Besides the function p, we need a function q whose purpose is not immediately clear; it soon turns out to interact usefully with p. Define

$$q := \prod_{\alpha>0} \big(e(\alpha/2) - e(-\alpha/2)\big).$$

A moment's thought shows that this lies in $\mathcal{X}$. Note that q can be rewritten using the fact that $\rho = \sum_\alpha \alpha/2$, or $e(\rho) = \prod_{\alpha>0} e(\alpha/2)$:

$$q = \prod_{\alpha>0} e(\alpha/2) * \big(e(0) - e(-\alpha)\big) = e(\rho) * \prod_{\alpha>0} \big(e(0) - e(-\alpha)\big).$$

This makes it obvious that $q \neq 0$: for example, $q(\rho) = 1$.

Exercise. Rewrite q as $e(-\rho) * \prod_{\alpha>0}(e(\alpha) - 1)$, using the fact that $e(-\rho) = \prod_{\alpha>0} e(-\alpha/2)$.

Now the special feature of q is that it is an **alternating** function relative to the action of W:

Lemma (B). *For all $w \in W$, we have $wq = (-1)^{\ell(w)}q$.*

Proof. In case $\ell(w) = 0$, $w = 1$ and there is nothing to prove. When $\ell(w) = 1$, $w = s_\alpha$ for some $\alpha \in \Delta$. Recall (0.3) that s_α sends α to $-\alpha$ but keeps all other positive roots in Φ^+. This shows from the rewritten formula for q that $s_\alpha q = -q$. Now continue by induction on $\ell(w)$. $\qquad\square$

It is easy now to see how p and q interact.

Lemma (C). *For arbitrary* $\lambda \in \mathfrak{h}^*$,

$$q * \operatorname{ch} M(\lambda) = q * p * e(\lambda) = e(\lambda + \rho).$$

Proof. We just have to prove the second equality, which is equivalent to $q * p = e(\rho)$. Combining Lemmas (A) and (B), we compute

$$q * p = \prod_{\alpha > 0} \big(e(0) - e(-\alpha)\big) * e(\rho) * p = \prod_{\alpha > 0} \big(\big(e(0) - e(-\alpha)\big) * f_\alpha\big) * e(\rho) = e(\rho).$$

2.4. Formulas of Weyl and Kostant

With these lemmas in hand it is easy to complete the proof of Weyl's Character Formula, in tandem with Kostant's Weight Multiplicity Theorem:

Theorem (Weyl). *Let* $\lambda \in \Lambda^+$, *so* $\dim L(\lambda) < \infty$. *Then*

$$q * \operatorname{ch} L(\lambda) = \sum_{w \in W} (-1)^{\ell(w)} e\big(w(\lambda + \rho)\big),$$

where $q = \prod_{\alpha > 0} \big(e(\alpha/2) - e(-\alpha/2)\big)$. *In particular, when* $\lambda = 0$,

$$q = \sum_{w \in W} (-1)^{\ell(w)} e(w\rho).$$

Proof. This is just a matter of putting the pieces together, starting as in 2.2 with

$$\operatorname{ch} L(\lambda) = \sum_{w \in W} b(\lambda, w) \operatorname{ch} M(w \cdot \lambda), \quad \text{where } b(\lambda, w) \in \mathbb{Z} \text{ and } b(\lambda, 1) = 1.$$

Writing $\operatorname{ch} M(w \cdot \lambda)$ as $p * e(w \cdot \lambda)$ and multiplying both sides of the equation by q, Lemma 2.3(C) yields

$$q * \operatorname{ch} L(\lambda) = \sum_{w \in W} b(\lambda, w) e(w \cdot \lambda + \rho).$$

Here $b(\lambda, 1) = 1$. We just have to show that $b(\lambda, w) = (-1)^{\ell(w)}$. Recalling from Lemma 2.3(B) that q is an alternating function under W, whereas characters of finite dimensional modules are W-invariant, the application to both sides of a simple reflection $s_\alpha \in W$ changes the sign of the left side while replacing $w \cdot \lambda + \rho$ on the right by $s_\alpha w \cdot \lambda + \rho$. Now compare the coefficients to get $b(\lambda, s_\alpha w) = -b(\lambda, w)$ for all w. Induction on length implies that $b(\lambda, w) = (-1)^{\ell(w)}$ as required.

Since $\operatorname{ch} L(0) = e(0)$ (Example 1.6), the last statement of the theorem follows immediately (and could in fact be proved independently). $\square$

In the course of proving the character formula, we have also proved Kostant's formula for the multiplicity of an individual weight μ of $L(\lambda)$. This was originally derived in [**195**] using Weyl's formula but was then observed to be essentially equivalent to it.

Corollary (Kostant). *If $\lambda \in \Lambda^+$ and $\mu \leq \lambda$, then*

$$
\begin{aligned}
\dim L(\lambda)_\mu &= \sum_{w \in W} (-1)^{\ell(w)} p(\mu - w \cdot \lambda) \\
&= \sum_{w \in W} (-1)^{\ell(w)} p\big((\mu + \rho) - w(\lambda + \rho)\big).
\end{aligned}
$$

While this formulation is quite elegant, it is usually not suitable for computations due to the size of the Weyl group. From the computational point of view, storing all data about W typically consumes an enormous amount of space. In practice, computer methods based on the older recursive method of Freudenthal are usually implemented.

Exercise. Fix $\nu \in \Gamma$. Prove that for $n \gg 0$, $\dim L(n\rho)_{n\rho - \nu} = p(-\nu)$.

Remarks. (1) By now there are a number of quite different settings for Weyl's formula, ranging from compact Lie groups and Haar integrals (which Weyl himself used) to complex or algebraic geometry related to flag varieties for semisimple algebraic groups. A special advantage of the BGG approach presented here is that it generalizes well to algebraic situations having an infinite dimensional flavor, especially the theory of Kac–Moody Lie algebras and its offshoots. (Carter [**60**, 19.3], Jantzen [**153**], Kac [**165**, Chap. 10], Moody–Pianzola [**223**, 6.4].) See 13.5–13.6 below.

(2) The equality of the multiplicative and additive formulas for the Weyl function q goes by the name **Weyl Denominator Formula** due to its role in the character formula. Here the sum is taken over W and the product over Φ^+. In the Kac–Moody setting, where W is typically an infinite Coxeter group, formal identities of this type involving infinite products and sums turn out to recapture many classical identities involving special functions such as the Dedekind η-function, e.g., Jacobi's Triple Product Identity. Large families of new identities emerge from this kind of representation theory.

(3) To emphasize the dot action of W, Weyl's Character Formula could be restated in the form:

$$
\operatorname{ch} L(\lambda) = \frac{\sum_{w \in W} (-1)^{\ell(w)} e(w \cdot \lambda)}{\sum_{w \in W} (-1)^{\ell(w)} e(w \cdot 0)}.
$$

2.5. Dimension Formula

As remarked earlier, Weyl's formula yields a practical closed formula for $\dim L(\lambda)$ when $\lambda \in \Lambda^+$. Though its derivation is unrelated to category $\mathcal{O}$ methods, the formula itself is an essential byproduct of Weyl's theory.

The initial idea is straightforward: thinking of the character of $L(\lambda)$ as the finite sum $\sum_\mu \dim L(\lambda)_\mu e(\mu)$ in $\mathbb{Z}\Lambda$, add the coefficients to get $\dim L(\lambda)$. In the setting of $\mathcal{X}$, this translates into adding up the values of $\operatorname{ch} \lambda$. The copy of $\mathbb{Z}\Lambda$ in $\mathcal{X}$ is the subring $\mathcal{Y}$ generated by all $e(\mu)$ with $\mu \in \Lambda$. Here one has a well-defined homomorphism $v : \mathcal{Y} \to \mathbb{Z}$ assigning to f the sum of its values. So the problem is to compute $v(\operatorname{ch} L(\lambda))$. Unfortunately, Weyl's formula expresses this character as a quotient of two alternating sums over W, on each of which v clearly takes the value 0: that is, applying v to both sides of $q * \operatorname{ch} L(\lambda) = \sum_{w \in W} e\big(w(\lambda + \rho)\big)$ fails to give any information about the value of v on the character.

For example, when $\mathfrak{g} = \mathfrak{sl}(2, \mathbb{C})$ and Λ is identified with $\mathbb{Z}$, we know (0.9) that when $\lambda \geq 0$,

$$\operatorname{ch} L(\lambda) = e(\lambda) + e(\lambda - 2) + \cdots + e(-\lambda) = \frac{e(\lambda + 1) - e(\lambda - 1)}{e(1) - e(-1)}.$$

So $\dim L(\lambda) = \lambda + 1$, but applying v to the numerator or denominator of the fraction produces $1 - 1 = 0$.

This is roughly analogous to the problem of evaluating at $t = 1$ the quotient $(t^{\lambda+1} - t^{-\lambda-1})/(t - t^{-1})$ (or an equivalent rational function in t). In elementary calculus one can often transform such a problem by replacing both the numerator and the denominator by their derivatives. Here we adopt a similar formal strategy. But in rank ℓ the formulas involve ℓ variables, corresponding to coordinates of a weight relative to fundamental weights. So we have to use formal analogues of partial differentiation. The end result is an elegant formula, which only requires products over the positive roots (rather than sums taken over W):

Theorem (Weyl). *If $\lambda \in \Lambda^+$, then*

$$\dim L(\lambda) = \frac{\prod_{\alpha > 0}\langle \lambda + \rho, \alpha^\vee \rangle}{\prod_{\alpha > 0}\langle \rho, \alpha^\vee \rangle}.$$

Proof. We outline the formal steps, leaving routine details to the reader.

- For each $\alpha > 0$, define an operator $\partial_\alpha : \mathcal{Y} \to \mathcal{X}_0$ by extending linearly the assignment $\partial_\alpha e(\mu) = \langle \mu, \alpha^\vee \rangle e(\mu)$ (where $\mu \in \Lambda$). It is routine to check that ∂_α is a *derivation* of $\mathcal{Y}$. These operators commute for different roots, so we can form $\partial := \prod_{\alpha > 0} \partial_\alpha$ and view it as a differential operator (no longer a derivation in rank > 1).

- Recall from 2.3 (and the exercise there), along with Theorem 2.4, that the Weyl function q can be written as either a sum or a product:

$$q = e(-\rho) * \prod_{\alpha > 0} \big(e(\alpha) - 1\big) = \sum_{w \in W} (-1)^{\ell(w)} e(w\rho).$$

- Using first the product form of q, apply ∂ to $q * \mathrm{ch}\, L(\lambda)$, followed by the homomorphism v. While this looks extremely complicated to write out in full, it simplifies considerably because $v(e(\alpha) - 1) = 0$. All that survives is $v(\partial q)\, v(\mathrm{ch}\, L(\lambda))$.

- Now compute $v(\partial q)$ using $q = \sum_{w \in W}(-1)^{\ell(w)} e(w\rho)$. By linearity, we only need to determine $v\big(\partial e(w\rho)\big)$. Clearly $\partial_\alpha e(\rho) = \langle \rho, \alpha^\vee \rangle e(\rho)$, which yields at once $v\big(\partial e(\rho)\big) = \prod_{\alpha > 0}\langle \rho, \alpha^\vee \rangle$. The result is similar for $w\rho$ in place of ρ, but here we get

$$\prod_{\alpha > 0} \langle w\rho, \alpha^\vee \rangle = \prod_{\alpha > 0} \langle \rho, w^{-1}\alpha^\vee \rangle = \prod_{\alpha > 0} \langle \rho, (w^{-1}\alpha)^\vee \rangle.$$

- The number of positive roots sent to negative roots by w^{-1} is $\ell(w^{-1}) = \ell(w)$ (see 0.3). Since $\langle \rho, -\alpha^\vee \rangle = -\langle \rho, \alpha^\vee \rangle$ when $\alpha > 0$, we can rewrite the previous step to conclude:

$$v(\partial q) = \sum_{w \in W} (-1)^{\ell(w)} v\big(\partial e(w\rho)\big) = \sum_{w \in W} \prod_{\alpha > 0} \langle \rho, \alpha^\vee \rangle = |W| \prod_{\alpha > 0} \langle \rho, \alpha^\vee \rangle.$$

- The same method, applied to the right side of the character formula (where $\lambda + \rho$ rather than ρ occurs), yields

$$v\Big(\partial \sum_{w \in W} (-1)^{\ell(w)} e\big(w(\lambda + \rho)\big)\Big) = |W| \prod_{\alpha > 0} \langle \lambda + \rho, \alpha^\vee \rangle.$$

Now the theorem follows by cancelling the factors $|W|$. $\square$

Exercise. If $k \in \mathbb{Z}^{>0}$, exhibit a simple module $L(\lambda)$ of dimension $k^{|\Phi^+|}$.

2.6. Maximal Submodule of $M(\lambda)$, $\lambda \in \Lambda^+$

It is an elementary fact that a Verma module or other highest weight module has a unique maximal submodule (Theorem 1.2(f)); the resulting quotient is $L(\lambda)$. But it is not so elementary to describe this maximal submodule explicitly. In the special case $\lambda \in \Lambda^+$, the maximal submodule $N(\lambda)$ of $M(\lambda)$ does turn out to have a simple description.

The proof of the theorem below requires a formal identity in $U(\mathfrak{g})$, which is valid more generally in arbitrary associative algebras over fields of characteristic 0.

Lemma. *Let A be an associative algebra over a field of characteristic 0. Let $k \in \mathbb{Z}^{>0}$ and write as usual $\binom{k}{i}$ for a binomial coefficient. If $a, b \in A$, then*

$$[a^k, b] = \binom{k}{1}[a, b]a^{k-1} + \binom{k}{2}[a, [a, b]]a^{k-2} + \cdots + [a, [a, \ldots, [a, b]\ldots].$$

Proof. In case $k = 1$, this is just the identity $[a, b] = [a, b]$. Then proceed by induction on k. $\qquad\square$

Theorem. *Enumerate the simple roots as $(\alpha_1, \ldots, \alpha_\ell)$, abbreviating $s_i := s_{\alpha_i}$. If $\lambda \in \Lambda^+$, the maximal submodule $N(\lambda)$ of $M(\lambda)$ is the sum of the submodules $M(s_i \cdot \lambda)$ for $1 \leq i \leq \ell$.*

Proof. In 1.3, we saw that $M(\lambda) \cong U(\mathfrak{g})/I$, where I is the left ideal generated by $\mathfrak{n}$ along with all $h - \lambda(h) \cdot 1$ ($h \in \mathfrak{h}$). Here I is precisely the annihilator in $U(\mathfrak{g})$ of a maximal vector v^+ of $M(\lambda)$.

In our situation $n_i := \langle \lambda, \alpha_i^\vee \rangle \in \mathbb{Z}^+$. Consider the left ideal J of $U(\mathfrak{g})$ generated by I together with all $y_i^{n_i+1}$ ($1 \leq i \leq \ell$). Thanks to Corollary 1.4, J annihilates a maximal vector of $L(\lambda)$. We claim it is the precise annihilator.

Indeed, set $M := U(\mathfrak{g})/J$, which has $L(\lambda)$ as a quotient. If we can show M is *finite dimensional*, our claim will follow from complete reducibility: a direct sum of more than one simple finite dimensional module clearly cannot be a highest weight module. As observed in Exercise 1.6, it will be enough to show that all y_i act locally nilpotently on M (which the x_i obviously do). In this situation, steps in the proof of Theorem 1.6 will imply the finite dimensionality based on the fact that M is a sum of finite dimensional $\mathfrak{s}_i$-submodules for each fixed i.

By construction a maximal vector (coset of 1) in M is killed by $y_i^{n_i+1}$ for each i. On the other hand, M is spanned by the cosets of all possible monomials $(*)\ y_{i_1} \cdots y_{i_t}$ with $i_j \in \{1, \ldots, \ell\}$. Now apply the above lemma, with $A = U(\mathfrak{g})$ and a, b root vectors y, z belonging to two negative roots. Since root strings have length at most 4 (0.2), $(\operatorname{ad} y)^4(z) = 0$. So if the coset of a monomial $(*)$ is killed by y_i^k, the identity in the lemma shows that the coset of the longer monomial $y_{i_0} y_{i_1} \cdots y_{i_t}$ is killed by y_i^{k+3}. Induction on t (starting with $t = 0$) implies the local nilpotence of the action of y_i on M.

To translate the conclusion that $U(\mathfrak{g})/J \cong L(\lambda)$ into the assertion made in the theorem, recall from 1.4 that the submodule $M(s_i \cdot \lambda)$ of $M(\lambda)$ is generated by $y_i^{n_i+1} \cdot v^+$. So the left ideal in $U(\mathfrak{g})$ annihilating the sum N of these submodules is precisely J. This forces $N = N(\lambda)$. $\qquad\square$

The theorem may be restated in a suggestive way. Recalling that the elements of length 1 in W are precisely the simple reflections, we see that the following sequence is exact whenever $\lambda \in \Lambda^+$:

$$\bigoplus_{w \in W,\, \ell(w)=1} M(w \cdot \lambda) \to M(\lambda) \to L(\lambda) \to 0.$$

The alternating sum format derived for $\operatorname{ch} L(\lambda)$ in terms of characters of Verma modules resembles an *Euler characteristic*, so it is tempting to extend this sequence to the left. The kth term in the sequence would be of the form

$$\bigoplus_{w \in W,\, \ell(w)=k} M(w \cdot \lambda),$$

ending with $\ell(w_\circ) = |\Phi^+|$, where $w_\circ$ denotes the longest element of W. Somewhat miraculously, this kind of resolution of $L(\lambda)$ does exist. But to develop it will require further study of maps between Verma modules in Chapter 4. The *BGG resolution* itself will then be taken up in Chapter 6.

2.7. Related Topics

There is a rich literature involving finite dimensional representations of $\mathfrak{g}$ and their applications in both mathematics and physics. Most of this is technically unrelated to category $\mathcal{O}$, but the reader should be aware of some of the possible directions for further study:

- When $\lambda, \mu \in \Lambda^+$, the decomposition into simple summands $L(\nu)$ of $L(\lambda) \otimes L(\mu)$ has been looked at from many angles. The multiplicities with which these summands occur can be computed in a variety of ways, especially in type A_ℓ or other classical types: Littlewood–Richardson Rule, etc. They turn out to be closely related to a number of apparently remote questions in algebraic geometry and combinatorics.

- The restriction of a simple module to a semisimple (or reductive) subalgebra of $\mathfrak{g}$ is a natural problem (leading to *branching rules*), which can often be studied together with the tensor product decomposition problem.

- Fundamental representations $L(\varpi_1), \ldots, L(\varpi_\ell)$ can be constructed explicitly for classical types, by exploiting the exterior powers of the "natural" representation of $\mathfrak{g}$ as a classical Lie algebra.

- In the study of multiplicities in $L(\lambda) \otimes L(\mu)$, a special role is played by the dominant weights which are W-conjugate to the weights $\lambda + w\mu$ (with $w \in W$): which of these occur as highest weights of summands, and then with what multiplicity? The work of Kumar and others on the PRV (Parasarathy, Ranga Rao, Varadarajan) Conjecture is a high point.

- In type A_ℓ (and other classical types), representations and multiplicities have been explored extensively using combinatorial methods, as just indicated for the tensor product problem. There are deep connections with representations of symmetric groups, classical invariant theory, and the like, going back to Schur and Weyl.

- While the algorithms for explicit computation of weight multiplicities $\dim L(\lambda)_\mu$ are lengthy in practice, there is a reasonable theoretical picture of the weight diagram of $L(\lambda)$ as a whole. Starting with the *extremal weights* $w\lambda$ ($w \in W$), take all other $\mu \in \Lambda^+$ satisfying $\mu \leq \lambda$, together with their W-conjugates. These are precisely the weights occurring with positive multiplicity. Equivalently, the weight diagram can be characterized as consisting of all points $\mu \in \Lambda$ which satisfy $\mu \leq \lambda$ and lie in the *convex hull* of the extremal weights.

- Taking $\mathfrak{b}^- = \mathfrak{h} \oplus \mathfrak{n}^-$, the $U(\mathfrak{b}^-)$-submodule of $L(\lambda)$ generated by one of its extremal weight vectors is called a *Demazure module*. Its formal character has been explicitly determined, following Demazure's conjecture; this module turns out to be intimately related to a Schubert variety for an associated semisimple group.

- Work of Lusztig and Kashiwara involving quantized enveloping algebras has revealed unexpectedly the existence of *canonical* bases for the finite dimensional modules $L(\lambda)$.

Exercise. Prove that every finite dimensional simple $U(\mathfrak{g})$-module occurs as a direct summand in a suitable tensor product of copies of the fundamental modules $L(\varpi_1), \ldots, L(\varpi_\ell)$.

Notes

The proof of Weyl's Character Formula given here follows the 1971 paper by BGG [25], who were inspired in part by Verma's 1966 thesis [251]. Previous algebraic treatments (such as Freudenthal's proof in [143]) were much less conceptual, relying on algebraic analogues of older analytic methods. A number of texts have followed the BGG argument, e.g., Bourbaki [46, VIII, §9], Carter [60, Chap. 12], Humphreys [125, §24].

Weyl's Dimension Formula (2.5) is derived by a standard algebraic variant of the analytic proof based on the expression of characters in terms of actual exponentials.

There are countless research papers and books treating aspects of finite dimensional representation theory for semisimple Lie algebras. For example, Fulton–Harris [97] emphasize the down-to-earth combinatorics involved in working out weight multiplicities and realizations of special representations.

Chapter 3

Category $\mathcal{O}$: Methods

Here we introduce some standard tools which will be applied throughout the following chapters:

- The functors Hom and Ext (3.1).
- A duality functor $M \mapsto M^\vee$ on $\mathcal{O}$ (3.2–3.3).
- Reflection groups $W_{[\lambda]}$ (3.4), dominant and antidominant weights (3.5).
- Tensoring Verma modules with finite dimensional modules (3.6).
- "Standard" filtrations having Verma modules as subquotients (3.7).
- Projective objects in $\mathcal{O}$ and BGG Reciprocity (3.8–3.13).
- "Contravariant" forms on modules (3.14–3.15).

3.1. Hom and Ext

A basic tool in the study of various objects in $\mathcal{O}$ such as Verma modules is the Hom functor. We already saw (Theorem 1.3) that $\dim \mathrm{Hom}_{\mathcal{O}}\left(L(\mu), L(\lambda)\right) = \delta_{\lambda\mu}$. At first glance it is far from obvious which computations are easy and which are not, for example the computation of $\mathrm{Hom}_{\mathcal{O}}\left(M(\mu), M(\lambda)\right)$. As the list of interesting objects grows, this sort of question will always be in the background.

It is well known that Hom is left exact but often fails to be right exact. This leads to right derived functors Ext^n, which will be studied in greater generality in Chapter 6. For the moment we limit ourselves to the use of $\mathrm{Ext}^0 = \mathrm{Hom}$ and Ext^1 (often just written Ext).

For arbitrary $U(\mathfrak{g})$-modules A, C, the vector space $\mathrm{Ext}_{U(\mathfrak{g})}(C, A)$ classifies up to a natural notion of equivalence the possible short exact sequences

$$(*) \qquad\qquad 0 \to A \to B \to C \to 0$$

of $U(\mathfrak{g})$-modules. The zero element corresponds to the class of split exact sequences. In case $A, C \in \mathcal{O}$, there is no need here for B to lie in $\mathcal{O}$. For example, there exist nonsplit extensions when $A = C$ is a Verma module; as shown below, these cannot be realized in $\mathcal{O}$. So it is essential to write $\mathrm{Ext}_{\mathcal{O}}$ when working in this category.

Exercise. Let $\mathfrak{g} = \mathfrak{sl}(2, \mathbb{C})$, and identify $\lambda \in \mathfrak{h}^*$ with a scalar as usual. Let N be a 2-dimensional $U(\mathfrak{b})$-module defined by letting x act as 0 and h act as $\begin{pmatrix} \lambda & 1 \\ 0 & \lambda \end{pmatrix}$. Show that the induced $U(\mathfrak{g})$-module $M := U(\mathfrak{g}) \otimes_{U(\mathfrak{b})} N$ fits into an exact sequence which fails to split:

$$0 \to M(\lambda) \to M \to M(\lambda) \to 0.$$

(Here M cannot lie in $\mathcal{O}$, thanks to part (a) of the proposition below.)

The most basic general tool available for the study of the derived functors Ext^n of Hom in a module category is the *long exact sequence* associated to a short exact sequence. This comes in two flavors, since Hom is contravariant when the second variable is fixed but covariant when the first variable is fixed. For example, taking D as the second variable, we get from $(*)$ above

$$0 \to \mathrm{Hom}(C, D) \to \mathrm{Hom}(B, D) \to \mathrm{Hom}(A, D)$$

$$\to \mathrm{Ext}^1(C, D) \to \mathrm{Ext}^1(B, D) \to \mathrm{Ext}^1(A, D) \to \cdots .$$

The sequence continues with Ext^2, etc.

In this chapter a number of specific computations will appear, but mainly of the easier sort. For example:

Proposition. *Let $\lambda, \mu \in \mathfrak{h}^*$.*

(a) *If M is a highest weight module of weight μ, with $\lambda \not\leq \mu$, then $\mathrm{Ext}_{\mathcal{O}}(M(\lambda), M) = 0$. In particular,*

$$\mathrm{Ext}_{\mathcal{O}}\big(M(\lambda), L(\lambda)\big) = 0 = \mathrm{Ext}_{\mathcal{O}}\big(M(\lambda), M(\lambda)\big).$$

(b) *If $\mu \leq \lambda$, then $\mathrm{Ext}_{\mathcal{O}}\big(M(\lambda), L(\mu)\big) = 0$.*

(c) *If $\mu < \lambda$ and $N(\lambda)$ is the maximal submodule of $M(\lambda)$, then*

$$\mathrm{Hom}_{\mathcal{O}}\big(N(\lambda), L(\mu)\big) \cong \mathrm{Ext}_{\mathcal{O}}\big(L(\lambda), L(\mu)\big).$$

(d) $\mathrm{Ext}_{\mathcal{O}}\big(L(\lambda), L(\lambda)\big) = 0.$

Proof. (a) Given an extension $0 \to M \to E \to M(\lambda) \to 0$ in $\mathcal{O}$, the hypothesis guarantees that any preimage in E_λ of a maximal vector of weight λ in $M(\lambda)$ is also a maximal vector. So it generates a submodule mapping isomorphically onto $M(\lambda)$, which splits the sequence.

(b) follows immediately from (a).

(c) Starting with the exact sequence $0 \to N(\lambda) \to M(\lambda) \to L(\lambda) \to 0$, one gets as part of a long exact sequence:

$$\cdots \to \mathrm{Hom}_{\mathcal{O}}\big(M(\lambda), L(\mu)\big) \to \mathrm{Hom}_{\mathcal{O}}\big(N(\lambda), L(\mu)\big)$$

$$\to \mathrm{Ext}_{\mathcal{O}}\big(L(\lambda), L(\mu)\big) \to \mathrm{Ext}_{\mathcal{O}}\big(M(\lambda), L(\mu)\big) \to \cdots .$$

Since $\mu < \lambda$, the first term vanishes because $L(\lambda)$ is the unique simple quotient of $M(\lambda)$ and the last term vanishes by part (b).

(d) Replace μ by λ in the proof of (c). Using (a) and the fact that $L(\lambda)$ is not a composition factor of $N(\lambda)$, the second and fourth terms of the resulting sequence vanish. $\qquad\square$

3.2. Duality in $\mathcal{O}$

There is a useful duality functor on $\mathcal{O}$, obtained by modifying the usual notion of dual module for $U(\mathfrak{g})$ (which arises naturally from the Hopf algebra structure of $U(\mathfrak{g})$). Recall that for any $U(\mathfrak{g})$-module M, the dual vector space M^* becomes a $U(\mathfrak{g})$-module if we define

$$(x \cdot f)(v) := -f(x \cdot v) \text{ for all } v \in M,\ f \in M^*,\ x \in \mathfrak{g}.$$

But if M lies in $\mathcal{O}$, it will usually not be the case that M^* also does. For one thing, M^* will be "too large" whenever $\dim M = \infty$; so it will be necessary to cut down to a suitable subspace in order to get a finitely generated module. Moreover, the standard action of $U(\mathfrak{g})$ creates intrinsic problems and must therefore be modified.

Twisting the action of $\mathfrak{g}$ improves the situation. Recall from 0.5 the transpose map τ on $\mathfrak{g}$, which fixes $\mathfrak{h}$ pointwise but sends $\mathfrak{g}_\alpha$ to $\mathfrak{g}_{-\alpha}$ for all roots α. A new action of $\mathfrak{g}$ on M^* is defined by $(x \cdot f)(v) := f(\tau(x) \cdot v)$. We always work with this action on dual vector spaces in what follows. In particular, $(h \cdot f)(v) = f(h \cdot v)$ for all $h \in \mathfrak{h}, f \in M^*, v \in M$ (where M can be any $U(\mathfrak{g})$-module).

To get started, consider the category $\mathcal{C}$ of $U(\mathfrak{g})$-modules which are weight modules with finite dimensional weight spaces. (This includes $\mathcal{O}$.) If $M \in \mathcal{C}$ and $\lambda \in \mathfrak{h}^*$, the dual space $(M_\lambda)^*$ of M_λ can be identified naturally with the set of those $f \in M^*$ which vanish on all summands M_μ of M for which $\mu \neq \lambda$. On the other hand, any $f \in M^*$ which has weight λ (under the twisted action of $\mathfrak{h}$) vanishes at all $v \in M_\mu$: whenever $\lambda(h) \neq \mu(h)$ we get

$$\mu(h)f(v) = f(\mu(h)v) = (h \cdot f)(v) = \lambda(h)f(v).$$

Thus $f \in (M_\lambda)^*$, identifying $(M^*)_\lambda$ with $(M_\lambda)^*$. This allows us to drop the parentheses. Now we can define

$$M^\vee := \bigoplus_{\lambda \in \mathfrak{h}^*} M_\lambda^* \text{ for any } M \in \mathcal{C}$$

and call this the **dual** in $\mathcal{C}$.

Since the notion of formal character makes sense for $\mathcal{C}$, it follows from this discussion that $\operatorname{ch} M^\vee = \operatorname{ch} M$. Moreover, $M \mapsto M^\vee$ defines an exact contravariant functor on $\mathcal{C}$: an exact sequence $0 \to L \to M \to N \to 0$ in $\mathcal{C}$ yields an exact sequence $0 \to N^\vee \to M^\vee \to L^\vee \to 0$. This implies that M is simple if $M^\vee$ is simple. It is also obvious that $(M \oplus N)^\vee \cong M^\vee \oplus N^\vee$. Finally, the functor sending $M \mapsto M^{\vee\vee}$ is naturally isomorphic to the identity functor on $\mathcal{C}$.

It still has to be checked that $\mathcal{O}$ is closed under duality: when $M \in \mathcal{O}$, is $M^\vee$ locally finite for $\mathfrak{n}$ as well as finitely generated? If $M \in \mathcal{O}$ and $f \in M_\mu^\vee$ with $\mu \in \mathfrak{h}^*$, the twisted action of $\mathfrak{g}$ on $M^\vee$ ensures that $x_\alpha \cdot f \in M_{\mu+\alpha}^\vee$. This implies that the action of $\mathfrak{n}$ on M is locally finite. Observe next that each $L(\lambda)^\vee$ is simple because its dual $L(\lambda)^{\vee\vee} \cong L(\lambda)$ is simple; its highest weight is λ because $\operatorname{ch} L(\lambda)^\vee = \operatorname{ch} L(\lambda)$. Therefore $L(\lambda)^\vee$ is isomorphic to $L(\lambda)$. Since M has finite length, we conclude from the exactness of duality that $M^\vee$ has finite length (with the same composition factors) and is therefore finitely generated. So $M^\vee \in \mathcal{O}$.

How does duality interact with the subcategories $\mathcal{O}_\chi$? Recall from Exercise 1.10 that τ fixes $Z(\mathfrak{g})$ pointwise: τ commutes with the Harish-Chandra homomorphism ξ (which is injective), while $U(\mathfrak{h})$ is commutative. If $M \in \mathcal{O}_\chi$, it follows from the definition of $M^\vee$ that $Z(\mathfrak{g})$ also acts on $M^\vee$ by the generalized character χ; in particular, $z \in Z(\mathfrak{g})$ acts on any maximal vector as scalar multiplication by $\chi(z)$.

To summarize:

Theorem. *The duality functor on $\mathcal{O}$ sending M to $M^\vee$ is exact and contravariant, with the properties:*

(a) *Duality induces a self-equivalence on category $\mathcal{O}$: its square is naturally isomorphic to the identity functor.*

(b) *For any $M \in \mathcal{O}$ and any central character χ, $(M^\vee)^\chi \cong (M^\chi)^\vee$; in particular, if $M \in \mathcal{O}_\chi$, then $M^\vee \in \mathcal{O}_\chi$.*

(c) *If $M \in \mathcal{O}$, then $\operatorname{ch} M = \operatorname{ch} M^\vee$; so M and $M^\vee$ have the same composition factor multiplicities and therefore define the same element of $K(\mathcal{O})$. In particular, $L(\lambda)^\vee \cong L(\lambda)$.*

(d) *If $M, N \in \mathcal{O}$, then $(M \oplus N)^\vee \cong M^\vee \oplus N^\vee$; so if M is indecomposable, $M^\vee$ is also indecomposable.*

(e) $\mathrm{Ext}_\mathcal{O}(M, N) \cong \mathrm{Ext}_\mathcal{O}(N^\vee, M^\vee)$ *for all $M, N \in \mathcal{O}$. In particular,* $\mathrm{Ext}_\mathcal{O}\big(L(\lambda), L(\mu)\big) \cong \mathrm{Ext}_\mathcal{O}\big(L(\mu), L(\lambda)\big).$ $\qquad\square$

Exercise. Show that $(L \otimes M)^\vee \cong L^\vee \otimes M^\vee$ if $M \in \mathcal{O}$ and $\dim L < \infty$.

3.3. Duals of Highest Weight Modules

It is easy to apply the duality functor in $\mathcal{O}$ to simple modules and Verma modules:

Theorem. *Let $\lambda, \mu \in \mathfrak{h}^*$.*

(a) $L(\lambda)^\vee \cong L(\lambda)$.

(b) $M(\lambda)^\vee$ *has $L(\lambda)$ as its unique simple submodule. Its other composition factors $L(\mu)$ satisfy $\mu < \lambda$.*

(c) *Any nonzero homomorphism $M(\mu) \to M(\lambda)^\vee$ has the simple submodule $L(\lambda)$ as its image. Thus $\dim \mathrm{Hom}_\mathcal{O}\big(M(\lambda), M(\lambda)^\vee\big) = 1$, while $\mathrm{Hom}_\mathcal{O}\big(M(\mu), M(\lambda)^\vee\big) = 0$ when $\mu \neq \lambda$.*

(d) $\mathrm{Ext}_\mathcal{O}\big(M(\mu), M(\lambda)^\vee\big) = 0$ *for all λ, μ.*

Proof. (a) This was already observed in 3.2.

(b) Applying duality to the short exact sequence

$$0 \to N(\lambda) \to M(\lambda) \to L(\lambda) \to 0$$

produces an exact sequence

$$0 \to L(\lambda)^\vee \to M(\lambda)^\vee \to N(\lambda)^\vee \to 0.$$

Thanks to (a), $L(\lambda)^\vee \cong L(\lambda)$ occurs as a simple submodule. No other simple submodule can occur in $M(\lambda)^\vee$, since its dual would have to be a simple quotient of $M(\lambda)$. Since $M(\lambda)^\vee$ has the same formal character as $M(\lambda)$, hence the same composition factor multiplicities, (b) follows.

(c) The image M in $M(\lambda)^\vee$ of a nonzero homomorphism $M(\mu) \to M(\lambda)^\vee$ is a highest weight module of weight μ; by (b), it has a submodule isomorphic to $L(\lambda)^\vee \cong L(\lambda)$. Therefore $\mu \geq \lambda$. On the other hand, (b) implies that $\mu \leq \lambda$. Thus $\mu = \lambda$. In this case, the universal property of $M(\lambda)$ provides a nonzero homomorphism $M(\lambda) \to L(\lambda) \hookrightarrow M(\lambda)^\vee$. Since $\dim M(\lambda)^\vee_\lambda = 1$, Exercise 1.3 forces $\dim \mathrm{Hom}_\mathcal{O}\big(M(\lambda), M(\lambda)^\vee\big) = 1$.

(d) We have to show that any short exact sequence

$$0 \to M(\lambda)^\vee \to M \to M(\mu) \to 0$$

splits. If μ happens to be a maximal weight of M, the pre-image of a maximal vector of weight μ in $M(\mu)$ will be a maximal vector in M; then the universal property of Verma modules will ensure a splitting. Otherwise we must have $\mu < \lambda$. But then we can dualize and use the fact that $M(\lambda)^{\vee\vee} \cong M(\lambda)$ to get another such short exact sequence

$$0 \to M(\mu)^{\vee} \to M^{\vee} \to M(\lambda) \to 0.$$

Since $\lambda > \mu$, this sequence splits; as a result, the original sequence splits. $\square$

Exercise. Does part (d) of the theorem remain true if $M(\mu)$ and $M(\lambda)^{\vee}$ are interchanged?

3.4. The Reflection Group $W_{[\lambda]}$

What do we know at this point about the possible composition factors $L(\mu)$ of a Verma module $M(\lambda)$? The most obvious condition is (1) $\mu \leq \lambda$; in particular, μ must lie in the same coset of $\mathfrak{h}^*/\Lambda_r$ as λ. From 1.7 we know that in addition, (2) $\mu = w \cdot \lambda$ for some $w \in W$. These two conditions narrow the choices for μ considerably, but turn out to be far from optimal. This will become clear in 5.1.

To focus the problem better when $\lambda \in \mathfrak{h}^*$ is not assumed to be *integral*, Jantzen [**147**, 1.3] introduced for each λ an associated subgroup of W which can replace W in (2). Since $s_\alpha\lambda = \lambda - \langle\lambda, \alpha^\vee\rangle\alpha$, the scalar $\langle\lambda, \alpha^\vee\rangle$ plays a crucial role here. Note that it lies in $\mathbb{Z}$ if and only if $\langle\lambda + \rho, \alpha^\vee\rangle$ does.

Start by defining

$$\Phi_{[\lambda]} := \{\alpha \in \Phi \mid \langle\lambda, \alpha^\vee\rangle \in \mathbb{Z}\}$$

and

$$W_{[\lambda]} := \{w \in W \mid w\lambda - \lambda \in \Lambda_r\}.$$

The notation $[\lambda]$ is motivated by the fact that

$$\Phi_{[\lambda]} = \Phi_{[\mu]} \text{ and } W_{[\lambda]} = W_{[\mu]} \text{ whenever } \lambda \equiv \mu \pmod{\Lambda}.$$

Since $\rho \in \Lambda$, one could equally well write

$$\Phi_{[\lambda]} = \{\alpha \in \Phi \mid \langle\lambda + \rho, \alpha^\vee\rangle \in \mathbb{Z}\}, \ W_{[\lambda]} = \{w \in W \mid w \cdot \lambda - \lambda \in \Lambda_r\}.$$

It follows that $\Phi_{[\lambda]} = \Phi_{[\mu]}$ and $W_{[\lambda]} = W_{[\mu]}$ whenever $\mu \in W_{[\lambda]} \cdot \lambda$.

It is clear that $\Lambda = \{\lambda \in \mathfrak{h}^* \mid \Phi_{[\lambda]} = \Phi\}$. Moreover,

$$\Phi_{[\lambda]} = \{\alpha \in \Phi \mid s_\alpha \in W_{[\lambda]}\},$$

thanks to the fact that the only scalar multiples of α in Λ_r are $\mathbb{Z}\alpha$ (a fact which is obvious for simple roots and hence for their W-conjugates).

In terms of this subgroup of W we can refine the above necessary conditions for $L(\mu)$ to occur as a composition factor of $M(\lambda)$:

(3) $\qquad\qquad \mu \leq \lambda \text{ and } \mu = w \cdot \lambda \text{ for some } w \in W_{[\lambda]}.$

In fact $W_{[\lambda]}$ also plays the role of a Weyl group, acting on a subspace of $E = \mathbb{R} \otimes_{\mathbb{Z}} \Lambda$:

Theorem. *Let $\lambda \in \mathfrak{h}^*$. Then:*

    (a) $\Phi_{[\lambda]}$ *is a root system in its $\mathbb{R}$-span $E(\lambda) \subset E$.*

    (b) $W_{[\lambda]}$ *is the Weyl group of the root system $\Phi_{[\lambda]}$. In particular, it is generated by the reflections s_α with $\alpha \in \Phi_{[\lambda]}$.*

For the proof we need some standard facts about root systems and reflection groups:

(A) Associated to W is the **affine Weyl group** W_a, which acts in the affine space A associated with E as a group generated by affine reflections. Abstractly W_a is the semidirect product $W \ltimes \Lambda_r$. (See Bourbaki [**45**, VI, §2], Humphreys [**129**, Chap. 3].) Here W acts naturally on Λ_r, regarded as an additive group of translations in E. It is a Coxeter group, generated by the *affine reflections*

$$s_{\alpha,n} : \mu \mapsto s_\alpha \mu + n\alpha \text{ with } \alpha \in \Phi,\ n \in \mathbb{Z},\ \mu \in E.$$

Now if $\lambda \in E$, its isotropy group in W is itself the Weyl group of the root system spanned by all $\alpha \in \Phi$ orthogonal to λ. In particular, it is a reflection group. Similarly, the isotropy group in W_a of a point in A is generated by the reflections $s_{\alpha,n}$ it contains. (See [**45**, V, §3, Prop. 1] or [**129**, 1.12, 4.8].)

(B) If $\Phi_1 \subset \Phi$ is a root system in the subspace E_1 of E which it spans, while $\Phi \cap E_1 = \Phi_1$, then there is a basis Δ of Φ for which $\Delta \cap \Phi_1$ is a basis of Φ_1. (See [**45**, VI, §1, Prop. 24].)

Remark. The basic idea in the proof of the theorem is suggested by Exercise 1 in Bourbaki [**45**, VI, §2], where the setting is the real euclidean space E. But Jantzen's proof in [**147**, 1.3] requires additional technical steps to pass from $\mathbb{R}$ to an arbitrary splitting field for $\mathfrak{g}$. We present a simplified version of his proof when the ground field is $\mathbb{C}$; here we may write $\mathfrak{h}^* = E + iE$ with $i^2 = -1$ in $\mathbb{C}$. (For another version see Bernstein–Gelfand [**24**, App. 1].)

Proof of Theorem. For (a), the only nonobvious point to check is that $s_\alpha \Phi_{[\lambda]} = \Phi_{[\lambda]}$ for all $\alpha \in \Phi_{[\lambda]}$. Let $\beta \in \Phi_{[\lambda]}$ and note that $(s_\alpha \beta)^\vee = s_\alpha \beta^\vee$ (since β and $s_\alpha \beta$ have the same length). Then

$$\langle \lambda, (s_\alpha \beta)^\vee \rangle = \langle \lambda, s_\alpha \beta^\vee \rangle = \langle s_\alpha \lambda, \beta^\vee \rangle = \langle \lambda, \beta^\vee \rangle - \langle \lambda, \alpha^\vee \rangle \langle \alpha, \beta^\vee \rangle \in \mathbb{Z},$$

since both α and β are assumed to lie in $\Phi_{[\lambda]}$.

The proof of (b) is more subtle. The basic trick is to interpret the defining condition for $W_{[\lambda]}$ as requiring that λ be fixed by a subgroup of the affine Weyl group W_a. Then (A) above applies. This can only be done directly in case $\lambda \in E(\lambda)$, the $\mathbb{R}$-span of $\Phi_{[\lambda]}$ in E. Now the condition that

$s_\alpha \in \Phi_{[\lambda]}$, or $s_\alpha\lambda - \lambda \in \Lambda_r$, translates into the condition that $s_{\alpha,n}\lambda = \lambda$ for some integer n. From (A) we conclude that the isotropy group of λ in W_a is generated by precisely these reflections. The natural projection of W_a onto W maps this isotropy group onto $W_{[\lambda]}$, which is therefore the reflection subgroup of W generated by the s_α with $\alpha \in \Phi_{[\lambda]}$.

It remains to consider the general case, with $\lambda \in \mathfrak{h}^*$ arbitrary. As remarked above, we assume that $\mathfrak{h}^* = E \oplus iE$. Write λ uniquely in the form $\lambda = \lambda_0 + i\lambda_1$, where $\lambda_k \in E$. Then $w\lambda - \lambda = (w\lambda_0 - \lambda_0) + i(w\lambda_1 - \lambda_1)$, with each expression in parentheses still lying in E. It follows that $w \in W_{[\lambda]}$ if and only if $w\lambda_0 - \lambda_0 \in \Lambda_r$ (forcing $\lambda_0 \in E(\lambda)$) and $w\lambda_1 = \lambda_1$. Denote by W_1 the isotropy group of λ_1 in W. Thanks to (A) above, this is itself the Weyl group of the root system Φ_1 (in the subspace of E it spans) consisting of all $\alpha \in \Phi$ orthogonal to λ_1. Thus $W_{[\lambda]} = W_{[\lambda_0]} \cap W_1$.

Any root in the $\mathbb{R}$-span E_1 of Φ_1 is orthogonal to λ_1 and hence lies in Φ_1. From (B) above it follows at once that $E_1 \cap \Lambda_r$ is precisely the root lattice of Φ_1. Thus $W_{[\lambda]}$ consists of those $w \in W_1$ for which $w\lambda - \lambda$ lies in the root lattice associated with W_1. Now we can apply to W_1 and λ_0 the earlier argument in the real case. $\qquad\square$

When dealing with nonintegral weights, these reflection subgroups $W_{[\lambda]}$ enable us to refine the study of a single linkage class, by partitioning it among various cosets of Λ_r in $\mathfrak{h}^*$. Each such subset is itself the linkage class of a weight λ under the reflection group $W_{[\lambda]}$. Eventually we will show in 4.9 that the subcategories corresponding to these orbits $W_{[\lambda]} \cdot \lambda$ are precisely the *blocks* of $\mathcal{O}$ in the sense of 1.13.

Exercise. Show that $\Phi_{[\lambda]} \cap \Phi^+$ is a positive system in the root system $\Phi_{[\lambda]}$; but the corresponding simple system (call it $\Delta_{[\lambda]}$) may be unrelated to Δ. For a concrete example, take Φ to be of type B_2 with short simple root α and long simple root β. If $\lambda := \alpha/2$, check that $\Phi_{[\lambda]}$ contains just the four short roots in Φ.

3.5. Dominant and Antidominant Weights

When $\lambda \in \Lambda$, we have $W_{[\lambda]} = W$. In that case, the linkage class of λ can be parametrized by any weight in the class: for example, a weight in $\Lambda^+ - \rho$. In the case of a nonintegral weight λ, it is not so obvious at first sight how to select an optimal representative of the orbit $W_{[\lambda]} \cdot \lambda$. It turns out that the *lowest* weight in the orbit is most convenient in the further study of Verma modules, while for other purposes the *highest* weight may be preferred.

Call $\lambda \in \mathfrak{h}^*$ **antidominant** if $\langle \lambda + \rho, \alpha^\vee \rangle \notin \mathbb{Z}^{>0}$ for all $\alpha \in \Phi^+$. For example, $-\rho$ is antidominant. Similarly, call λ **dominant** if $\langle \lambda + \rho, \alpha^\vee \rangle \notin \mathbb{Z}^{<0}$ for all $\alpha \in \Phi^+$. *Warning:* This differs from the usual notion of dominance in

Λ: now the set of "dominant" weights in Λ becomes $\Lambda^+ - \rho$, with $-\rho$ being both dominant and antidominant. To avoid confusion we usually emphasize just the notion of antidominance for $\mathfrak{h}^*$, the other case being symmetric.

Notice that any linkage class of W contains at least one antidominant weight: If a weight μ in the class is minimal relative to the standard partial ordering, but some $\langle \mu + \rho, \alpha^\vee \rangle \in \mathbb{Z}^{>0}$, then the linked weight $s_\alpha \cdot \mu < \mu$ contradicts the minimality. Similarly, for any $\lambda \in \mathfrak{h}^*$, each dot-orbit of $W_{[\lambda]}$ contains at least one antidominant weight.

Proposition. *Let $\lambda \in \mathfrak{h}^*$, with corresponding root system $\Phi_{[\lambda]}$ and Weyl group $W_{[\lambda]}$. Let $\Delta_{[\lambda]}$ be the simple system corresponding to the positive system $\Phi_{[\lambda]} \cap \Phi^+$ in $\Phi_{[\lambda]}$. Then λ is antidominant if and only if one of the following three equivalent conditions holds:*

(a) $\langle \lambda + \rho, \alpha^\vee \rangle \leq 0$ for all $\alpha \in \Delta_{[\lambda]}$.

(b) $\lambda \leq s_\alpha \cdot \lambda$ for all $\alpha \in \Delta_{[\lambda]}$.

(c) $\lambda \leq w \cdot \lambda$ for all $w \in W_{[\lambda]}$.

Therefore there is a unique antidominant weight in the orbit $W_{[\lambda]} \cdot \lambda$.

Proof. Observe first that λ is antidominant if and only if (a) holds. The "only if" part is obvious. In the other direction, assume (a). If $\langle \lambda + \rho, \alpha^\vee \rangle$ lies in $\mathbb{Z}$ for some $\alpha \in \Phi^+$, then $\alpha \in \Phi_{[\lambda]} \cap \Phi^+$ and is therefore a $\mathbb{Z}^+$-linear combination of $\Delta_{[\lambda]}$. It follows from (a) that $\langle \lambda + \rho, \alpha^\vee \rangle \leq 0$. So λ must be antidominant.

For any root α we have

$$s_\alpha \cdot \lambda = s_\alpha(\lambda + \rho) - \rho = \lambda + \rho - \langle \lambda + \rho, \alpha^\vee \rangle \alpha - \rho = \lambda - \langle \lambda + \rho, \alpha^\vee \rangle \alpha.$$

From this the equivalence of (a) and (b) is clear.

Since (c) obviously implies (b), it remains to show that (b) implies (c). Here we use induction on length in $W_{[\lambda]}$, relative to the simple system $\Delta_{[\lambda]}$. When $\ell(w) = 0$, we have $w = 1$ and there is nothing to prove. When $\ell(w) > 0$, write $w = w' s_\alpha$ in $W_{[\lambda]}$, with $\alpha \in \Delta_{[\lambda]}$ and $\ell(w') = \ell(w) - 1$. From 0.3(4) we get $w\alpha < 0$. Now

$$\lambda - w \cdot \lambda = (\lambda - w' \cdot \lambda) + w' \cdot (\lambda - s_\alpha \cdot \lambda).$$

By induction, $\lambda - w' \cdot \lambda \leq 0$. On the other hand,

$$w' \cdot (\lambda - s_\alpha \cdot \lambda) = w(s_\alpha \cdot \lambda - \lambda) = -\langle \lambda + \rho, \alpha^\vee \rangle w\alpha \leq 0,$$

using the equivalence of (a) and (b) along with the fact that $w\alpha < 0$. $\qquad \square$

Exercise. If $\lambda \in \mathfrak{h}^*$ is antidominant, prove that $M(\lambda) = L(\lambda)$. [This and its converse will be discussed further in Chapter 4.]

Remark. Recall from Exercise 1.13 the refined decomposition of $\mathcal{O}_\chi$ when χ is a central character belonging to a linkage class of *nonintegral* weights. This decomposition is indexed by the nonempty intersections of the linkage class with cosets in $\mathfrak{h}^*/\Lambda_r$. From now on we can use the antidominant weight λ in such an intersection to parametrize the corresponding subcategory of $\mathcal{O}_\chi$. Call this subcategory $\mathcal{O}_\lambda$. (This conflicts with the special notation $\mathcal{O}_0$ often used for the principal block, introduced in 1.13. But since $\lambda = 0$ is never antidominant, there should be no confusion in practice.) In 4.9 we will be able to show that the $\mathcal{O}_\lambda$ are precisely the blocks of $\mathcal{O}$.

Notice that the various subsets $W_{[\lambda]} \cdot \lambda$ of the given linkage class (with λ antidominant and lying in one coset modulo Λ_r) have the same cardinality. Indeed, if μ is another such antidominant weight and $w \cdot \mu = \lambda$, then $W_{[\mu]} = w^{-1} W_{[\lambda]} w$. Moreover, in each case the structure of partially ordered set is the same, with a unique maximal (dominant) and a unique minimal (antidominant) element.

3.6. Tensoring Verma Modules with Finite Dimensional Modules

As we saw in Theorem 1.1(d), category $\mathcal{O}$ is closed under tensoring with finite dimensional modules. As a tool in the study of projective modules, and in preparation for the introduction of "translation functors" in Chapter 7, we examine here an interesting special situation.

Theorem. *Let M be a finite dimensional $U(\mathfrak{g})$-module. For any $\lambda \in \mathfrak{h}^*$, the tensor product $T := M(\lambda) \otimes M$ has a finite filtration with quotients isomorphic to Verma modules of the form $M(\lambda + \mu)$. Here μ ranges over the weights of M, each occurring $\dim M_\mu$ times in the filtration.*

Proof. A direct computational proof is outlined in the exercise below, requiring some bookkeeping with formal characters; but here we place the theorem in a more general context which is useful elsewhere.

Let M be a $U(\mathfrak{g})$-module and L a $U(\mathfrak{b})$-module. Then $L \otimes M$ is also a $U(\mathfrak{b})$-module (with the extra $U(\mathfrak{g})$-module structure on M forgotten). The **Tensor Identity** provides a natural isomorphism

$$\big(U(\mathfrak{g}) \otimes_{U(\mathfrak{b})} L\big) \otimes M \cong U(\mathfrak{g}) \otimes_{U(\mathfrak{b})} \big(L \otimes M\big),$$

where the left side is a tensor product of $U(\mathfrak{g})$-modules and the right side is an induced module viewed as left $U(\mathfrak{g})$-module. This is valid in a general setting for Lie algebras and subalgebras (see Knapp [**188**, Prop. 6.5]) and is implicated in ideas such as Frobenius Reciprocity in representation theory.

As pointed out in Remark 1.3, $N \mapsto U(\mathfrak{g}) \otimes_{U(\mathfrak{b})} N$ is an *exact* functor from $U(\mathfrak{b})$-modules to $U(\mathfrak{g})$-modules whenever $\dim N < \infty$.

Now assume that L and M are finite dimensional, and set $N := L \otimes M$. Order a basis of N consisting of weight vectors $v_1, \ldots, v_n$ (with weights $\nu_1, \ldots, \nu_n$) so that $i \leq j$ whenever $\nu_i \leq \nu_j$. This yields a filtration $0 \subset N_n \subset \cdots \subset N_1 = N$, with N_k the $U(\mathfrak{b})$-submodule spanned by $v_k, \ldots, v_n$. An easy induction on $\dim N$ then shows that the induced module on the right side of the Tensor Identity has a corresponding filtration with quotients which are Verma modules of highest weights equal to the weights of N.

In our situation the $U(\mathfrak{b})$-module L is $\mathbb{C}_\lambda$, so induction yields $M(\lambda)$. On the other hand, $L \otimes M$ has the same dimension as M, but λ is added to all weights μ of M. Thus we can identify the tensor product T with the induced module $U(\mathfrak{g}) \otimes_{U(\mathfrak{b})} (L \otimes M)$. By the argument above, this module has the asserted type of filtration by Verma modules. $\qquad\square$

Example. A special case of the induced module construction above will be needed in Chapter 6. There one takes M to be the kth exterior power of the vector space $\mathfrak{g}/\mathfrak{b}$, viewed as a $U(\mathfrak{b})$-module. (What are its weights?)

Exercise. Without invoking the Tensor Identity, give a direct proof of the theorem by filling in these steps:

(a) Check that the formal character of T agrees with the sum of formal characters of the indicated Verma modules: Write $\operatorname{ch} M = \sum_\mu m_\mu e(\mu)$, with $m_\mu := \dim M_\mu$. Then Proposition 1.16 implies that

$$
\begin{aligned}
\operatorname{ch} T \;&=\; \operatorname{ch} M(\lambda) * \operatorname{ch} M = p * e(\lambda) * \sum_\mu m_\mu e(\mu) \\
&=\; \sum_\mu m_\mu \big(p * e(\lambda + \mu)\big) = \sum_\mu m_\mu \operatorname{ch} M(\lambda + \mu).
\end{aligned}
$$

(b) To construct the desired filtration, denote by v^+ a maximal vector in $M(\lambda)$ and by $v_1, \ldots, v_n$ a basis of M consisting of vectors having respective weights $\mu_1, \ldots, \mu_n$ (with repetitions allowed). Order this basis in such a way that $i \leq j$ whenever $\mu_i \leq \mu_j$. Then set $t_i := v^+ \otimes v_i$ in T and let T_i be the submodule generated by $t_i, \ldots, t_n$. So $0 \subset T_n \subset T_{n-1} \subset \cdots \subset T_1 \subset T$.

(c) Prove that $T = T_1$, using induction with respect to the PBW filtration of $U(\mathfrak{g})$. This starts with the calculation

$$
(x \cdot v^+) \otimes v = x \cdot (v^+ \otimes v) - v^+ \otimes x \cdot v \quad \text{for } x \in \mathfrak{g}, \; v \in M.
$$

Evidently the vector on the right side lies in T_1.

(d) The vector t_i has weight $\lambda + \mu_i$, as does its image $\overline{t_i}$ in T_i/T_{i+1}, which clearly generates the quotient module. Using the chosen ordering of weights, show that $\overline{t_i}$ is either 0 or else a maximal vector of weight $\lambda + \mu_i$. In either case the universal property of Verma modules gives a surjection $\varphi_i : M(\lambda + \mu_i) \to T_i/T_{i+1}$.

(e) Each φ_i is an isomorphism, by the formal character comparison above.

Remark. Either proof of the theorem shows more precisely that T has a submodule isomorphic to $M(\lambda + \mu)$, where μ is any maximal weight of M. Similarly, T has a quotient isomorphic to $M(\lambda + \nu)$, where ν is any minimal weight of M.

While it is easy to keep track of formal characters, the detailed submodule structure of the tensor product in the theorem is not so easy to study. But the decomposition of $\mathcal{O}$ into the direct sum of subcategories $\mathcal{O}_\chi$ does provide some further information due to the fact that each Verma module corresponds to a single central character. It is instructive to look at the rank 1 case, though even here the full picture requires development of some further theory. Let $\mathfrak{g} = \mathfrak{sl}(2, \mathbb{C})$ and identify $\mathfrak{h}^*$ with $\mathbb{C}$. The reader should be able to describe for each $\lambda \in \mathbb{Z}^+$ the decomposition of $M(0) \otimes L(\lambda)$ relative to central characters χ.

3.7. Standard Filtrations

The type of filtration described in Theorem 3.6 plays a prominent role in the study of projective modules in $\mathcal{O}$ as well as the later study of "tilting" modules (Chapter 11). We say that $M \in \mathcal{O}$ has a **standard filtration** (also sometimes called a *Verma flag*) if there is a sequence of submodules $0 = M_0 \subset M_1 \subset M_2 \subset \cdots \subset M_n = M$ for which each $M^i := M_i/M_{i-1}$ ($1 \le i \le n$) is isomorphic to a Verma module. While the M_i are not in general unique, comparison of formal characters shows that the **filtration length** n is well defined, as is the multiplicity with which each Verma module occurs as a subquotient. By convention, the 0 module may also be said to have a standard filtration of length 0.

Denote the multiplicity with which each Verma module $M(\lambda)$ occurs in a standard filtration of M by $\big(M : M(\lambda)\big)$, reserving the earlier notation $[M : L(\lambda)]$ for the multiplicity of $L(\lambda)$ in a Jordan–Hölder series of M. (The notation has to be handled with care, for example when $M(\lambda) = L(\lambda)$.) It is clear that a direct sum $M \oplus N$ of modules with standard filtrations also has such a filtration; by comparing formal characters, we get $\big(M \oplus N : M(\lambda)\big) = \big(M : M(\lambda)\big) + \big(N : M(\lambda)\big)$.

Here are some less obvious, but still elementary, observations about standard filtrations:

Proposition. *Let $M \in \mathcal{O}$ have a standard filtration.*

(a) *If λ is maximal among the weights of M, then M has a submodule isomorphic to $M(\lambda)$ and $M/M(\lambda)$ has a standard filtration.*

(b) *If $M = M' \oplus M''$ in $\mathcal{O}$, then M' and M'' have standard filtrations.*

(c) *M is free as a $U(\mathfrak{n}^-)$-module.*

Proof. (a) The assumption on λ implies that M has a maximal vector of weight λ, which (by the universal property of Verma modules) ensures the existence of a nonzero homomorphism $\varphi : M(\lambda) \to M$. We claim it is injective. In the given standard filtration, let i be the smallest index for which $\varphi\big(M(\lambda)\big) \subset M_i$. Thus the induced map $\psi : M(\lambda) \to M^i = M_i/M_{i-1}$ is nonzero. But $M_i/M_{i-1} \cong M(\mu)$ for some μ, which forces $\lambda \leq \mu$. By the maximality of λ, we have $\lambda = \mu$. This forces ψ to be an isomorphism. In turn, φ itself is injective. Write briefly $M(\lambda) \subset M$. Now $M(\lambda) \cap M_{i-1} = \operatorname{Ker}\psi = 0$, resulting in an induced short exact sequence

$$0 \to M_{i-1} \to M/M(\lambda) \to M/M_i \to 0.$$

Both M_{i-1} and M/M_i have standard filtrations, which combine to produce a standard filtration for $M/M(\lambda)$.

(b) There is nothing to prove if M is a Verma module, so we can proceed by induction on the filtration length of M. Let λ be maximal among the weights of M; without loss of generality, say $M'_\lambda \neq 0$. As before, there is a maximal vector in M' of weight λ and an associated homomorphism $M(\lambda) \to M' \hookrightarrow M$. Applying part (a) to the composite map, we see that $M(\lambda) \to M'$ is injective, while $M/M(\lambda) \cong \big(M'/M(\lambda)\big) \oplus M''$ admits a standard filtration. The induction hypothesis, applied to $M/M(\lambda)$, shows that $M'/M(\lambda)$ and M'' admit standard filtrations (so M' also does).

(c) Suppose M has a standard filtration. To show that M is $U(\mathfrak{n}^-)$-free, use induction on the standard filtration length starting with the obvious case when $M \cong M(\lambda)$. For the induction step, use part (a) to find a submodule N isomorphic to some $M(\lambda)$ for which M/N has a standard filtration. This module is $U(\mathfrak{n}^-)$-free by induction, so M is also $U(\mathfrak{n}^-)$-free. $\qquad\square$

Exercise. (a) If a module M has a standard filtration and there exists an epimorphism $\varphi : M \to M(\lambda)$, prove that $\operatorname{Ker}\varphi$ admits a standard filtration.

(b) Show by example when $\mathfrak{g} = \mathfrak{sl}(2, \mathbb{C})$ that the existence of a monomorphism $\varphi : M(\lambda) \to M$ when M has a standard filtration fails to imply that $\operatorname{Coker}\varphi$ also has a standard filtration.

For a module with a standard filtration, we can characterize the resulting filtration multiplicities in a useful way (needed in 3.11 below):

Theorem. *If M has a standard filtration, then for all $\lambda \in \mathfrak{h}^*$ we have*

$$\big(M : M(\lambda)\big) = \dim \operatorname{Hom}_{\mathcal{O}}\big(M, M(\lambda)^\vee\big).$$

Proof. The proof is by induction on the filtration length of M. In the case of length 1, combine the obvious fact that $\big(M(\mu) : M(\lambda)\big) = \delta_{\lambda\mu}$ with the earlier computation $\dim \operatorname{Hom}_{\mathcal{O}}\big(M(\mu), M(\lambda)^{\vee}\big) = \delta_{\lambda\mu}$ in Theorem 3.3(c).

For the induction step, look at the beginning of the long exact sequence associated with a short exact sequence $0 \to N \to M \to M(\mu) \to 0$:

$$0 \to \operatorname{Hom}_{\mathcal{O}}\big(M(\mu), M(\lambda)^{\vee}\big) \to \operatorname{Hom}_{\mathcal{O}}\big(M, M(\lambda)^{\vee}\big)$$

$$\to \operatorname{Hom}_{\mathcal{O}}\big(N, M(\lambda)^{\vee}\big) \to \operatorname{Ext}_{\mathcal{O}}\big(M(\mu), M(\lambda)^{\vee}\big) \to \cdots.$$

The Ext term vanishes by Theorem 3.3(d), while $\dim \operatorname{Hom}_{\mathcal{O}}\big(M(\mu), M(\lambda)^{\vee}\big)$ is $\delta_{\lambda\mu}$ as just recalled. By induction, $\big(N : M(\lambda)\big)$ is the dimension of the associated Hom space. Since $\big(M : M(\lambda)\big) = \big(N : M(\lambda)\big) + \delta_{\lambda\mu}$ by the choice of μ, the desired value of $\big(M : M(\lambda)\big)$ follows. $\qquad\square$

3.8. Projectives in $\mathcal{O}$

In order to make effective use of homological methods, we want to show that $\mathcal{O}$ has enough projectives: for each $M \in \mathcal{O}$, there is a projective object $P \in \mathcal{O}$ and an epimorphism $P \to M$. Recall that an object P in an abelian category is called **projective** if the left exact functor $\operatorname{Hom}(P, ?)$ is also right exact. Equivalently, given an epimorphism $\pi : M \to N$ and any morphism $\varphi : P \to N$, there is a lifting $\psi : P \to M$ such that $\pi\psi = \varphi$. Dually, an object Q is **injective** if the right exact functor $\operatorname{Hom}(?, Q)$ is also left exact. To say that $\mathcal{O}$ has enough injectives is to say that for each $M \in \mathcal{O}$, there is an injective object Q and a monomorphism $M \to Q$.

In category $\mathcal{O}$, the existence of enough projectives will automatically imply the existence of enough injectives: given a projective P, it follows immediately from the definitions that the duality functor (which is exact and contravariant) sends P to an injective $P^{\vee}$. So we focus here on projectives. Some examples are easy to find. Recall from 3.5 that $\lambda \in \mathfrak{h}^*$ is called "dominant" if $-\lambda$ is antidominant.

Proposition. *Category $\mathcal{O}$ contains some projective objects:*

(a) *Suppose $\lambda \in \mathfrak{h}^*$ is dominant, hence maximal in its $W_{[\lambda]}$-orbit, for example $\lambda \in \Lambda^+ - \rho$. Then $M(\lambda)$ is projective in $\mathcal{O}$.*

(b) *If $P \in \mathcal{O}$ is projective, while $\dim L < \infty$, then $P \otimes L$ is projective in $\mathcal{O}$.*

Proof. (a) Given an epimorphism $\pi : M \to N$ in $\mathcal{O}$ and a homomorphism $\varphi : M(\lambda) \to N$ (which may be assumed to be nonzero), we need to find a lifting $\psi : M(\lambda) \to M$. Since $M(\lambda) \in \mathcal{O}_{\chi}$ (where $\chi = \chi_\lambda$), the image of φ lies in N^{χ}. So we may as well assume that $M, N \in \mathcal{O}_{\chi}$.

Now if v^+ is a maximal vector in $M(\lambda)$, its image $\varphi(v^+)$ is a maximal vector in N of weight λ. Since π is surjective, there exists a vector v of weight λ in M such that $\varphi(v^+) = \pi(v)$. The $\mathfrak{n}$-submodule of M generated by v must contain a maximal vector having a weight linked to λ. But by assumption there is no such weight $> \lambda$. This forces $\mathfrak{n} \cdot v = 0$. In turn, the universal property of $M(\lambda)$ implies the existence of the desired homomorphism ψ sending v^+ to v.

(b) It has to be shown that $\mathrm{Hom}_{\mathcal{O}}(P \otimes L, ?)$ is right exact. In general, we have

$$\mathrm{Hom}_{\mathcal{O}}(P \otimes L, M) \cong \mathrm{Hom}_{\mathcal{O}}\left(P, \mathrm{Hom}(L, M)\right).$$

This starts with the standard vector space isomorphism, which is then easily checked to be compatible with the $U(\mathfrak{g})$-actions. Here the vector space $\mathrm{Hom}(L, M)$ has the usual structure of $U(\mathfrak{g})$-module. Since $\dim L < \infty$, we can further replace $\mathrm{Hom}(L, M)$ by $L^* \otimes M$, where again the $U(\mathfrak{g})$-actions are compatible. Since P is projective, while tensoring with L^* is exact in $\mathcal{O}$, this shows that $P \otimes L$ is also projective. $\qquad\square$

Part (a) of the proposition shows that at least some of the Verma modules have categorical meaning. In a similar but more artificial vein, one can use the same argument to show that an arbitrary $M(\lambda)$ is a projective object in the full subcategory of $\mathcal{O}$ whose objects are the modules with all weights bounded above by λ.

Example. In case $\lambda = -\rho$, we know already from Example 1.11 that $M(-\rho) = L(-\rho)$. Now part (a) of the proposition shows that $M(-\rho)$ is projective.

Exercise. If Q is injective in $\mathcal{O}$, while $\dim L < \infty$, prove that $Q \otimes L$ is injective in $\mathcal{O}$.

By combining the two parts of the proposition, we can construct a large collection of projectives. Although these are typically hard to analyze in detail, they do ensure the existence of enough projectives:

Theorem. *Category $\mathcal{O}$ has enough projectives (and enough injectives).*

Proof. The first goal is to find for each $\lambda \in \mathfrak{h}^*$ a projective object in $\mathcal{O}$ mapping onto $L(\lambda)$. It is clear that $\mu := \lambda + n\rho$ is *dominant* for sufficiently large n. Therefore $M(\mu)$ is projective, by the proposition. In turn, $n\rho \in \Lambda^+$, so $\dim L(n\rho) < \infty$ and $P := M(\mu) \otimes L(n\rho)$ is projective. Obviously the lowest weight of $L(n\rho)$ is $-n\rho = w_\circ(n\rho)$. So Remark 3.6 shows that the tensor product has a quotient isomorphic to $M(\mu - n\rho) = M(\lambda)$. This shows that $M(\lambda)$ and hence also $L(\lambda)$ is the quotient of a projective module.

For use in 3.10 below, note that Theorem 3.6 ensures the existence of a standard filtration in P. The subquotients are the $M(\mu + \nu)$, where ν runs over the weights of $L(n\rho)$ (counting multiplicity). In particular, $M(\lambda)$ occurs just once, with all other $\mu + \nu > \lambda$.

Now an induction based on length will complete the proof that an arbitrary nonzero $M \in \mathcal{O}$ is the quotient of a projective module. Assuming M has length > 1, it has a simple submodule $L(\lambda)$, giving rise to a short exact sequence $0 \to L(\lambda) \to M \to N \to 0$. Here the quotient N has smaller length than M, so by induction there is a surjection $Q \to N$ for some projective Q in $\mathcal{O}$. Lifting this map to $Q \to M$ by the projective property, either $Q \to M$ is surjective (and we are done) or else the map gives a splitting $M \cong L(\lambda) \oplus N$. In the latter case, a direct sum of projectives mapping respectively onto $L(\lambda)$ and N will map onto M. $\qquad\square$

3.9. Indecomposable Projectives

As in other artinian module categories, it follows from the existence of enough projectives that each $M \in \mathcal{O}$ has a **projective cover** $\pi : P_M \to M$. Here π is an epimorphism and is *essential*, meaning that no proper submodule of the projective module P_M is mapped onto M. Up to isomorphism, the module P_M is then the unique projective having this property. (See for example Curtis–Reiner [**74**, §6C].)

In particular, for each $\lambda \in \mathfrak{h}^*$ denote by $\pi_\lambda : P(\lambda) \to L(\lambda)$ a fixed projective cover. If N were a maximal submodule of $P(\lambda)$ different from the maximal submodule $\operatorname{Ker} \pi_\lambda$, then N would be mapped onto $L(\lambda)$ by π_λ contrary to the map being essential. Thus $P(\lambda)$ has a unique maximal submodule and is therefore *indecomposable*. Moreover, thanks to the projective property, one actually has epimorphisms $P(\lambda) \to M(\lambda) \to L(\lambda)$. (This in turn implies that $P(\lambda)$ *is a projective cover* of $M(\lambda)$.)

Are there any other indecomposable projectives in $\mathcal{O}$? If P is one of these, it has at least one simple quotient $L(\lambda)$, which defines an epimorphism $\varphi : P \to L(\lambda)$. The projective property of P then yields a homomorphism $\psi : P \to P(\lambda)$ for which $\pi_\lambda \psi = \varphi$. Since π_λ is essential, ψ must be surjective. In turn the projective property of $P(\lambda)$ yields a map $\psi' : P(\lambda) \to P$ for which $\psi\psi'$ is the identity on $P(\lambda)$. Thus $P(\lambda)$ is isomorphic to a direct summand of P. Since P is indecomposable, $P \cong P(\lambda)$. (This argument makes no use of the special properties of $\mathcal{O}$.)

We have proved part (a) of the following theorem.

Theorem. *The modules $P(\lambda)$ with $\lambda \in \mathfrak{h}^*$ satisfy:*

(a) *In category $\mathcal{O}$, every indecomposable projective module is isomorphic to some projective cover $P(\lambda)$.*

    (b) *When a projective module $P \in \mathcal{O}$ is written as a direct sum of indecomposables, the number of those isomorphic to $P(\lambda)$ is equal to $\dim \operatorname{Hom}_{\mathcal{O}}\big(P, L(\lambda)\big)$.*

    (c) *For all $M \in \mathcal{O}$, $\dim \operatorname{Hom}_{\mathcal{O}}(P(\lambda), M) = [M : L(\lambda)]$. In particular, $\dim \operatorname{End}_{\mathcal{O}} P(\lambda) = [P(\lambda) : L(\lambda)]$.*

Proof. (b) Each projective in $\mathcal{O}$ is a direct sum of copies of various $P(\lambda)$, while $P(\lambda)$ has $L(\lambda)$ as its unique simple quotient. Since $\dim \operatorname{End}_{\mathcal{O}} L(\lambda) = 1$ (Theorem 1.2(g)), we see that $\dim \operatorname{Hom}_{\mathcal{O}}\big(P, L(\lambda)\big)$ is the number of summands isomorphic to $P(\lambda)$ in any decomposition of P into indecomposables.

(c) Comparison of formal characters shows that the right side is "additive" in M: given a short exact sequence $0 \to M_1 \to M \to M_2 \to 0$,

$$[M : L(\lambda)] = [M_1 : L(\lambda)] + [M_2 : L(\lambda)].$$

But since $\operatorname{Hom}_{\mathcal{O}}(P(\lambda), ?)$ is exact, we get

$$0 \to \operatorname{Hom}_{\mathcal{O}}(P(\lambda), M_1) \to \operatorname{Hom}_{\mathcal{O}}(P(\lambda), M) \to \operatorname{Hom}_{\mathcal{O}}(P(\lambda), M_2) \to 0.$$

Thus the dimension on the left side is also additive in M. Using additivity, the desired equality now follows by induction on the length of M, starting with the case when $M = L(\mu)$: $\dim \operatorname{Hom}_{\mathcal{O}}(P(\lambda), M) = \delta_{\lambda\mu} = [M : L(\lambda)]$. This completes the proof. $\square$

Applying Theorem 3.2(d), we see that the indecomposable injectives in $\mathcal{O}$ are precisely the modules $Q(\lambda) := P(\lambda)^{\vee}$. Here $Q(\lambda)$ is the **injective hull** of $L(\lambda) \cong L(\lambda)^{\vee}$.

Remark. Before taking up the main results of BGG [**27**], we pause to illustrate some of the techniques developed so far by proving a special result which will be useful in Chapter 10. The reader is encouraged to work out the proof independently as an exercise; but for completeness we give the details: *Let $\lambda, \mu \in \mathfrak{h}^*$, with λ dominant, and let $\dim L < \infty$. Then*

$$(*) \qquad \dim \operatorname{Hom}_{\mathcal{O}}\big(M(\lambda) \otimes L, M(\mu)\big) = \dim L_{\mu - \lambda}.$$

Recalling the discussion in the proof of Proposition 3.8(b), we can use the assumption $\dim L < \infty$ to replace the Hom space on the left by

$$\operatorname{Hom}_{\mathcal{O}}\big(M(\lambda), L^* \otimes M(\mu)\big).$$

Since λ is dominant, Proposition 3.8(a) says that $M(\lambda)$ is projective; thus $M(\lambda) = P(\lambda)$. In turn, part (c) of the above theorem shows that

$$\dim \operatorname{Hom}_{\mathcal{O}}\big(M(\lambda), L^* \otimes M(\mu)\big) = [L^* \otimes M(\mu) : L(\lambda)].$$

Recall from 1.6 that the weights of L^* are just the negatives of the weights of L. By Theorem 3.6, $L^* \otimes M(\mu)$ has a standard filtration, with each quotient $M(-\nu + \mu)$ occurring $\dim L_\nu$ times. So $L(\lambda)$ occurs as a composition factor of

$L^* \otimes M(\mu)$ at least $\dim L_{\mu-\lambda}$ times. On the other hand, since λ is dominant, $L(\lambda)$ cannot occur as a composition factor of any other Verma module in the filtration.

3.10. Standard Filtrations of Projectives

Having found enough projectives in $\mathcal{O}$, our next goal is to show that each such module has a standard filtration (3.7).

Theorem. *Each projective module in $\mathcal{O}$ has a standard filtration. In a standard filtration of $P(\lambda)$, the multiplicity $\big(P(\lambda) : M(\mu)\big)$ is nonzero only if $\mu \geq \lambda$ while $\big(P(\lambda) : M(\lambda)\big) = 1$.*

Proof. The idea is to exploit the tensor product construction in Proposition 3.8(b). We can extract $P(\lambda)$ as a direct summand of a projective of the form $M(\mu) \otimes L$ with μ dominant and $\dim L < \infty$. As noted in the proof of Theorem 3.8 (based on Theorem 3.6), this tensor product has a standard filtration involving $M(\lambda)$ once along with other Verma modules of highest weight $> \lambda$. Thanks to Proposition 3.7(b), the direct summand $P(\lambda)$ inherits the same type of standard filtration. $\qquad\square$

Corollary. *Each projective module $P \in \mathcal{O}$ is determined up to isomorphism by its formal character: if P' is also projective and $\operatorname{ch} P = \operatorname{ch} P'$, then $P \cong P'$.*

Proof. Thanks to Theorem 3.9(b), P determines uniquely integers $d_\lambda > 0$ for which $P \cong \bigoplus_\lambda d_\lambda P(\lambda)$. So it suffices to show that $\operatorname{ch} P$ determines the d_λ. Use induction on the length of P, the length 1 case $P = P(\lambda) = L(\lambda)$ being obvious.

By Exercise 1.16, $\operatorname{ch} P = \sum_\mu c_\mu \operatorname{ch} M(\mu)$ with the $c_\mu \in \mathbb{Z}$ determined by P. The theorem shows that all $c_\mu \geq 0$. Let λ be minimal among the weights for which $c_\lambda > 0$. Then the theorem implies that $P(\lambda)$ occurs precisely c_λ times in the direct sum decomposition of P: that is, $c_\lambda = d_\lambda$. After discarding these summands, the induction hypothesis takes over for the remaining summand. $\qquad\square$

Exercise. Extend Exercise 1.16 by showing that the $[P(\lambda)]$ form a $\mathbb{Z}$-basis of the subgroup they span in $K(\mathcal{O})$.

Remark. Note that if $P(\lambda) \rightarrow M(\lambda)$ is the canonical map, the existence of a standard filtration in $P(\lambda)$ implies that the kernel of this map also has a standard filtration: see Exercise 3.7.

3.11. BGG Reciprocity

At this point we have attached to each $\lambda \in \mathfrak{h}^*$ three indecomposable modules related by epimorphisms: $P(\lambda) \to M(\lambda) \to L(\lambda)$. Though the Verma modules have no obvious categorical meaning, they play some kind of intermediate role here. This is made precise in the following fundamental result discovered by Bernstein–Gelfand–Gelfand [27]:

Theorem (BGG Reciprocity). *Let* $\lambda, \mu \in \mathfrak{h}^*$. *Then*

$$\big(P(\lambda) : M(\mu)\big) = [M(\mu) : L(\lambda)],$$

which is the same as $[M(\mu)^\vee : L(\lambda)]$.

For example, in the special case when λ is *maximal* in its linkage class, we have $P(\lambda) = M(\lambda)$ thanks to Proposition 3.8(a); here both sides in the theorem are equal to $\delta_{\lambda\mu}$.

Proof. Unless λ and μ are linked (and also lie in the same coset of $\mathfrak{h}^*$ modulo Λ_r) there is nothing to prove. Fix a linkage class and corresponding central character χ. Most of the groundwork for the proof has already been laid in the previous sections.

Since $\operatorname{ch} M(\mu)^\vee = \operatorname{ch} M(\mu)$, both have the same composition factor multiplicities. For technical reasons it is better to work here with the dual. The strategy is to show separately that each side of the equality in the statement of the theorem agrees with $\dim \operatorname{Hom}_{\mathcal{O}}\big(P(\lambda), M(\mu)^\vee\big)$.

(a) The equality $\dim \operatorname{Hom}_{\mathcal{O}}\big(P(\lambda), M(\mu)^\vee\big) = [M(\mu)^\vee : L(\lambda)]$ is true more generally when $M(\mu)^\vee$ is replaced by an arbitrary $M \in \mathcal{O}$, thanks to Theorem 3.9(c).

(b) It remains to check that $\big(P(\lambda) : M(\mu)\big) = \dim \operatorname{Hom}_{\mathcal{O}}\big(P(\lambda), M(\mu)^\vee\big)$. Since $P(\lambda)$ has a standard filtration, this is a special case of Theorem 3.7. $\square$

Exercise. Prove that $M(\lambda)$ is projective only when $\lambda \in \mathfrak{h}^*$ is dominant.

Remark. The original proof of the theorem in [27] used the same strategy of comparing multiplicities with dimensions of a Hom space; but the details were more complicated, going back to a somewhat elaborate indirect construction of projective objects in $\mathcal{O}$. The modern proof, which relies essentially on the duality functor, is developed by Irving [132, §4] in an unpublished manuscript.

Irving emphasizes the module structure in explaining why BGG Reciprocity is natural. Roughly speaking, each occurrence of $L(\lambda)$ in a composition series of $M(\mu)$ yields a quotient M of $M(\mu)$ with head $L(\mu)$ and socle $L(\lambda)$. The resulting dual module $M^\vee$ is then a homomorphic image of $P(\lambda)$. The "pre-image" of the socle $L(\mu)$ of $M^\vee$ in $P(\lambda)$ is then shown to be the head of a copy of $M(\mu)$ in a suitable standard filtration of $P(\lambda)$. This leads to the desired equality of multiplicities in the theorem.

By now we have seen the main features of projectives in $\mathcal{O}$. However, many fine points are yet to be investigated, some requiring fairly sophisticated techniques:

- Which simple modules occur in $\operatorname{Soc} P(\lambda)$, and with what multiplicity? More generally, one can ask for a detailed description of the socle or radical series of $P(\lambda)$.

- For which λ is $P(\lambda)$ self-dual (isomorphic to the injective hull $Q(\lambda)$ of $L(\lambda)$)?

- What is the structure of the algebra $\operatorname{End}_{\mathcal{O}} P(\lambda)$? For example, when is this algebra commutative?

3.12. Example: $\mathfrak{sl}(2, \mathbb{C})$

By now we have enough theory in hand to sort out the indecomposable modules in $\mathcal{O}$ when $\mathfrak{g} = \mathfrak{sl}(2, \mathbb{C})$. As usual $\mathfrak{h}^*$ is identified with $\mathbb{C}$, while Λ is identified with $\mathbb{Z}$. Some initial steps were already outlined in 1.5.

Since nonintegral weights (as well as $-\rho$) are linked to no comparable weights in the natural ordering, the only interesting subcategories are those of the form $\mathcal{O}_\chi$ with $\chi = \chi_\lambda$ and $\lambda \in \mathbb{Z}$ (but not equal to $-\rho$). So consider a typical linked pair $\lambda, \mu = -\lambda - 2$ with $\lambda \geq 0$. Because no lower weight is linked to μ, we must have $L(\mu) = M(\mu)$. On the other hand, $\dim L(\lambda) < \infty$, while there exists a maximal vector of weight μ in $M(\lambda)$ (by Proposition 1.4). It must be unique (up to scalars) and generate a submodule isomorphic to $L(\mu)$ in view of the fact that all weight spaces in $M(\lambda)$ and $L(\mu)$ are 1-dimensional. So $M(\lambda)$ has just these two composition factors.

By Proposition 3.8(a), $M(\lambda) = P(\lambda)$ is projective. Its dual is the injective module $Q(\lambda)$, having socle $L(\lambda)$ and head $L(\mu)$. What is the structure of $P(\mu)$? It has a standard filtration with $M(\mu) = L(\mu)$ as top quotient (Theorem 3.10). Thanks to BGG Reciprocity (3.11) and the fact that $L(\mu)$ occurs just once as a composition factor of $M(\lambda)$ and $M(\mu)$, any standard filtration involves these Verma modules with multiplicity 1. Therefore we must have an embedding $M(\lambda) \subset P(\mu)$, forcing $L(\mu) = \operatorname{Soc} P(\mu)$. In turn, the extension $0 \to M(\lambda) \to P(\mu) \to M(\mu) \to 0$ is nonsplit. Moreover, $P(\mu) \cong Q(\mu)$ is *self-dual*.

Exercise. Use the formulas in 0.9 to compute explicitly how the Casimir element $z := h^2 + 2h + 4yx \in Z(\mathfrak{g})$ acts on these modules. In particular, z acts on each of $M(\lambda)$ and $M(\mu)$ by the scalar $\chi_\lambda(z) = \lambda^2 + 2\lambda$ (call it c), but $z - c \neq 0$ on $P(\mu)$; instead $(z - c)^2 = 0$. What is the structure of the algebra $\mathrm{End}_{\mathcal{O}}\, P(\mu)$?

So far we have found five nonisomorphic indecomposable modules in $\mathcal{O}_\chi$:

$$L(\lambda),\ L(\mu) = M(\mu),\ M(\lambda) = P(\lambda),\ M(\lambda)^\vee = Q(\lambda),\ P(\mu) = Q(\mu).$$

Are there any others?

Proposition. *If* $\mathfrak{g} = \mathfrak{sl}(2,\mathbb{C})$ *and* $\{\lambda, \mu\}$ *is a linkage class of integral weights with* $\lambda \geq 0, \mu = -\lambda - 2$, *and* $\chi = \chi_\lambda$, *then every indecomposable module in* $\mathcal{O}_\chi$ *is isomorphic to one of the five modules listed above.*

Proof. We use a mixture of general and *ad hoc* reasoning. Recall that $\mathrm{Ext}_{\mathcal{O}}\big(M(\nu), M(\nu)\big) = 0 = \mathrm{Ext}_{\mathcal{O}}\big(L(\nu), L(\nu)\big)$ for all ν (Proposition 3.1). In our situation, we can also show at this point that $\mathrm{Ext}_{\mathcal{O}}\big(L(\lambda), M(\lambda)\big) = 0$, by inspecting part of a long exact sequence:

$$\cdots \to \mathrm{Hom}_{\mathcal{O}}\big(L(\lambda), M(\lambda)\big) \to \mathrm{Hom}_{\mathcal{O}}\big(M(\lambda), M(\lambda)\big) \to$$

$$\mathrm{Hom}_{\mathcal{O}}\big(N(\lambda), M(\lambda)\big) \to \mathrm{Ext}_{\mathcal{O}}\big(L(\lambda), M(\lambda)\big) \to \mathrm{Ext}_{\mathcal{O}}\big(M(\lambda), M(\lambda)\big) \to \cdots .$$

The first term is clearly 0, while the second is $\mathbb{C}$ since $\dim M(\lambda)_\lambda = 1$. In the third term, the maximal submodule $N(\lambda) \cong L(\mu)$ and $\mathrm{Hom}_{\mathcal{O}}\big(L(\mu), M(\lambda)\big) \cong \mathrm{Hom}_{\mathcal{O}}\big(L(\mu), L(\mu)\big) = \mathbb{C}$. As just noted, the fifth term is 0, which forces the fourth term to be 0 as claimed.

Now suppose $M \in \mathcal{O}_\chi$ is an indecomposable module but not isomorphic to any of those listed. Its only composition factors are $L(\lambda)$ and $L(\mu)$. Clearly each nonzero vector in M_λ is a maximal vector and therefore generates a submodule isomorphic to $L(\lambda)$ or $M(\lambda)$. Using the above vanishing results for Ext, together with the standard fact that Ext commutes with direct sums in either variable, we see by induction that the submodule $N \subset M$ generated by M_λ is a direct sum of copies of $L(\lambda)$ and/or $M(\lambda)$. So N must be a proper submodule of M. In turn, M/N is a direct sum of copies of $L(\mu)$. At least one $L(\mu)$ must extend N nontrivially, lest M split into a direct sum of known indecomposables. So $L(\mu)$ must extend some summand $L(\lambda)$ or $M(\lambda)$ nontrivially. In these respective cases, we get a copy of the injective module $Q(\lambda)$ or $Q(\mu)$ as a submodule of M. By the injective property, it splits off and must therefore be all of M, a contradiction. $\square$

Beyond rank 1, the study of indecomposable modules in categories $\mathcal{O}_\chi$ is far more complicated. This becomes especially evident when some of the Verma modules have multiple composition factors and hence infinitely many different submodules.

3.13. Projective Generators and Finite Dimensional Algebras

Enough features of the subcategories $\mathcal{O}_\chi$ are now visible to permit a comparison with the representation theory of certain finite dimensional algebras. The latter subject is well developed and sometimes suggests methods which are useful in the study of category $\mathcal{O}$.

If $\chi = \chi_\lambda$ is a central character corresponding to the linkage class of an antidominant weight $\lambda \in \mathfrak{h}^*$, consider a projective module $P :=$ $\bigoplus_w n_w P(w \cdot \lambda)$. Here the sum is taken over coset representatives in W of the isotropy group of λ, while all $n_w > 0$. The special property of such a module is that a large enough direct sum $P \oplus \cdots \oplus P$ covers any given $M \in \mathcal{O}_\chi$. This is an obvious consequence of the fact that the $P(w \cdot \lambda)$ represent all isomorphism classes of indecomposable projectives in $\mathcal{O}_\chi$. Call P a **projective generator** of the category $\mathcal{O}_\chi$. In turn, define $A := \mathrm{End}_{\mathcal{O}} P$. This is a finite dimensional algebra (Theorem 1.11).

Proposition. *Let P as above be a projective generator of $\mathcal{O}_\chi$, with $A = \mathrm{End}_{\mathcal{O}} P$. Then the functor $\mathrm{Hom}_{\mathcal{O}}(P, ?)$ defines a category equivalence between $\mathcal{O}_\chi$ and the category of finite dimensional right A-modules.*

Proof. It is easy to get started. Given $M \in \mathcal{O}_\chi$, the right A-module structure on $\mathrm{Hom}_{\mathcal{O}}(P, M)$ is the obvious one induced by composing $\eta : P \to P$ with $\varphi : P \to M$. The further details of the argument are standard, as given for example in Bass [**14**, Thm. II.1.3] in the general setting of an abelian category with suitable finiteness conditions. $\qquad\square$

While the choice of P (and hence of A) is not unique, in practice this makes little difference. Comparison of $\mathcal{O}_\chi$ with the module category over a finite dimensional algebra illuminates some of the categorical features and suggests further problems. On the other hand, this broader viewpoint is not immediately helpful in working out the combinatorial aspects of $\mathcal{O}$, for example the formal characters of simple modules.

3.14. Contravariant Forms

Following work of Jantzen and Shapovalov (see the Notes), we introduce here a basic tool in the study of category $\mathcal{O}$. This uses the transpose map τ of $U(\mathfrak{g})$ (0.5). If M is any $U(\mathfrak{g})$-module, a symmetric bilinear form $(v, v')_M$ on M is called **contravariant** if

$$(u \cdot v, v')_M = (v, \tau(u) \cdot v')_M \text{ for all } u \in U(\mathfrak{g}),\ v, v' \in M.$$

Obviously a scalar multiple of the form will have the same property. It turns out that nonzero contravariant forms always exist for highest weight modules (and some others) in $\mathcal{O}$. But first we look at some elementary consequences of the definition.

Proposition. *Contravariant forms have the following properties:*

(a) *If the $U(\mathfrak{g})$-module M has a contravariant form, then $(M_\lambda, M_\mu)_M = 0$ whenever $\lambda \neq \mu$ in $\mathfrak{h}^*$.*

(b) *Suppose $M = U(\mathfrak{g}) \cdot v^+$ is a highest weight module generated by a maximal vector v^+ of weight λ. If M has a nonzero contravariant form, then the form is uniquely determined up to a scalar multiple by the (nonzero!) value $(v^+, v^+)_M$.*

(c) *If $U(\mathfrak{g})$-modules M_1, M_2 have contravariant forms $(v, v')_{M_1}$ and $(w, w')_{M_2}$, then $M := M_1 \otimes M_2$ also has a contravariant form, given by $(v \otimes w, v' \otimes w')_M := (v, v')_{M_1} (w, w')_{M_2}$. In case both of the forms are nondegenerate, so is the product form.*

(d) *If M has a contravariant form and N is a submodule, the orthogonal space $N^\perp := \{v \in M \mid (v, v')_M = 0 \text{ for all } v' \in N\}$ is also a submodule.*

(e) *If $M \in \mathcal{O}$ has a contravariant form, then the summands M^χ for distinct central characters χ are orthogonal.*

Proof. (a) Given weight vectors v, v' of respective weights λ, μ, use the fact that $\tau(h) = h$ for $h \in \mathfrak{h}$:

$$\lambda(h)\,(v, v')_M = (h \cdot v, v')_M = (v, h \cdot v')_M = (v, \mu(h)v')_M = \mu(h)(v, v')_M.$$

When $\lambda \neq \mu$ this forces $(v, v')_M = 0$.

(b) In view of (a) it suffices to look at values of the form on a weight space M_μ. Typical vectors $v, v' \in M_\mu$ can be written as $u \cdot v^+$, $u' \cdot v^+$ for suitable $u, u' \in U(\mathfrak{n}^-)$. Note that since u takes M_λ into M_μ, the element $\tau(u) \in U(\mathfrak{n})$ takes M_μ into M_λ (which is spanned by v^+). Then $(v, v')_M = (u \cdot v^+, u' \cdot v^+)_M = (v^+, \tau(u)u' \cdot v^+)_M$, which is a scalar multiple of $(v^+, v^+)_M$ depending just on the action of $U(\mathfrak{n}^-)$ and not on the choice of the form.

(c) and (d) follow readily from the definitions. Note that the nondegeneracy in (c) is a general fact unrelated to contravariance. To prove it directly, suppose that $\sum_{i,j} a_{ij}\, v_i \otimes w_j \neq 0$ lies in the radical of the form on M; take the $\{v_i\}$ and $\{w_j\}$ to be independent sets, so for fixed i or j some $a_{ij} \neq 0$. Then for each fixed j, set $v_0^{(j)} := \sum_i a_{ij} v_i \neq 0$ and find $v^{(j)}$ for which $c_j := (v_0^{(j)}, v^{(j)})_{M_1} \neq 0$. This forces $\sum_j c_j w_j \neq 0$ to lie in the radical of the form on M_2, a contradiction.

(e) Abbreviate by writing (v, w) in place of $(v, w)_M$. Let χ_1, χ_2 be distinct central characters. Thanks to (a), weight spaces of M belonging to distinct weights are orthogonal; so it is enough to show for each weight λ of

M that $(M_\lambda^{\chi_1}, M_\lambda^{\chi_2}) = 0$. Fix $z \in Z(\mathfrak{g})$ for which $\chi_1(z) = c_1 \neq c_2 = \chi_2(z)$. In the special case when $v, w \in M$ are both eigenvectors for z, we get (as in a familiar linear algebra argument):

$$c_1(v, w) = (z \cdot v, w) = (v, z \cdot w) = c_2(v, w),$$

using the contravariance and the fact that $\tau(z) = z$ (Exercise 1.10). This forces $(v, w) = 0$. In general, to $v \in M_\lambda^{\chi_1}$ corresponds the least $e \in \mathbb{Z}^+$ such that $(z - c_1)^e \cdot v = 0$, and similarly for $w \in M_\lambda^{\chi_2}$. Use induction on the sum of these two exponents. The induction hypothesis applies to each of the pairs $(z - c_1) \cdot v$, w and v, $(z - c_2)w$. This yields

$$c_1(v, w) = (z \cdot v, w) = (v, z \cdot w) = c_2(v, w),$$

forcing $(v, w) = 0$ as before. $\qquad\qquad\qquad\qquad\qquad\qquad\qquad\square$

Exercise. Show that any contravariant form on a highest weight module must induce the zero form on its maximal submodule. In case $\mathfrak{g} = \mathfrak{sl}(2, \mathbb{C})$, use the explicit formulas for the action of $\mathfrak{g}$ on $L(\lambda)$ in 0.9 (or 1.5) to construct a nondegenerate contravariant form on $L(\lambda)$.

3.15. Universal Construction

We still have to show the existence of (nonzero!) contravariant forms on special modules in $\mathcal{O}$ such as highest weight modules.

Rather than look immediately at $M(\lambda)$, we start with a sort of universal construction involving $U(\mathfrak{g})$; this will later play an essential role in the study of Verma modules (5.8). Let $\varepsilon^+ : U(\mathfrak{n}) \to \mathbb{C}$ be the augmentation map sending all nonconstant PBW basis elements to 0 (and define ε^- similarly). Using the PBW Theorem, consider the linear map $\varphi := \varepsilon^- \otimes \mathrm{id} \otimes \varepsilon^+ :$ $U(\mathfrak{g}) \cong U(\mathfrak{n}^-) \otimes U(\mathfrak{h}) \otimes U(\mathfrak{n}) \to U(\mathfrak{h})$. Next define a bilinear form on $U(\mathfrak{g})$ by the recipe $C(u, u') := \varphi(\tau(u)u')$. Since τ is an anti-automorphism and interchanges $U(\mathfrak{n})$ and $U(\mathfrak{n}^-)$, it is clear that $\varphi(\tau(u)) = \varphi(u)$ for all u. This implies that our form is *symmetric*:

$$C(u, u') = \varphi(\tau(u)u') = \varphi\big(\tau(\tau(u)u')\big) = \varphi\big(\tau(u')\tau^2(u)\big) = C(u', u).$$

Moreover, $C(1, 1) = 1$. A further property, related to contravariance, follows immediately from the definition and the anti-automorphism property of τ:

$$(*) \qquad\qquad C(u_\circ u, u') = C(u, \tau(u_\circ)u') \text{ for all } u_\circ, u, u' \in U(\mathfrak{g}).$$

This universal form yields $\mathbb{C}$-valued forms as follows. Each $\lambda \in \mathfrak{h}^*$ defines a representation of the abelian Lie algebra $\mathfrak{h}$ and thus induces an algebra homomorphism $\lambda : U(\mathfrak{h}) \to \mathbb{C}$. (Since $U(\mathfrak{h}) = S(\mathfrak{h})$, we can also view this as evaluation of polynomial functions on $\mathfrak{h}^*$.) Now compose λ with φ to get a linear map $\varphi_\lambda : U(\mathfrak{g}) \to \mathbb{C}$. This defines in turn a $\mathbb{C}$-valued symmetric bilinear form $C^\lambda(u, u')$ on $U(\mathfrak{g})$.

With this form in hand, it is easy to construct a contravariant form on a highest weight module M of weight λ generated by a maximal vector v^+. Here a typical vector can be written (in general non-uniquely) as $u \cdot v^+$ with $u \in U(\mathfrak{n}^-)$. Set $M' = \sum_{\mu \neq \lambda} M_\mu$ and let N be the unique maximal submodule of M (so $N \subset M'$). Writing $u \in U(\mathfrak{g})$ in the above PBW format, observe that $u \cdot v^+ \equiv \varphi_\lambda(u)v^+ \pmod{M'}$. From this it is easy to verify the equivalence of the following statements:

$$(1) \quad u \cdot v^+ \in N,$$
$$(2) \quad U(\mathfrak{g})u \cdot v^+ \subset N,$$
$$(3) \quad \varphi_\lambda(U(\mathfrak{g})u) = 0,$$
$$(4) \quad C^\lambda(U(\mathfrak{g}), u) = 0,$$
$$(5) \quad u \text{ lies in the radical of the form } C^\lambda.$$

In particular, if $u_1, u_2 \in U(\mathfrak{n}^-)$ satisfy $u_1 \cdot v^+ = u_2 \cdot v^+$, then $u_1 - u_2$ is in the radical of C^λ. This allows us to transfer the form to M itself, by defining (unambiguously) $(v, v')_M := C^\lambda(u, u')$ whenever $v = u \cdot v^+$ and $v' = u' \cdot v^+$ with $u, u' \in U(\mathfrak{n}^-)$. From the above equivalent statements we then conclude that N is the radical of the resulting form.

It still has to be checked that the form on M is *contravariant*:

$$(u_\circ u \cdot v^+, u' \cdot v^+)_M = (u \cdot v^+, \tau(u_\circ)u' \cdot v^+)_M$$

for all $u_\circ \in U(\mathfrak{g})$ and all $u, u' \in U(\mathfrak{n}^-)$. This follows immediately from $(*)$ above. In view of Proposition 3.14, we have shown:

Theorem. *If M is a highest weight module of weight λ, there exists a (nonzero) contravariant form $(v, v')_M$ on M. The form is defined uniquely up to scalar multiples by the value $(v^+, v^+)_M$. Its radical is the unique maximal submodule of M. In particular, the form is nondegenerate if and only if $M \cong L(\lambda)$.* $\qquad\square$

In principle the theorem affords a strategy for deciding when $M(\lambda) = L(\lambda)$. But for this one has to be able to determine the radical of the form, which is not easy to do. We return to the simplicity question in Chapter 4.

Exercise. Recall that $U(\mathfrak{g})$ is the graded direct sum of subspaces $U(\mathfrak{g})_\nu$ with $\nu \in \Lambda_r$ (0.5). In particular, restriction of C to the finite dimensional space $U(\mathfrak{n}^-)_{-\nu}$ with $\nu \in \Gamma$ defines a $U(\mathfrak{h})$-valued bilinear form C_ν. Prove that this form is *nondegenerate*, so its determinant relative to a PBW basis is nonzero. [Use the fact that $M(-\rho)$ is simple (Example 1.11), so its contravariant form defined above is nondegenerate on each weight space.]

Notes

(3.3) In his study of parallel module categories, Donkin has popularized an alternative notation: $\Delta(\lambda)$ in place of $M(\lambda)$ and $\nabla(\lambda)$ in place of $M(\lambda)^\vee$.

Proposition 3.5 is based on Jantzen [**147**, 1.8].

Proposition 3.7(c) was suggested by Kåhrström (based on lectures of Mazorchuk).

The basic source for the results in 3.8–3.11 is Bernstein–Gelfand–Gelfand [**27**], but the proofs have been simplified over the years. We have drawn on some of the arguments in Irving [**132**, §4] and Jantzen [**148**, 4.6–4.9] as well as Irving [**137**] (where the proof of BGG Reciprocity is formalized in the setting of "BGG algebras": see 13.7 below).

The ideas in 3.13 are formulated by BGG [**27**, Thm. 3], with the same reference to Bass [**14**] given here.

(3.14–3.15) Contravariant forms arose in work of Steinberg [**243**, 228–229] as well as work of Burgoyne and Wong in 1971–72: see Humphreys [**131**, 2.4, 3.8, 4.2]. In the present setting the results are mainly due to Jantzen [**144, 145, 146, 147**], with independent contributions by Shapovalov [**231**], as will be seen in Chapters 4–5.

Chapter 4

Highest Weight Modules I

In this and the next chapter we focus on Verma modules and their simple quotients. A number of natural questions arise, which turn out to be of widely varying difficulty:

- When is $M(\lambda)$ simple?

- What are the composition factors of $M(\lambda)$, apart from multiplicity?

- What are the precise multiplicities $[M(\lambda) : L(\mu)]$ for all $\lambda \in \mathfrak{h}^*$? (For example, when is the multiplicity of $L(\mu)$ equal to 1?) The answer would give a recursive method to compute $\operatorname{ch} L(\lambda)$ for all $\lambda \in \mathfrak{h}^*$.

- What is the submodule structure of $M(\lambda)$? For example, what are its simple submodules?

- What can be said about $\operatorname{Hom}_{\mathcal{O}}(M(\mu), M(\lambda))$ for arbitrary λ, μ?

Underlying everything here is the question: How much depends just on the Weyl group, as opposed to the specific weights involved? In the case of composition factor multiplicities in Verma modules, this line of inquiry eventually culminates in the Kazhdan–Lusztig Conjecture (Chapter 8).

Our first goal is a necessary and sufficient condition for $[M(\lambda) : L(\mu)]$ to be nonzero, which requires the treatment of some of the other questions above as well. Along the way we will bring into play a variety of methods for the study of highest weight modules: contravariant forms, Jantzen filtrations, translation functors, etc. We start by sorting out some of the easier matters, then move on to the more challenging ones.

Several factors tend to complicate the logic of the exposition: (1) It is more straightforward to deal with the case when the weights involved are *integral*, which is richer than the general case and more often needed in applications. As a result we sometimes postpone treatment of the nonintegral case. (2) The proofs of some of the main theorems are intertwined with other theorems and therefore require temporary detours. (3) Alternative approaches to some theorems have by now been developed, requiring us to make choices. (See the Notes for comments on the history.)

Features of Verma modules discussed in this chapter include: simple submodules (4.1), homomorphisms (4.2), simplicity criterion (4.4, 4.8), and the existence of embeddings $M(s_\alpha \cdot \lambda) \hookrightarrow M(\lambda)$ (4.5–4.7, 4.12–4.13).

4.1. Simple Submodules of Verma Modules

It is quite easy to show that an arbitrary Verma module has a unique simple submodule, by taking advantage of the $U(\mathfrak{n}^-)$-module isomorphism between $U(\mathfrak{n}^-)$ and $M(\lambda)$. The absence of zero-divisors in the noetherian ring $U(\mathfrak{n}^-)$ (see 0.5) can be exploited via the following elementary lemma:

Lemma. *Let R be a left noetherian ring. If $x \in R$ is not a right zero-divisor, then the left ideal Rx meets every nonzero left ideal of R nontrivially. In particular, if R has no right zero-divisors, then any two nonzero left ideals intersect nontrivially.*

Proof. Suppose I is a nonzero left ideal of R, but $I \cap Rx = 0$. Construct an ascending chain of left ideals:

$$0 \subset I \subset I + Ix \subset I + Ix + Ix^2 \subset \cdots .$$

We claim that each sum here is *direct*; this ensures that the inclusions are proper, contrary to the noetherian hypothesis on R. Indeed, suppose that $a_0 + a_1 x + \cdots + a_n x^n = 0$ for some $a_i \in I$. Then

$$a_0 = -(a_1 + a_2 x + \cdots + a_n x^{n-1})x \in I \cap Rx = 0,$$

forcing $a_0 = 0$ and $a_1 + a_2 x + \cdots + a_n x^{n-1} = 0$ (because x is not a right zero-divisor). Induction on n shows that all $a_i = 0$. $\square$

Proposition. *For any $\lambda \in \mathfrak{h}^*$, the module $M(\lambda)$ has a unique simple submodule, which is therefore its socle.*

Proof. If $M(\lambda)$ has distinct simple submodules L and L', these must intersect trivially. On the other hand, in a fixed isomorphism of $M(\lambda)$ with $U(\mathfrak{n}^-)$ the submodules must correspond to distinct nonzero left ideals, which intersect nontrivially because of the lemma. This is impossible. $\square$

Thanks to 1.3, the simple submodule of $M(\lambda)$ is isomorphic to some $L(\mu)$ with $\mu \leq \lambda$. Moreover, $\mu = w \cdot \lambda$ for some $w \in W_{[\lambda]}$ (3.4). In order to say anything more precise about $L(\mu)$, we need some further theory.

Exercise. Let M be a nonzero submodule of a Verma module $M(\lambda)$. If M has a nondegenerate contravariant form (3.14), prove that M is the unique simple submodule of $M(\lambda)$.

4.2. Homomorphisms between Verma Modules

Consider next $\mathrm{Hom}_{\mathcal{O}}(M(\mu), M(\lambda))$, which carries some (though as it turns out not all) information about the possible occurrences of $L(\mu)$ as a composition factor of $M(\lambda)$. As in the previous section, we identify $U(\mathfrak{n}^-)$ as a left $U(\mathfrak{n}^-)$-module with either of these Verma modules. Given $\varphi : M(\mu) \to M(\lambda)$, denote respective maximal vectors by v_μ^+ and v_λ^+. Then $\varphi(v_\mu^+) = u \cdot v_\lambda^+$ for a unique $u \in U(\mathfrak{n}^-)$. In turn, $\varphi(u' \cdot v_\mu^+) = u'u \cdot v_\lambda^+$ for all $u' \in U(\mathfrak{n}^-)$. This means that φ is uniquely determined by u.

Theorem. *Let* $\lambda, \mu \in \mathfrak{h}^*$.

(a) *Any nonzero homomorphism* $\varphi : M(\mu) \to M(\lambda)$ *is injective.*

(b) *In all cases,* $\dim \mathrm{Hom}_{\mathcal{O}}(M(\mu), M(\lambda)) \leq 1$.

(c) *The unique simple submodule* $L(\mu)$ *of* $M(\lambda)$ *in Proposition 4.1 is itself a Verma module.*

Proof. (a) As in 4.1, we use the fact that $U(\mathfrak{n}^-)$ has no zero-divisors. We just saw that φ corresponds to the map $U(\mathfrak{n}^-) \to U(\mathfrak{n}^-)$ given by $u' \mapsto u'u$ for some fixed $u \neq 0$. Since this algebra has no zero-divisors, $\mathrm{Ker}\,\varphi = 0$.

(b) Suppose φ_1 and φ_2 are nonzero homomorphisms $M(\mu) \to M(\lambda)$, therefore injective by part (a). Thanks to Proposition 4.1, $M(\mu)$ has a unique simple submodule L. Then φ_1 and φ_2 map L isomorphically onto respective simple submodules L_1 and L_2 of $M(\lambda)$. Again by Proposition 4.1, $L_1 = L_2$. This means that for some nonzero $c \in \mathbb{C}$, the module homomorphism $\varphi_1 - c\varphi_2$ takes L to 0. From (a) we conclude that $\varphi_1 - c\varphi_2 = 0$, forcing φ_1 and φ_2 to be proportional.

(c) By the universal property of Verma modules, $L(\mu)$ is the image of a homomorphism $M(\mu) \to M(\lambda)$, which by part (a) must be injective. $\quad\square$

Whenever $\mathrm{Hom}_{\mathcal{O}}(M(\mu), M(\lambda)) \neq 0$, the theorem allows us to write unambiguously $M(\mu) \subset M(\lambda)$. In the following sections we investigate when such embeddings exist. Along the way, we can also determine which Verma modules are simple.

4.3. Special Case: Dominant Integral Weights

We pause to look at the important special case when $\lambda \in \Lambda^+ - \rho$ (so λ is "dominant" in the general sense introduced in 3.5). An extreme case is $\lambda = -\rho$, but typically there are many other weights in $\Lambda^+ - \rho$ having a mix of negative and nonnegative coordinates relative to fundamental weights.

In this case the inclusions among Verma modules embedded in $M(\lambda)$ are transparent:

Proposition. *Suppose $\lambda + \rho \in \Lambda^+$. Then $M(w \cdot \lambda) \subset M(\lambda)$ for all $w \in W$; thus all $[M(\lambda) : L(w \cdot \lambda)] > 0$. More precisely, if $w = s_n \cdots s_1$ is a reduced expression for w, with s_i the reflection relative to the simple root α_i, there is a sequence of embeddings*

$$M(w \cdot \lambda) = M(\lambda_n) \subset M(\lambda_{n-1}) \subset \cdots \subset M(\lambda_0) = M(\lambda).$$

Here

$$\lambda_0 := \lambda, \ \lambda_k := s_k \cdot \lambda_{k-1} = (s_k \cdots s_1) \cdot \lambda, \ so \ \lambda_n = s_n \cdot \lambda_{n-1} = w \cdot \lambda.$$

In particular, $w \cdot \lambda = \lambda_n \leq \lambda_{n-1} \leq \cdots \leq \lambda_0 = \lambda$, with $\langle \lambda_k + \rho, \alpha_{k+1}^\vee \rangle \in \mathbb{Z}^+$ for $k = 0, \ldots, n-1$.

Proof. The proof uses induction on $\ell(w)$ and relies on some bookkeeping with indices along with repeated applications of Proposition 1.4 as reformulated in 1.8.

The case $\ell(w) = 0$ is obvious. For the inductive step, fix $0 < k < n$ and write $w' := s_k \cdots s_1$. From 0.3 we know that $(w')^{-1} \alpha_{k+1} > 0$. Since $\lambda + \rho \in \Lambda^+$, we have:

$$\langle \lambda_k + \rho, \alpha_{k+1}^\vee \rangle = \langle w' \cdot \lambda + \rho, \alpha_{k+1}^\vee \rangle = \langle w'(\lambda + \rho), \alpha_{k+1}^\vee \rangle$$
$$= \langle \lambda + \rho, (w')^{-1} \alpha_{k+1}^\vee \rangle = \langle \lambda + \rho, \big((w')^{-1} \alpha_{k+1}\big)^\vee \rangle \in \mathbb{Z}^+.$$

Applying the reformulated Proposition 1.4 to this situation, we get a homomorphism (embedding!) $M(\lambda_{k+1}) \subset M(\lambda_k)$, with $\lambda_{k+1} \leq \lambda_k$. $\qquad\square$

Exercise. Assume $\lambda + \rho \in \Lambda^+$.

(a) Prove that the unique simple submodule of $M(\lambda)$ is isomorphic to $M(w_\circ \cdot \lambda)$, where $w_\circ$ is the longest element of W.

(b) In case $\lambda \in \Lambda^+$, show that the inclusions obtained in the proposition are all proper.

For arbitrary $\lambda \in \mathfrak{h}^*$ (even for $\lambda \in \Lambda$), it is not so easy at this point to specify those $\mu = w \cdot \lambda$ for which an embedding $M(\mu) \subset M(\lambda)$ exists. A basic obstacle is the failure of Proposition 1.4 to carry over to nonsimple positive roots: even when $s_\alpha \cdot \lambda < \lambda$, there is no obvious way to construct an explicit embedding $M(s_\alpha \cdot \lambda) \subset M(\lambda)$. (This is explored further in 4.5 below.)

Example. To illustrate the potential problem here, let $\mathfrak{g} = \mathfrak{sl}(3, \mathbb{C})$, with $\Delta = \{\alpha, \beta\}$. Fix $\lambda \in \Lambda^+$ and set $\mu := s_\alpha \cdot \lambda$. If $\gamma = \alpha + \beta$, then $s_\gamma \cdot \mu < \mu$. Since the method of Proposition 1.4 does not apply here, it is nontrivial to find an embedding $M(s_\gamma \cdot \mu) \hookrightarrow M(\mu)$. On the other hand, there is a reduced expression $s_\gamma = s_\alpha s_\beta s_\alpha$, so $s_\gamma s_\alpha = s_\alpha s_\beta$ and $s_\gamma \cdot \mu = s_\alpha s_\beta \cdot \lambda$. From the above proposition one gets embeddings $M(s_\gamma \cdot \mu) \hookrightarrow M(s_\beta \cdot \lambda) \hookrightarrow M(\lambda)$. But no similar chain of embeddings passes through $M(\mu)$.

4.4. Simplicity Criterion: Integral Case

There is a straightforward necessary and sufficient condition for $M(\lambda)$ to be simple. To state it, recall from 3.5 what it means for $\lambda \in \mathfrak{h}^*$ to be *antidominant*:
$$\langle \lambda + \rho, \alpha^\vee \rangle \notin \mathbb{Z}^{>0} \text{ for all } \alpha \in \Phi^+.$$

Theorem. *Let $\lambda \in \mathfrak{h}^*$. Then $M(\lambda) = L(\lambda)$ if and only if λ is antidominant.*

While this is easy to state in general, the proof requires more preparation in case λ fails to lie in Λ. So we start with the integral case and return in 4.8 to the general case.

Proof (Integral Case). Assume $\lambda \in \Lambda$.

(a) Suppose $M(\lambda)$ is simple, i.e., its maximal submodule $N(\lambda)$ is 0. Recall that in the integral case, antidominance requires only that $\langle \lambda + \rho, \alpha^\vee \rangle \leq 0$ for all *simple* roots α. If this fails for some $\alpha \in \Delta$, then $s_\alpha \cdot \lambda < \lambda$ and Proposition 1.4 provides a nonzero homomorphism $M(s_\alpha \cdot \lambda) \to M(\lambda)$ (which we now know to be an embedding). But this forces $M(s_\alpha \cdot \lambda) \subset N(\lambda)$, which is absurd.

(b) Conversely, suppose λ is antidominant. Thanks to 3.5, $\lambda \leq w \cdot \lambda$ for all $w \in W$. Since all composition factors of $M(\lambda)$ are of the form $L(w \cdot \lambda)$ with $w \cdot \lambda \leq \lambda$, it follows that only $L(\lambda)$ can occur as a composition factor. But it occurs just once, so $M(\lambda) = L(\lambda)$. $\square$

Exercise. If $\lambda \in \Lambda$ is antidominant, prove that the socle of any projective module $P(w \cdot \lambda)$ with $w \in W$ is a direct sum of copies of $L(\lambda)$.

Remark. Observe how part (b) of the proof of the theorem can be extended to arbitrary λ, by exploiting the reflection group $W_{[\lambda]}$ studied in 3.4: The only candidates for highest weights of composition factors of $M(\lambda)$ are of the form $w \cdot \lambda \leq \lambda$ with $w \in W_{[\lambda]}$. Since λ is assumed to be antidominant, Proposition 3.5 ensures that $\lambda \leq w \cdot \lambda$ for all such w. To extend the first part of the proof is more challenging, since we now have to consider arbitrary positive roots; here Proposition 1.4 is not applicable. So we turn next to the study of embeddings $M(s_\alpha \cdot \lambda) \hookrightarrow M(\lambda)$ when $\alpha \in \Phi^+$. This starts again with the easier case when $\lambda \in \Lambda$ and then uses a Zariski density argument.

4.5. Existence of Embeddings: Preliminaries

Assuming that $\lambda \in \Lambda$ while $\mu = s_\alpha \cdot \lambda \leq \lambda$ for some $\alpha \in \Phi^+$, we want to prove the existence of an embedding $M(\mu) \subset M(\lambda)$ without actually exhibiting a maximal vector of weight μ in $M(\lambda)$. Of course, we can assume $\mu < \lambda$. The proof of Proposition 4.3 might suggest a step-by-step approach using a reduced expression for s_α. But without the dominance assumption there we cannot guarantee at each step that successive weights are lower in the partial ordering (Example 4.3).

Instead we proceed indirectly, applying Proposition 4.3 to an element $w \in W$ which links μ to a dominant weight. The proof requires a couple of preparatory results, after which it is mainly a matter of careful bookkeeping. First we record an elementary commutation property, which is needed in the proposition below:

Lemma. *Let $\mathfrak{a}$ be a nilpotent Lie algebra, with $x \in \mathfrak{a}$ and $u \in U(\mathfrak{a})$. Given a positive integer n, there exists an integer t depending on x and u such that $x^t u \in U(\mathfrak{a})x^n$.*

Proof. Recall from 0.5 that the adjoint action of $\mathfrak{a}$ on itself extends naturally to an action by derivations on $U(\mathfrak{a})$ (which in this case are nilpotent). If l_x and r_x denote respectively the left and right multiplication by x on $U(\mathfrak{a})$, then $\operatorname{ad} x = l_x - r_x$ (with all these operators commuting).

Since $\operatorname{ad} x$ is nilpotent on $U(\mathfrak{a})$, there exists $q > 0$ such that $(\operatorname{ad} x)^q u = 0$. Now choose any $t \geq q + n$ and compute:

$$x^t u = l_x^t u = (r_x + \operatorname{ad} x)^t u = \sum_{i=0}^{t} \binom{t}{i} r_x^{t-i} (\operatorname{ad} x)^i u$$

$$= \sum_{i=0}^{q} \binom{t}{i} \left((\operatorname{ad} x)^i u \right) x^{t-i} \in U(\mathfrak{a})x^{t-q}.$$

By the choice of t, the final exponent is $\geq n$ as required. $\qquad\square$

The following proposition contains the key technical step in the proof to be given in the next section, but without the distracting subscripts needed there. For its proof we need a standard identity valid in any associative algebra with elements $x, y, h := [xy]$ satisfying the commutation relations for $\mathfrak{sl}(2, \mathbb{C})$; this is easily proved by induction on t (as in Lemma 1.4(c)):

$$(1) \qquad\qquad [xy^t] = ty^{t-1}(h - t + 1).$$

Proposition. *Let $\lambda, \mu \in \mathfrak{h}^*$ and $\alpha \in \Delta$, with $n := \langle \lambda + \rho, \alpha^\vee \rangle \in \mathbb{Z}$. Assume that*

$$M(s_\alpha \cdot \mu) \subset M(\mu) \subset M(\lambda).$$

Then there are two possibilities for the position of $M(s_\alpha \cdot \lambda)$:

  (a) *If $n \leq 0$, then $M(\lambda) \subset M(s_\alpha \cdot \lambda)$.*
  (b) *If $n > 0$, then $M(s_\alpha \cdot \mu) \subset M(s_\alpha \cdot \lambda) \subset M(\lambda)$.*

Proof. (a) The case $n \leq 0$ is straightforward:

$$\langle s_\alpha \cdot \lambda + \rho, \alpha^\vee \rangle = \langle s_\alpha(\lambda + \rho), \alpha^\vee \rangle = \langle \lambda + \rho, -\alpha^\vee \rangle = -n \geq 0.$$

Since α is simple, Proposition 1.4 implies that $M(\lambda) \subset M(s_\alpha \cdot \lambda)$.

(b) Since $n > 0$, Proposition 1.4 shows that $M(s_\alpha \cdot \lambda) \subset M(\lambda)$. The proof of the remaining inclusion requires much more work. To keep track of the various maximal vectors involved, denote by v_λ^+ a maximal vector of weight λ in $M(\lambda)$, so $y_\alpha^n \cdot v_\lambda^+$ is a maximal vector in $M(s_\alpha \cdot \lambda)$. Similarly, let v_μ^+ be a maximal vector in $M(\mu)$. The assumption $M(s_\alpha \cdot \mu) \subset M(\mu)$ implies that $y_\alpha^s \cdot v_\mu^+$ is a maximal vector of $M(s_\alpha \cdot \mu)$, where $s := \langle \mu + \rho, \alpha^\vee \rangle \in \mathbb{Z}^+$.

With this notation, the assumption $M(\mu) \subset M(\lambda)$ ensures the existence of $u \in U(\mathfrak{n}^-)$ for which $u \cdot v_\lambda^+ = v_\mu^+$. In this situation, we can apply the above lemma: there exists $t > 0$ such that $y_\alpha^t u \in U(\mathfrak{n}^-)y_\alpha^n$. Thus

$$y_\alpha^t \cdot v_\mu^+ = y_\alpha^t u \cdot v_\lambda^+ \in U(\mathfrak{n}^-)y_\alpha^n \cdot v_\lambda^+ \subset M(s_\alpha \cdot \lambda).$$

Increasing t if needed, we can assume $t \geq s$. If $t > s$, apply the above identity (1) to get:

$$(2) \qquad [x_\alpha y_\alpha^t] \cdot v_\mu^+ = t y_\alpha^{t-1}(h_\alpha - t + 1) \cdot v_\mu^+ = (s - t)t y_\alpha^{t-1} \cdot v_\mu^+,$$

thanks to $\mu(h_\alpha) = \langle \mu, \alpha^\vee \rangle = s - 1$. Then expand the bracket on the left:

$$(3) \qquad x_\alpha y_\alpha^t \cdot v_\mu^+ - y_\alpha^t x_\alpha \cdot v_\mu^+ = x_\alpha y_\alpha^t \cdot v_\mu^+ \in M(s_\alpha \cdot \lambda).$$

Comparison of (2) and (3) shows that $y_\alpha^{t-1} \cdot v_\mu^+ \in M(s_\alpha \cdot \lambda)$. Reducing t stepwise in this fashion (always with the restriction $t > s$), we arrive at $y_\alpha^s \cdot v_\mu^+ \in M(s_\alpha \cdot \lambda)$. Since this vector generates $M(s_\alpha \cdot \mu)$, the desired embedding follows. $\square$

4.6. Existence of Embeddings: Integral Case

With Proposition 4.5 in hand, the proof of the following theorem in the special case $\lambda \in \Lambda$ mainly requires keeping track of many subscripts. However, further ideas will be needed to handle arbitrary λ in 4.7 below.

Theorem (Verma). *Let $\lambda \in \mathfrak{h}^*$. Given $\alpha > 0$, suppose $\mu := s_\alpha \cdot \lambda \leq \lambda$. Then there exists an embedding $M(\mu) \subset M(\lambda)$.*

Proof (Integral Case). Assume $\lambda \in \Lambda$, so all weights involved here are integral. We proceed in steps.

(1) Since $\mu \in \Lambda$, there exists $w \in W$ such that $\mu' := w^{-1} \cdot \mu$ lies in $\Lambda^+ - \rho$. Thanks to Proposition 4.3, there is a sequence of (not necessarily proper) embeddings relative to a reduced expression $w = s_n \cdots s_1$ (where s_i is the reflection relative to a simple root α_i):

$$M(\mu') = M(\mu_0) \supset M(\mu_1) \supset \cdots \supset M(\mu_n) = M(\mu),$$

where $\mu_k := s_k \cdot \mu_{k-1} = (s_k \cdots s_1) \cdot \mu'$ for $k = 1, \ldots, n$. Moreover, $\mu' \geq \mu_1 \geq \cdots \geq \mu$.

(2) Next define a parallel list of weights associated to $\lambda' := w^{-1} \cdot \lambda$, starting with $\lambda_0 := \lambda'$. For $k = 1, \ldots, n - 1$, define $\lambda_k := s_k \cdot \lambda_{k-1} = (s_k \cdots s_1) \cdot \lambda$, so that $\lambda_n = \lambda$.

(3) Without loss of generality, we can assume that $\mu < \lambda$. In turn, this forces all $\mu_k \neq \lambda_k$. How are these λ_k related to the μ_k? Since μ is linked to λ by the reflection s_α, a quick calculation (left to the reader) shows that $\mu_k = s_{\beta_k} \cdot \lambda_k$ for some root β_k (with $\beta_n = \alpha$). The reflection here is the conjugate $w_k^{-1} s_\alpha w_k$, where $w_k := s_n \cdots s_{k+1}$ (setting $w_n := 1$). In particular, $\mu_k - \lambda_k$ is an integral multiple of β_k.

(4) Although all the λ_k and μ_k lie in the linkage class of λ, we have little control over the inequalities among these weights. We do know that $\mu_k \geq \mu_{k+1}$ for all k. We also know that $\mu < \lambda$, whereas $\mu' > \lambda'$ because μ' is the unique dominant weight in this linkage class. Therefore there must exist a least index k for which $\mu_k > \lambda_k$ but $\mu_{k+1} < \lambda_{k+1}$.

(5) The objective now is to prove step-by-step that there exist embeddings $M(\mu_{k+1}) \subset M(\lambda_{k+1})$, $M(\mu_{k+2}) \subset M(\lambda_{k+2}) \cdots$, culminating in the desired $M(\mu) \subset M(\lambda)$.

(6) Rewrite the definitions (recalling Exercise 1.8): $\mu_{k+1} - \lambda_{k+1} = s_{k+1}(\mu_k - \lambda_k)$. By the choice of k, the left side is a negative multiple of β_{k+1}, whereas $\mu_k - \lambda_k$ is a positive multiple of β_k. Since s_{k+1} is a simple reflection, the only positive root which it takes to a negative root is α_{k+1}, forcing β_k (and β_{k+1}) to equal α_{k+1}. Now Proposition 1.4 yields $M(\mu_{k+1}) \subset M(\lambda_{k+1})$.

(7) Combining the previous steps, we obtain embeddings:

$$M(\mu_{k+2}) = M(s_{k+2} \cdot \mu_{k+1}) \subset M(\mu_{k+1}) \subset M(\lambda_{k+1}).$$

This puts us precisely in the situation of Proposition 4.5, where either of the alternatives (a) and (b) yields $M(\mu_{k+2}) \subset M(s_{k+2} \cdot \lambda_{k+1}) = M(\lambda_{k+2})$.

(8) Iterate the argument to get $M(\mu) = M(\mu_n) \subset M(\lambda_n) = M(\lambda)$. $\square$

Exercise. Work through the steps of this argument in the special case discussed in Example 4.3.

4.7. Existence of Embeddings: General Case

Somewhat in the spirit of the density argument in 1.9, we can complete the proof of Theorem 4.6 for arbitrary $\lambda \in \mathfrak{h}^*$. The idea here is to fix a root $\alpha > 0$ and $n \in \mathbb{Z}^+$, then consider the set $X = X_{\alpha,n} \subset \mathfrak{h}^*$ consisting of all λ for which an embedding $M(\lambda - n\alpha) \subset M(\lambda)$ exists. Write $\nu := n\alpha$. (More generally, we could replace ν in the argument by any element of Γ, as the reader can check.) Since an embedding of this type can exist only if $\lambda - n\alpha$ is linked to λ, the set X is contained in the affine hyperplane $H = H_{\alpha,n} \subset \mathfrak{h}^*$ defined by the condition $\langle \lambda + \rho, \alpha^\vee \rangle = n$.

We have already proved that $\Lambda \cap H \subset X$. On the other hand, an argument like that in 1.9 shows that $\Lambda \cap H$ is *Zariski dense* in H. To conclude that $X = H$ it will therefore suffice to prove that X is *Zariski closed*, i.e., the set of common zeros of certain polynomial functions on H. To make the situation concrete, we use the fundamental weights $\varpi_1, \dots, \varpi_\ell$ as a basis for $\mathfrak{h}^*$ and express polynomial functions in terms of the coordinates of λ. For $n\alpha$, this requires us to express α first as a $\mathbb{Z}$-linear combination of the ϖ_i. Fortunately, we do not need to write down explicit polynomials, only to show their existence.

If v^+ is a maximal vector of weight λ in $M(\lambda)$, any maximal vector of weight $\lambda - \nu$ must be annihilated by $\mathfrak{n}$ and have the form $v := u \cdot v^+$ for a nonzero $u \in N := U(\mathfrak{n}^-)_{-\nu}$. This is the subspace of $U(\mathfrak{n}^-)$ having as basis the PBW monomials $y_1^{r_1} \cdots y_m^{r_m}$ which correspond to positive roots $\alpha_1, \dots, \alpha_m$ for which $\nu = \sum_i r_i \alpha_i$. Here $\dim N = \mathcal{P}(\nu)$, where $\mathcal{P}$ is Kostant's partition function.

The idea now is to define linear maps $g^\lambda : N \to U(\mathfrak{n}^-)^\ell$ (the direct sum of ℓ copies of the vector space $U(\mathfrak{n}^-)$) for all $\lambda \in \mathfrak{h}^*$ in such a way that λ belongs to X if and only if $\operatorname{rank} g^\lambda < \dim N$ (that is, g^λ has nonzero nullity). If the matrices of such linear maps are written in terms of the weight coordinates of λ, we know from linear algebra that the rank inequality translates into the fact that certain minors of the matrix for g^λ have determinant 0 (a polynomial condition on the entries). In turn these minors depend polynomially on λ.

Take standard bases (h_i, x_i, y_i) for copies $\mathfrak{s}_i$ of $\mathfrak{sl}(2, \mathbb{C})$ corresponding to simple roots. Then for $u \in U(\mathfrak{n}^-)$, the commutation relations show that $[x_i\, u] = u_i + u_i' h_i$, where $u_i, u_i' \in U(\mathfrak{n}^-)$ depend linearly on u. Then each $\lambda \in \mathfrak{h}^*$ defines linear maps $f_i^\lambda : N \to U(\mathfrak{n}^-)$, where $f_i^\lambda(u) = u_i + \lambda(h_i) u_i'$. These combine to give a single linear map $g^\lambda : N \to U(\mathfrak{n}^-)^\ell$, as promised.

Now the condition above that $v = u \cdot v^+$ be a maximal vector of weight $\lambda - \nu$ translates successively into equivalent conditions on u:

(a) $\mathfrak{n} \cdot v = \mathfrak{n} u \cdot v^+ = 0$.

(b) $x_1 u \cdot v^+ = x_2 u \cdot v^+ = \cdots = x_\ell u \cdot v^+ = 0$.

(c) $[x_1 u] \cdot v^+ = \cdots = [x_\ell u] \cdot v^+ = 0$ (since all $x_i \cdot v^+ = 0$).

(d) $f_1^\lambda(u) \cdot v^+ = \cdots = f_\ell^\lambda(u) \cdot v^+ = 0$.

(e) all $f_i^\lambda(u) = 0$ (since the action of $U(\mathfrak{n}^-)$ on $M(\lambda)$ is free).

Condition (e) amounts to the single requirement that $\operatorname{rank} g^\lambda < \dim N$.

This completes the proof of Theorem 4.6 in general. $\qquad\square$

Unlike the more limited formulation in which α is required to be simple, the theorem allows one to iterate the procedure and construct a sequence of embeddings $M(\lambda) \supset M(s_\alpha \cdot \lambda) \supset M(s_\beta s_\alpha \cdot \lambda) \supset \cdots$ provided the weights involved decrease in the partial ordering. This yields in particular a more precise *sufficient* condition for $L(\mu)$ to occur as a composition factor of $M(\lambda)$ (with multiplicity at least 1). In 5.1 we will go on to discuss the necessity of this condition.

4.8. Simplicity Criterion: General Case

Having obtained the analogue of Proposition 1.4 for an arbitrary positive root (and arbitrary highest weight), we can now fill in the proof of the criterion stated earlier for simplicity of a Verma module (4.4) but proved there only for integral weights:

Theorem. *Let $\lambda \in \mathfrak{h}^*$. Then $M(\lambda) = L(\lambda)$ if and only if λ is antidominant.*

Proof. (a) Suppose $M(\lambda)$ is simple. Going back to the original definition of "antidominant", we have to show for all $\alpha \in \Phi^+$ that $\langle \lambda + \rho, \alpha^\vee \rangle \notin \mathbb{Z}^{>0}$. Suppose this fails for some α. Since $s_\alpha \cdot \lambda = \lambda - \langle \lambda + \rho, \alpha^\vee \rangle \alpha$, we have $s_\alpha \cdot \lambda < \lambda$. Thanks to Theorem 4.6, there is an embedding $M(s_\alpha \cdot \lambda) \subset M(\lambda)$, which is proper because of the strict inequality of weights. This contradicts the simplicity of $M(\lambda)$.

(b) Suppose λ is antidominant. As we already pointed out after the proof of the integral case in 4.4, use of the group $W_{[\lambda]}$ allows that proof to be carried over to the general case: since λ is minimal in its $W_{[\lambda]}$ dot-orbit, Harish-Chandra's Theorem ensures that $M(\lambda)$ is simple. $\qquad\square$

Corollary. *Let $\lambda \in \mathfrak{h}^*$ be antidominant. Then for all $w \in W_{[\lambda]}$, the socle of $P(w \cdot \lambda)$ is a direct sum of copies of $L(\lambda)$.*

Proof. By Theorem 3.10, there is a standard filtration

$$0 = P_0 \subset P_1 \subset \cdots \subset P_n = P(w \cdot \lambda),$$

with each P_i/P_{i-1} isomorphic to some $M(w' \cdot \lambda)$. If L is a simple summand of $\operatorname{Soc} P(w \cdot \lambda)$, let i be the least index for which $L \subset P_i$ (so $L \cap P_{i-1} = 0$).

Thus L embeds in some $M(w' \cdot \lambda)$. Combining the above theorem with Theorem 4.2(c), $L \cong L(\lambda)$. $\qquad\square$

It is not so easy to determine how many copies of $L(\lambda)$ occur in this socle. Using more sophisticated arguments (see 13.14 below), Stroppel [**246**, Thm. 8.1] shows (for integral weights) that $\operatorname{Soc} P(w \cdot \lambda)$ is a direct sum of r copies of $L(\lambda)$, where $r = (P(w \cdot \lambda) : M(w_\circ \cdot \lambda))$. By BGG Reciprocity, this filtration multiplicity is the same as $[M(w_\circ \cdot \lambda) : L(w \cdot \lambda)]$, the computation of which in general requires Kazhdan–Lusztig theory (8.4).

The corollary has an interesting further consequence, which will be refined in 4.10 below for integral weights and in 7.16 for arbitrary weights:

Exercise. Let $\lambda \in \mathfrak{h}^*$. If $P(\lambda) \cong P(\lambda)^{\vee}$ is self-dual, i.e., $P(\lambda) \cong Q(\lambda)$, prove that λ must be antidominant. What can you say about the converse?

4.9. Blocks of $\mathcal{O}$ Revisited

In 1.13 we formulated the notion of "block" in category $\mathcal{O}$ and showed by elementary arguments that the blocks involving *integral* weights are the subcategories $\mathcal{O}_{\chi_\lambda}$ containing the modules whose composition factors have highest weights linked to $\lambda \in \Lambda$. Now we can formulate a more definitive result using the subgroups $W_{[\lambda]}$ of W introduced in 3.4:

Theorem. *The blocks of $\mathcal{O}$ are precisely the subcategories consisting of modules whose composition factors all have highest weights linked by $W_{[\lambda]}$ to an antidominant weight λ. Thus the blocks are in natural bijection with antidominant (or alternatively, dominant) weights.*

Proof. Start with an arbitrary weight μ. Thanks to Proposition 4.1 and Theorem 4.2(c), $M(\mu)$ has a unique simple submodule $L(\lambda) = M(\lambda)$. But the highest weight of a simple Verma module must be antidominant, by Theorem 4.8. Thus all composition factors of $M(\mu)$ including $L(\mu)$ lie in the block of $L(\lambda)$. The highest weights involved are then linked to λ by $W_{[\lambda]}$. In the other direction, $L(\lambda)$ is the unique simple submodule of any $M(w \cdot \lambda)$ with $w \in W_{[\lambda]}$. $\qquad\square$

As in Remark 3.5, we denote the block associated to an antidominant weight by $\mathcal{O}_\lambda$ (keeping in mind the mild inconsistency in our use of $\mathcal{O}_0$ for the block corresponding to the linkage class of the dominant integral weight 0). It was observed there that each of the $\mathcal{O}_\lambda$ belonging to a fixed central character involves the same number of linked weights (hence simple modules). This is consistent with the deeper result that all such blocks are equivalent as categories: see Mathieu [**206**, Lemma A.3], where the proof requires a localization technique using nonintegral powers of root vectors. (Is there a more direct proof of the equivalence?)

Exercise. Refine Proposition 3.14(e) as follows: If $M \in \mathcal{O}$ has a contravariant form, then its block summands in distinct blocks $\mathcal{O}_\lambda, \mathcal{O}_\mu$ are orthogonal.

4.10. Example: Antidominant Projectives

Here we examine an interesting special case which expands on Exercise 4.8 and makes use of much of the theory developed so far. It illustrates well what can and cannot be said based on what we know at this point.

The goal here is to describe the projective cover of a simple module whose highest weight is *antidominant* and *integral*. Start with a weight $\lambda \in \Lambda^+ - \rho$. Let $\chi = \chi_\lambda$ be the corresponding central character. Now $\dim L(\lambda + \rho) < \infty$ (1.6), so tensoring it with a projective produces another projective (Proposition 3.8(b)).

We know that $P(-\rho) = M(-\rho) = L(-\rho)$ (Example 3.8); being self-dual, this module is also injective. Now set $T := M(-\rho) \otimes L(\lambda + \rho)$. This is projective (as well as injective) and has a standard filtration by Verma modules with highest weights $-\rho + \mu$, where μ ranges over the weights of $L(\lambda + \rho)$ (3.6). The same is true of its direct summands. In particular, the extremal weights $w(\lambda + \rho)$, with $w \in W$, yield Verma modules $M(w \cdot \lambda)$ as filtration quotients (with multiplicity 1) of T^χ. These are clearly the only weights μ of $L(\lambda + \rho)$ for which $-\rho + \mu$ is linked to λ. Thus the block summand T^χ is filtered only by the $M(w \cdot \lambda)$.

Thanks to Remark 3.6, T^χ has $M(w_\circ \cdot \lambda)$ as a quotient. This Verma module is actually simple, since $w_\circ \cdot \lambda$ is antidominant (4.4). Since T^χ is projective, the projective cover $P(w_\circ \cdot \lambda)$ must occur as a summand. In particular, a standard filtration of $P(w_\circ \cdot \lambda)$ involves various modules $M(w \cdot \lambda)$, each occurring only once. By BGG Reciprocity (3.11), we get $[M(w \cdot \lambda) : L(w_\circ \cdot \lambda)] \leq 1$ for all $w \in W$. On the other hand, every $M(w \cdot \lambda)$ has $L(w_\circ \cdot \lambda)$ as its unique simple submodule, by Theorems 4.2(c) and 4.4. This forces $[M(w \cdot \lambda) : L(w_\circ \cdot \lambda)] \geq 1$. Again invoking BGG Reciprocity, we conclude that $T^\chi = P(w_\circ \cdot \lambda)$.

Now Remark 3.6 shows that $M(\lambda)$ occurs as a submodule of T^χ, hence also its unique simple submodule $L(w_\circ \cdot \lambda)$. Since T^χ is injective (and indecomposable), it must be the injective envelope of this simple module. In other words, in this special case the projective cover is actually *self-dual*. (This can happen only in case the highest weight of the simple module is *antidominant*, as noted in Exercise 4.8.)

To summarize:

Theorem. *Let $\lambda + \rho \in \Lambda^+$, so $w_\circ \cdot \lambda$ is antidominant and integral. Then a standard filtration of $P(w_\circ \cdot \lambda)$ involves all of the distinct $M(w \cdot \lambda)$ exactly* *once. Thus $[M(w \cdot \lambda) : L(w_\circ \cdot \lambda)] = 1$ for all $w \in W$. Moreover, $P(w_\circ \cdot \lambda)$ is self-dual and equal to the injective envelope of $L(w_\circ \cdot \lambda)$.* $\qquad\square$

In the language of translation functors (7.1), we have shown in particular that translation from $-\rho$ to $w_\circ \cdot \lambda$ takes $M(-\rho)$ to $P(w_\circ \cdot \lambda)$. (See the more general formulation in 7.13.)

What can be said if $w_\circ \cdot \lambda$ is replaced by an arbitrary antidominant weight? Although the method of proof we have used here fails to generalize, the essential features of the proposition remain true in general. For the proof it seems necessary to work explicitly with translation functors: see 7.16.

Remark. Self-dual projectives P turn out to be of considerable theoretical interest, as discussed later in 13.11. A natural but rather subtle problem is to describe the structure of the algebra $\mathrm{End}_{\mathcal{O}}\, P$. For the easy case $\mathfrak{g} = \mathfrak{sl}(2,\mathbb{C})$, see Exercise 3.12.

Exercise. Under the hypothesis of the theorem, what is $\dim \mathrm{End}_{\mathcal{O}}\, P(w_\circ \cdot \lambda)$?

4.11. Application to $\mathfrak{sl}(3,\mathbb{C})$

It requires a lot of new ideas (going as far as Chapter 8) to work out the composition factor multiplicities of arbitrary Verma modules. Along the way it is worth keeping track of what can be said in the simplest nontrivial case: $\mathfrak{g} = \mathfrak{sl}(3,\mathbb{C})$. Denote the two simple roots by α, β. Here

$$W = \{1, s_\alpha, s_\beta, s_\alpha s_\beta, s_\beta s_\alpha, w_\circ\}.$$

Consider first a linkage class $\{w \cdot \lambda \mid w \in W\}$, where λ is regular, integral, and antidominant. Because λ is antidominant, Theorem 4.4 shows that $M(\lambda) = L(\lambda)$, while no other $M(w \cdot \lambda)$ can be simple. Moreover, Theorem 4.2(c) shows that $L(\lambda)$ is the unique simple submodule of each $M(w \cdot \lambda)$. Even more is true: by Theorem 4.10, $[M(w \cdot \lambda) : L(\lambda)] = 1$ for all w.

This makes it easy to find the composition factors (or formal character) of $M(s_\alpha \cdot \lambda)$: The only linked weight strictly below $s_\alpha \cdot \lambda$ is λ. Since both $L(s_\alpha \cdot \lambda)$ and $L(\lambda)$ occur just once as composition factors, it follows that $\mathrm{ch}\, L(s_\alpha \cdot \lambda) = \mathrm{ch}\, M(s_\alpha \cdot \lambda) - \mathrm{ch}\, M(\lambda)$. (Of course, α can be replaced by β here.) The remaining three Verma modules require more theory to handle: see 5.4 below.

Exercise. In the case of $\mathfrak{sl}(3,\mathbb{C})$, what can be said at this point about Verma modules with a singular integral highest weight? Leaving aside the trivial case of $-\rho$, a typical linkage class has three elements: for example, if λ lies just in the α-hyperplane and is antidominant, the linked weights are $\lambda, s_\beta \cdot \lambda, s_\alpha s_\beta \cdot \lambda$.

4.12. Shapovalov Elements

Before discussing the further implications of Verma's Theorem 4.6, we take up the natural question of whether it is possible for arbitrary $\gamma > 0$ to construct an explicit embedding $M(s_\gamma \cdot \lambda) \hookrightarrow M(\lambda)$ whenever $s_\gamma \cdot \lambda < \lambda$ (as done for simple roots in Proposition 1.4). This amounts to finding an element $u \in U(\mathbf{n}^-)$ such that $u \cdot v^+$ is a maximal vector of weight $s_\gamma \cdot \lambda$ in $M(\lambda)$. Theorem 4.6 guarantees the existence of u, which is unique up to nonzero scalar multiples (Theorem 4.2).

Though it is sometimes possible to write down u in a concrete case, as in the exercise below, this is usually a daunting task. Indeed, if $\nu = \lambda - s_\gamma \cdot \lambda = r\gamma$, where $r = \langle \lambda + \rho, \gamma^\vee \rangle \in \mathbb{Z}^{>0}$, then u lies in $U(\mathbf{n}^-)_{-\nu}$. This space has dimension $\mathcal{P}(\nu)$: a basis consists of the various PBW monomials $y_1^{r_1} \cdots y_m^{r_m}$ such that $\sum_i r_i \alpha_i = \gamma$ if the positive roots are enumerated as $\alpha_1, \ldots, \alpha_m$. Unless γ is simple, this space is usually too large for explicit computations.

Moreover, one wants to understand the construction conceptually. For example, how does the choice of u depend on λ? It is instructive to look closely at the easiest nontrivial case, where computations can be done by hand:

Exercise. Let $\mathfrak{g} = \mathfrak{sl}(3, \mathbb{C})$, with $\Phi^+ = \{\alpha, \beta, \gamma\}$ (where $\gamma = \alpha + \beta$) and standard basis

$$(x_\alpha, x_\beta, x_\gamma, h_\alpha, h_\beta, y_\alpha, y_\beta, y_\gamma).$$

  (a) Use the commutation relations in $\mathfrak{g}$ to work out products of basis elements in the PBW ordering, such as $x_\alpha y_\gamma = y_\gamma x_\alpha - y_\beta$.

  (b) Setting $\nu := \gamma \ (= \rho)$, the elements $y_\alpha y_\beta$ and y_γ form a basis of $U(\mathbf{n}^-)_{-\nu}$. Rewrite the left multiple of each of these by x_α and by x_β in the PBW ordering:

$$\begin{aligned}
x_\alpha y_\alpha y_\beta &= y_\alpha y_\beta x_\alpha + y_\beta h_\alpha + y_\beta, \\
x_\beta y_\alpha y_\beta &= y_\alpha y_\beta x_\beta + y_\alpha h_\beta, \\
x_\alpha y_\gamma &= y_\gamma x_\alpha - y_\beta, \\
x_\beta y_\gamma &= y_\gamma x_\beta + y_\alpha.
\end{aligned}$$

Similarly, rewrite $x_\alpha y_\gamma h_\beta$ and $x_\beta y_\gamma h_\beta$.

  (c) Write $\lambda \in \mathfrak{h}^*$ as $a\varpi_\alpha + b\varpi_\beta$. Then $\langle \lambda + \rho, \gamma^\vee \rangle = 1$ precisely when $a + b = -1$ (defining an affine hyperplane H in $\mathfrak{h}^* \cong \mathbb{C}^2$). Whenever λ lies in H, $s_\gamma \cdot \lambda = \lambda - \gamma$. From now on, assume that λ satisfies this condition.

  (d) If $r, s \in \mathbb{C}$, set $u := r y_\alpha y_\beta + s y_\gamma \in U(\mathbf{n}^-)$, with not both r and s zero. If v^+ is a maximal vector in $M(\lambda)$, suppose that $u \cdot v^+$ is a maximal vector (of weight $\lambda - \gamma$). Show that $r(a + 1) - s = 0$ and

$br + s = 0$, forcing $r \neq 0$. But the choice of r and s (which is unique up to a nonzero scalar multiple) depends on λ. For example, take $\lambda = -\varpi_\alpha$ with $r = 1, s = 0$ or $\lambda = -\varpi_\beta$ with $r = 1, s = 1$.

(e) Now set $u := y_\alpha y_\beta - y_\gamma h_\beta \in U(\mathfrak{b}^-)_{-\gamma}$. Show that $u \cdot v^+$ is always a maximal vector in $M(\lambda)$, independent of the choice of λ in H.

This small example already illustrates fairly well a general idea due to Shapovalov [**231**], which will be essential later in the proof of Theorem 5.8. The basic data here include a positive but not necessarily simple root γ and an integer $r > 0$ (here $r = 1$). On the one hand, an element $u \in U(\mathfrak{n}^-)_{-\gamma}$ which produces a maximal vector $u \cdot v^+$ in $M(\lambda)$ of weight $\lambda - \gamma$ is unique (up to a nonzero scalar multiple) but typically depends on λ. The relevant λ form an affine hyperplane in $\mathfrak{h}^*$, defined by $\langle \lambda + \rho, \gamma^\vee \rangle = 1$. On the other hand, an element of $U(\mathfrak{b}^-)$ might serve the same purpose but be independent of λ. This element is no longer unique up to scalars, only unique modulo the left ideal in $U(\mathfrak{b}^-)$ generated by $h_\gamma + \rho(h_\gamma) - 1 \, (= h_\gamma + 1)$. In our example, we also see that the monomial $y_\alpha y_\beta$ occurs without a factor from $U(\mathfrak{h})$.

To formulate the general result we start with a standard basis of $\mathfrak{g}$ and resulting PBW basis of $U(\mathfrak{g})$. As usual, $y_1, \ldots, y_m$ correspond to the negative roots $-\alpha_1, \ldots, -\alpha_m$, where $\alpha_1, \ldots, \alpha_\ell$ are simple. By the "degree" of a PBW monomial $y_1^{r_1} \cdots y_m^{r_m}$ we mean its degree when viewed as a polynomial in m variables, namely $\sum_i r_i$. This contrasts with the grading of $U(\mathfrak{g})$ by the root lattice.

Theorem (Shapovalov). *Fix $\gamma \in \Phi^+$ and an integer $r > 0$. There exists an element $\theta_{\gamma,r} \in U(\mathfrak{b}^-)_{-r\gamma}$ having the following properties:*

(a) *For each root $\beta > 0$, the commutator $[x_\beta, \theta_{\gamma,r}]$ lies in the left ideal*
$$I_{\gamma,r} := U(\mathfrak{g})(h_\gamma + \rho(h_\gamma) - r) + U(\mathfrak{g})\mathfrak{n}.$$

(b) *If $\gamma = \sum_{i=1}^\ell a_i \alpha_i$, we can write*

$$\theta_{\gamma,r} = \prod_{i=1}^\ell y_i^{r a_i} + \sum_j p_j q_j,$$

with $p_j \in U(\mathfrak{n}^-)_{-r\gamma}$, $q_j \in U(\mathfrak{h})$, and $\deg p_j < r \sum_i a_i$.

Moreover, $\theta_{\gamma,r}$ is unique (up to a nonzero scalar multiple) modulo the left ideal $J_{\gamma,r} := U(\mathfrak{b}^-)(h_\gamma + \rho(h_\gamma) - r)$.

In (b) note that the highest degree term does not depend on the ordering of the simple roots $\alpha_1, \ldots, \alpha_\ell$. Condition (b) contributes to the proof of uniqueness by removing ambiguity as to scalar multiples; at the same time it helps in an inductive step of the proof and makes the description of $\theta_{\gamma,r}$ more precise. The theorem will be proved in the next section, but first we emphasize that it is difficult to devise a constructive method for obtaining $\theta_{\gamma,r}$ explicitly. Instead we settle for a somewhat roundabout approach to the existence.

4.13. Proof of Shapovalov's Theorem

It is easy to get started. We fix $r > 0$ once and for all. If γ is *simple*, then thanks to Proposition 1.4 we can just set $\theta_{\gamma,r} := y_\gamma^r$. This suggests an induction based on $\operatorname{ht}\gamma$. Assuming $\gamma \notin \Delta$, there exists $\alpha \in \Delta$ such that $p := \langle \gamma, \alpha^\vee \rangle > 0$ (0.2). We keep γ and α fixed from now on. Thus $\beta := s_\alpha\gamma = \gamma - p\alpha > 0$ with $\operatorname{ht}\beta < \operatorname{ht}\gamma$. The induction hypothesis provides an element $\theta_{\beta,r}$ with the desired properties.

Unfortunately, there is no obvious way to produce $\theta_{\gamma,r}$ from this data. The enveloping algebra setting is too abstract for this purpose, so we try to exploit some of the Verma module theory already developed. Given $\lambda \in \mathfrak{h}^*$ and a known element $\theta \in U(\mathfrak{b}^-)$ (such as $\theta_{\beta,r}$), write $\theta(\lambda)$ for the unique element of $U(\mathfrak{n}^-)$ congruent to θ modulo the left ideal I. In other words, given a maximal vector v^+ of weight λ in $M(\lambda)$, the element $\theta(\lambda)$ is the unique element $u \in U(\mathfrak{n}^-)$ such that $\theta \cdot v^+ = u \cdot v^+$. (This is independent of the choice of v^+.) The hope is to reconstruct the unknown $\theta_{\gamma,r}$ from such specializations.

As in some earlier proofs, a Zariski-density argument will enable us to focus just on selected weights. Since we are only interested here in those λ for which $M(\lambda)$ has a submodule isomorphic to $M(\lambda - r\gamma)$, we work in the hyperplane $H_{\gamma,r} \subset \mathfrak{h}^*$ defined by

$$H_{\gamma,r} := \{\lambda \in \mathfrak{h}^* \mid \langle \lambda + \rho, \gamma^\vee \rangle = r\}.$$

We know that $\Lambda \cap H_{\gamma,r}$ is Zariski-dense in $H_{\gamma,r}$. Intersecting this set with the half-space

$$H_\alpha := \{\lambda \in \mathfrak{h}^* \mid \langle \lambda + \rho, \alpha^\vee \rangle < 0\}$$

still yields a dense subset of the hyperplane; call it Θ.

The strategy now is to assign to each $\lambda \in \Theta$ an element $\theta_{\gamma,r}(\lambda)$ in $U(\mathfrak{n}^-)_{-r\gamma}$ satisfying analogues of (a) and (b) in the theorem together with a polynomial condition on its coordinates relative to a PBW basis of $U(\mathfrak{n}^-)$. This will allow the construction to be extended to all $\lambda \in H_{\gamma,r}$ by using the density of Θ.

Recall that I denotes the annihilator in $U(\mathfrak{g})$ of a maximal vector v^+ in $M(\lambda)$. We require $\theta_{\gamma,r}(\lambda)$ to satisfy:

(a') $[x_\gamma, \theta_{\gamma,r}(\lambda)] \in I$.

(b′) Independent of the choice of $\lambda \in \Theta$, the highest degree term in $\theta_{\gamma,r}(\lambda)$ when written in a standard PBW basis is $\prod_i y_i^{ra_i}$. (This condition is independent of the ordering of roots.)

(c′) The coefficients of $\theta_{\gamma,r}(\lambda)$ in the PBW basis (hence in any other basis of $U(\mathfrak{n}^-)$) depend polynomially on λ.

What (c′) means is this: relative to the fixed basis of $U(\mathfrak{n}^-)$, there are elements $p_j \in U(\mathfrak{n}^-)_{-r\gamma}$ and $q_j \in U(\mathfrak{h})$ (independent of λ) so that $\theta_{\gamma,r}(\lambda) = \sum_j p_j q_j(\lambda)$. Here the q_j are viewed as polynomial functions on $\mathfrak{h}^*$. According to (b′), $q_j = 1$ whenever $p_j = \prod_i y_i^{ra_i}$. Since Θ is Zariski-dense in $H_{\gamma,r}$, parts (a) and (b) of the theorem will follow if we set $\theta_{\gamma,r} := \sum_j p_j q_j \in U(\mathfrak{b}^-)$. In view of (b), it follows from the proof that $\theta_{\gamma,r}$ is unique modulo the left ideal $J_{\gamma,r}$ of $U(\mathfrak{b}^-)$, which specializes at each λ to the annihilator in $U(\mathfrak{b}^-)$ of a maximal vector generating $M(\lambda)$.

Recall the data established at the outset: $r > 0$ and $\gamma > 0$ of height > 1, together with $\alpha \in \Delta$ such that $p = \langle \gamma, \alpha^\vee \rangle > 0$ and $\beta = s_\alpha \gamma = \gamma - p\alpha > 0$ of lower height. The induction hypothesis provided $\theta_{\beta,r} \in U(\mathfrak{n}^-)$ satisfying the conditions of the theorem; in particular, all possible specializations $\theta_{\beta,r}(\mu)$ are defined for μ in the hyperplane $H_{\beta,r}$. Note that $\beta = \sum_{i=1}^{\ell} b_i \alpha_i$ with $b_i = a_i$ except when $\alpha_i = \alpha$, in which case $b_i = a_i - p$. Thus the highest degree term of $\theta_{\beta,r}$ is $\prod_i y_i^{rb_i}$.

Now consider $\lambda \in \Theta$, so $\langle \lambda + \rho, \alpha^\vee \rangle = -q < 0$ with $q \in \mathbb{Z}^{>0}$. Define $\mu := s_\alpha \cdot \lambda = \lambda + q\alpha > \lambda$ and observe that $\mu \in H_{\beta,r}$. A quick calculation shows that $s_\alpha \cdot (\mu - r\beta) = \lambda - r\gamma$, with $\langle \mu - r\beta + \rho, \alpha^\vee \rangle = q + rp$ (call this positive integer n). So we have an embedding $M(\lambda - r\gamma) \hookrightarrow M(\mu - r\beta)$. Finally, we can embed $M(\mu - r\beta) \hookrightarrow M(\mu)$ because $r > 0$. The resulting inclusions of Verma modules are pictured in Figure 1.

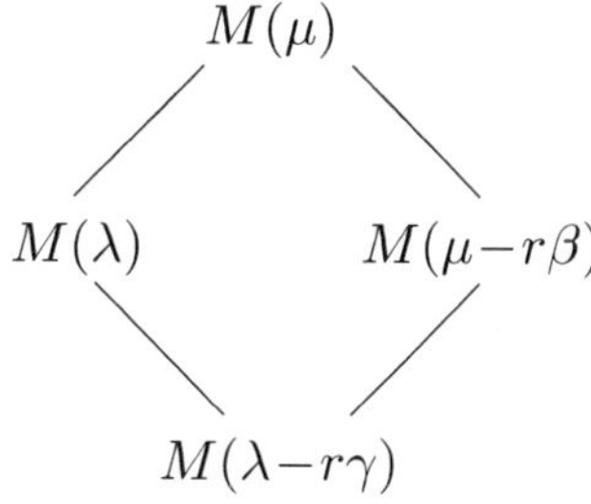

Figure 1. Inclusions of Verma modules

Thanks to the induction hypothesis, in three of the four cases we have in hand an explicit element in $U(\mathfrak{n}^-)$ taking a given maximal vector of the Verma module to a maximal vector for the embedded submodule: y_α^q, y_α^n, and the specialization $\theta_{\beta,r}(\mu)$. Since such elements are unique up to scalars, the embedding $M(\lambda - r\gamma) \hookrightarrow M(\lambda)$ is induced by a unique element $\theta_{\gamma,r}(\lambda) \in U(\mathfrak{n}^-)$ satisfying:

$$(1) \qquad \theta_{\gamma,r}(\lambda)y_\alpha^q = y_\alpha^n \theta_{\beta,r}(\mu), \ \text{with } n = q + rp.$$

It is clear that this new element $\theta_{\gamma,r}(\lambda)$ satisfies (a′) above. To check (b′), compare the two sides of (1): On the right, the highest degree of a monomial is $n + r\sum_i b_i = q + rp + r\sum_i b_i = q + r\sum_i a_i$, since $b_i = a_i$ unless $\alpha_i = \alpha$ (in which case $b_i = a_i - p$). Thus the monomial of highest degree in $\theta_{\gamma,r}(\lambda)$ must be $\prod_i y_i^{ra_i}$, as required.

For (c′), use an alternative PBW basis for $U(\mathfrak{n}^-)$ in which the power of y_α in each monomial occurs at the right end. When the right side of (1) is rewritten in this way, comparison with the left side shows that y_α^q must be a factor in every monomial. Since $U(\mathfrak{n}^-)$ has no zero divisors, we can then cancel y_α^q from both sides to get an explicit expression for $\theta_{\gamma,r}(\lambda)$. The rewriting process just involves commutation relations in $\mathfrak{g}$ (independent of the weights). Now property (c′) for $\theta_{\beta,r}(\mu)$, in tandem with the fact that λ depends linearly on μ, ensures that all coefficients of $\theta_{\gamma,r}(\lambda)$ (for any basis of $U(\mathfrak{n}^-)$) depend polynomially on λ.

4.14. A Look Back at Verma's Thesis

Verma's 1966 thesis [**251**] (and the announcement [**252**]) stimulated the work of BGG and others in the 1970s. Although his results and methods have since been greatly improved and extended, the thesis itself merits some concluding comments in this chapter. We indicate briefly where Verma's main ideas fit into our treatment (using our notation and terminology rather than his).

His results Lemma 2.8 and Theorem 3.10 on homomorphisms between Verma modules are equivalent to our Theorem 4.2. However, the dimension bound on the Hom space was originally proved by studying the growth in dimensions of weight spaces using the partition function (Appendix B).

His Theorem 3.1 is our 1.4, while his Theorem 3.2 is our 4.3.

Verma made one crucial error: he thought he had proved that all submodules of $M(\lambda)$ are generated by maximal vectors; in turn, all composition factors $L(\mu)$ would result from embeddings $M(\mu) \hookrightarrow M(\lambda)$ and thus occur with multiplicity 1. The source is a subtle gap in the last paragraph of the proof of his Lemma 3.7 (which propagates to Theorem 3.8 and eventually Lemma 7.1, Theorem 7.2). Counterexamples to his assertion emerged in the work of BGG [**25**] and of Conze–Dixmier [**72**, Ex. 2]. The latter paper gives an example in type D_4 of a Verma module $M(\lambda)$ having infinitely many submodules not generated by their Verma submodules: here $\lambda + \rho$ is taken to be the fundamental weight corresponding to the vertex of the Dynkin diagram connected to all other vertices. A full understanding of the multiplicity problem only emerged with the formulation of the Kazhdan–Lusztig Conjecture in 1979 (as explained in Chapter 8 below).

Verma's §5, which is unaffected by the error, yields our Theorem 4.6 on embeddings $M(s_\alpha \cdot \lambda) \hookrightarrow M(\lambda)$.

In §6 of the thesis there are a number of interesting examples and counterexamples, leading to the elaboration in §7 of the conjecture later proved in a stronger form by BGG (5.1). Here the role of the Bruhat ordering is first made explicit.

Notes

The results in this chapter are mainly due to Verma [**251**], BGG [**25, 26**], and Shapovalov [**231**], with further refinements by Dixmier [**84**].

When reading BGG or Dixmier, it has to be kept in mind that their notational conventions differ from ours in one essential respect: their $M(\lambda)$ denotes the Verma module with highest weight $\lambda - \rho$, while $M(w\lambda)$ is the Verma module with highest weight linked to $\lambda - \rho$ by w. Some improvements were incorporated in the English translation of Dixmier's book, later reprinted by the AMS. (But this translation contains a large number of uncorrected misprints, most of which are not in the short list given at the end of the AMS reprint.)

For Lemma 4.1 see Dixmier [**84**, 3.6.9]. [This is incorrectly cited as 3.6.13 in the simplified proof due to W. Borho of [**84**, 7.6.3] in the English translation.] In Proposition 4.1 we follow Borho's line of argument.

Theorem 4.6 is due originally to Verma [**251**, §5]; here we mainly follow the proof in Dixmier [**84**, 7.6.9–7.6.13].

Following the work of Shapovalov [**231**], a number of people have studied further the embeddings of Verma modules $M(s_\alpha \cdot \lambda) \hookrightarrow M(\lambda)$. Lutsyuk [**204**] develops a recursive formula for the elements of $U(\mathfrak{n}^-)$ inducing such embeddings (but claims erroneously that these must coincide with Shapovalov's elements). Franklin [**96**] shows how to construct the Shapovalov elements in a $\mathbb{Z}$-form of $U(\mathfrak{g})$ which is well suited to reduction modulo a prime. This is inspired by work of Carter on raising and lowering operators. Computational methods explored more recently by de Graaf [**115**] admit natural generalizations to Kac–Moody algebras.

In the proof of Theorem 4.10 the multiplicity 1 property is a crucial point. For nonintegral weights this remains valid (7.16) and can be approached in several ways, none quite as direct as the argument given here and typically requiring more information than we have yet developed. See Borho–Jantzen [**42**, 2.10], Conze–Dixmier [**72**, Lemme 1], and Jantzen [**147**,

2.23]. In the setting of Chapter 8, the multiplicity 1 in 4.10 corresponds to the fact that a related Kazhdan–Lusztig polynomial is equal to 1 (see for example [**129**, Exercise 7.14]).

Chapter 5

Highest Weight Modules II

Now we turn to developments after 1970 due to Bernstein–Gelfand–Gelfand and Shapovalov, along with important refinements due to Jantzen (to be studied further in Chapter 7). A central result is the BGG Theorem on composition factors of Verma modules (5.1), which can be reformulated for regular weights in terms of the Bruhat ordering (5.2). We emphasize the alternative proof of the theorem by Jantzen in the framework of Jantzen filtrations (5.3–5.7). In the proof given here, the determinant of a nondegenerate contravariant form on a weight space plays an essential role. It was computed independently by Shapovalov and Jantzen (5.8–5.9).

5.1. BGG Theorem

In Chapter 4 we raised the question of determining which simple modules occur as composition factors of a given Verma module $M(\lambda)$. Leaving aside the much more delicate question of their exact multiplicities, we can now state the definitive answer which was conjectured (in an oversimplified form) by Verma following his proof of Theorem 4.6 and ultimately proved by Bernstein–Gelfand–Gelfand [**25, 26**]. For this a special notation will be useful: Let $\lambda, \mu \in \mathfrak{h}^*$ and write $\mu \uparrow \lambda$ if $\mu = \lambda$ or there is a root $\alpha > 0$ such that $\mu = s_\alpha \cdot \lambda < \lambda$; in other words, $\langle \lambda + \rho, \alpha^\vee \rangle \in \mathbb{Z}^{>0}$. More generally, if $\mu = \lambda$ or there exist $\alpha_1, \ldots, \alpha_r \in \Phi^+$ such that

$$\mu = (s_{\alpha_1} \cdots s_{\alpha_r}) \cdot \lambda \uparrow (s_{\alpha_2} \cdots s_{\alpha_r}) \cdot \lambda \uparrow \cdots \uparrow s_{\alpha_r} \cdot \lambda \uparrow \lambda,$$

we say that μ is **strongly linked** to λ and write $\mu \uparrow \lambda$.

Theorem. *Let* $\lambda, \mu \in \mathfrak{h}^*$.

(a) **(Verma)** *If* μ *is strongly linked to* λ, *then* $M(\mu) \hookrightarrow M(\lambda)$; *in particular,* $[M(\lambda) : L(\mu)] \neq 0$.

(b) **(BGG)** *If* $[M(\lambda) : L(\mu)] \neq 0$, *then* μ *is strongly linked to* λ.

By iterating Theorem 4.6, we immediately obtain part (a).

Part (b) of the theorem lies somewhat deeper. Rather than present the original proof by BGG [**25**, **26**] (which is exposed thoroughly by Dixmier [**84**, Chap. 7]), we shall develop the proof using methods of Jantzen [**147**] which are of independent interest: the Jantzen filtration and contravariant forms.

Example. As an application of the theorem, Proposition 4.3 can be extended to cover the case when $\lambda \in \mathfrak{h}^*$ is *dominant* in the sense of 3.5 but not necessarily integral: *For all* $w \in W_{[\lambda]}$, *there is an embedding* $M(w \cdot \lambda) \subset M(\lambda)$; *in particular,* $[M(\lambda) : L(w \cdot \lambda)] > 0$ *for all* $w \in W_{[\lambda]}$. The observation before the statement of Proposition 4.3 remains valid when $\alpha \in \Delta_{[\lambda]}$ and $w \in W_{[\lambda]}$; but here α need not lie in Δ (Exercise 3.4). So the above theorem replaces Proposition 1.4 in the proof of 4.3, ensuring the existence of the relevant embeddings of Verma modules.

Remarks. (1) Taken together, the two parts of the theorem imply the following equivalence:

$(*)$ $\qquad$ Let $\lambda, \mu \in \mathfrak{h}^*$. Then $[M(\lambda) : L(\mu)] \neq 0 \Leftrightarrow M(\mu) \hookrightarrow M(\lambda)$.

This can be proved independently of the BGG Theorem, by some inductive arguments involving BGG Reciprocity and the Ext functor: see Moody–Pianzola [**223**, 2.11]. The subtle point, however, is that not every occurrence of $L(\mu)$ as a composition factor of $M(\lambda)$ need be accounted for by a straightforward homomorphism $M(\mu) \rightarrow M(\lambda)$. Multiplicities bigger than 1 actually begin to show up in rank 3.

(2) Verma only conjectured that strong linkage should be a necessary condition for $M(\mu) \hookrightarrow M(\lambda)$, believing that he had already proved $(*)$ (as discussed in 4.14).

Exercise. Let $\lambda \in \mathfrak{h}^*$. If all $[M(\lambda) : L(\mu)] = 1$, use $(*)$ to prove that the maximal submodule $N(\lambda)$ of $M(\lambda)$ is the sum of the corresponding embedded Verma modules $M(\mu)$. (Examples show however that the converse is false.)

5.2. Bruhat Ordering

It is useful to reformulate Theorem 5.1 in terms of the Weyl group alone. For this we have to invoke some basic facts about the Chevalley–Bruhat ordering of W: see 0.4. To make the transition from weights to elements of W, we consider only *regular integral* weights (1.8), meaning "dot-regular". For example, 0 is regular, but $-\rho$ is not. The linkage class of such a weight has size $|W|$ and is indexed by its lowest weight λ, which is *antidominant*: this means that $\langle \lambda + \rho, \alpha^\vee \rangle \notin \mathbb{Z}^{>0}$ for all $\alpha > 0$ (3.5).

Now suppose $\alpha > 0$ and $s_\alpha \cdot (w \cdot \lambda) < w \cdot \lambda$ for some $w \in W$, which according to Theorem 4.6 ensures that $M(s_\alpha w \cdot \lambda) \hookrightarrow M(w \cdot \lambda)$. The assumption is equivalent to $\langle w \cdot \lambda + \rho, \alpha^\vee \rangle \in \mathbb{Z}^{>0}$. This pairing can be rewritten as

$$\langle w(\lambda + \rho), \alpha^\vee \rangle = \langle \lambda + \rho, w^{-1}\alpha^\vee \rangle = \langle \lambda + \rho, (w^{-1}\alpha)^\vee \rangle.$$

Since λ is antidominant, the positivity assumption forces $w^{-1}\alpha < 0$, or $(w^{-1}s_\alpha)\alpha > 0$. Writing $w' := s_\alpha w$, this implies

$$\ell(w) = \ell(s_\alpha w') = \ell((w')^{-1}s_\alpha) > \ell((w')^{-1}) = \ell(w')$$

(see 0.3 for the relationship between length and positivity). This means by definition (0.4) that $w' \xrightarrow{s_\alpha} w$. Scrutiny of the steps shows that they are reversible, so we conclude that

$$(s_\alpha w) \cdot \lambda < w \cdot \lambda \iff s_\alpha w < w \text{ in the Bruhat ordering.}$$

Iterating, we get for all $w', w \in W$:

$$w' \cdot \lambda < w \cdot \lambda \iff w' < w.$$

Since the right hand side is independent of which (regular, antidominant) λ we choose, the theorem implies:

Corollary. *Let $\lambda \in \Lambda$ be antidominant and regular, and let $w, w' \in W$. Then $[M(w \cdot \lambda) : L(w' \cdot \lambda)] \neq 0$ if and only if $w' \leq w$.* $\qquad\square$

This consequence of the BGG Theorem raises a natural question: Can the composition factor multiplicities here be specified purely in terms of W? (The answer requires some completely new ideas: see Chapter 8.)

5.3. Jantzen Filtration

Given an arbitrary Verma module $M(\lambda)$ with $\lambda \in \mathfrak{g}^*$, Jantzen introduced a remarkable filtration of $M(\lambda)$ which leads to a fairly conceptual proof of the BGG Theorem. It also leads to new information about the multiplicities of composition factors of $M(\lambda)$ in special cases. We first state his theorem; filling in the details of the proof will occupy the following sections.

Theorem (Jantzen). *Let $\lambda \in \mathfrak{h}^*$ be arbitrary. Then $M(\lambda)$ has a filtration by submodules*

$$M(\lambda) = M(\lambda)^0 \supset M(\lambda)^1 \supset M(\lambda)^2 \supset \cdots$$

with $M(\lambda)^i = 0$ for large enough i, satisfying the following conditions.

(a) *Each nonzero quotient $M(\lambda)^i/M(\lambda)^{i+1}$ has a nondegenerate contravariant form in the sense of 3.14.*

(b) $M(\lambda)^1 = N(\lambda)$, *the unique maximal submodule of $M(\lambda)$.*

(c) *The formal characters satisfy*

$$\sum_{i>0} \operatorname{ch} M(\lambda)^i = \sum_{\alpha>0,\, s_\alpha\cdot\lambda<\lambda} \operatorname{ch} M(s_\alpha \cdot \lambda).$$

The filtration itself is known as the **Jantzen filtration**, while the formula in part (c) is called the **Jantzen Sum Formula**.   Denote by $M(\lambda)_i := M(\lambda)^i/M(\lambda)^{i+1}$ the ith filtration layer.

In part (c), the summation on the right side is over a set of positive roots Φ_λ^+. It can be characterized in terms of the subset $\Phi_{[\lambda]}$ of roots α for which $\langle \lambda + \rho, \alpha^\vee \rangle \in \mathbb{Z}$ (3.4): this is a root system, with a positive system $\Phi_{[\lambda]} \cap \Phi^+$. In turn, write Φ_λ^+ for the subset of those α satisfying $s_\alpha \cdot \lambda < \lambda$, i.e., $\langle \lambda+\rho, \alpha^\vee \rangle \in \mathbb{Z}^{>0}$. Note that this set depends on λ, not just on its coset modulo Λ. (Jantzen [**147**, 5.3] writes $R_+(\lambda)$.)

If $M(\lambda)^n \neq 0$ but $M(\lambda)^{n+1} = 0$, the Sum Formula shows that $n = |\Phi_\lambda^+|$ provided one knows that the unique simple submodule $L(\mu)$ of $M(\lambda)$ occurs just once as a composition factor of $M(\lambda)$ (hence also of each submodule $M(s_\alpha \cdot \lambda)$). This is a general fact, as proved for $\lambda \in \Lambda$ in 4.10 and in 7.16 below for arbitrary λ.

Exercise. Let λ be regular, antidominant, and integral. In the Jantzen filtration of $M(w \cdot \lambda)$, prove that the number n above is just $\ell(w)$, so there are $\ell(w) + 1$ nonzero layers in the filtration (as illustrated in 5.4 below by $\mathfrak{sl}(3, \mathbb{C})$). [Use 0.3(2) to describe $\Phi_{w\cdot\lambda}^+$.]

More generally, if $\lambda \in \mathfrak{h}^*$ is regular and antidominant, while $w \in W_{[\lambda]}$, we have $|\Phi_{w\cdot\lambda}^+| = \ell_\lambda(w)$. Here ℓ_λ is the length function on $W_{[\lambda]}$ determined by $\Delta_{[\lambda]}$.

A number of natural questions arise from Jantzen's theorem. Not all have yet been completely answered, while in some cases the answers turn out to be quite deep: see Chapter 8.

(1) Is the Jantzen filtration of $M(\lambda)$ unique relative to properties such as the Sum Formula and the existence of nondegenerate contravariant forms on the quotients?

(2) What are the composition factor multiplicities in each filtration layer $M(\lambda)_i$?

(3) Are the filtration layers semisimple? If so, does the filtration coincide with one of the standard module filtrations (radical or socle) having semisimple quotients (see 8.15)? This would provide an

intrinsic definition not requiring change of base as in 5.7 below. (In any case, the Sum Formula would require further subtle arguments.)

(4) When $M(\mu) \hookrightarrow M(\lambda)$, how do the respective Jantzen filtrations behave relative to the essentially unique homomorphism involved? More precisely, suppose $\mu \uparrow \lambda$ and set $r := |\Phi_\lambda^+| - |\Phi_\mu^+|$. One might expect that $M(\mu) \subset M(\lambda)^i$ if $i \leq r$ while $M(\mu) \cap M(\lambda)^i = M(\mu)^{i-r}$ if $i \geq r$.

Most of these questions were raised by Jantzen in [**147**, 5.17]. All the evidence he compiled in low ranks encouraged the hope that the filtrations would behave reasonably. In particular, the hereditary property in (4) became known as the **Jantzen Conjecture**. This eventually turns out to be true, but has been proved only using deep geometric ideas in the framework of Kazhdan–Lusztig theory (8.12–8.13).

5.4. Example: $\mathfrak{sl}(3,\mathbb{C})$

Let $\mathfrak{g} = \mathfrak{sl}(3,\mathbb{C})$, with the notation of 4.11. Taking λ to be regular, antidominant, and integral, the Sum Formula makes it easy to work out the Jantzen filtration for each $M(w \cdot \lambda)$. From this in turn we can deduce that all composition factors occur with multiplicity one. This is already clear when $w = 1, s_\alpha, s_\beta$: the unique simple submodule $M(\lambda) = L(\lambda)$ always occurs just once as a composition factor of $M(w \cdot \lambda)$ as does $L(w \cdot \lambda)$.

Suppose $w = s_\alpha s_\beta$, so the four weights $w' \cdot \lambda \leq w \cdot \lambda$ correspond to $w' = 1, s_\alpha, s_\beta, w$ (which happen to be below w in the Bruhat ordering). We know that $L(w \cdot \lambda)$ and $L(\lambda)$ each occurs once as a composition factor of $M(w \cdot \lambda)$. On the other hand, the Sum Formula gives

$$\sum_{i>0} \mathrm{ch}\, M(w \cdot \lambda)^i = \mathrm{ch}\, M(s_\alpha \cdot \lambda) + \mathrm{ch}\, M(s_\beta \cdot \lambda)$$

$$= \mathrm{ch}\, L(s_\alpha \cdot \lambda) + \mathrm{ch}\, L(s_\beta \cdot \lambda) + 2\,\mathrm{ch}\, L(\lambda).$$

It follows quickly that $M(w \cdot \lambda)^2 = L(\lambda)$ and $M(w \cdot \lambda)^i = 0$ for $i > 2$. This implies that $M(w \cdot \lambda)$ has four composition factors, each with multiplicity one. Similar reasoning then applies to the cases $w = s_\beta s_\alpha$ and $w = w_\circ$.

Question: What can be said if λ is no longer regular or integral?

Exercise. Consider the principal block $\mathcal{O}_0$ for $\mathfrak{sl}(3,\mathbb{C})$. It has six simple modules $L_w := L(w \cdot (-2\rho))$ and corresponding Verma modules M_w. Here the character of $L_1 = M_1$ is given by the partition function p. When $w \neq 1$, compute $\mathrm{ch}\, L_w$ and show that all weight spaces have dimension one.

5.5. Application to BGG Theorem

As promised, we can deduce from Jantzen's theorem the BGG Theorem 5.1:

If $[M(\lambda) : L(\mu)] \neq 0$, then μ is strongly linked to λ.

Proof of BGG Theorem. Given $\lambda \in \mathfrak{h}^*$, proceed by induction on the number of linked weights $\mu \leq \lambda$. In case λ is minimal in its linkage class, $M(\lambda) = L(\lambda)$ and there is nothing to prove.

For the induction step, suppose $[M(\lambda) : L(\mu)] > 0$, with $\mu < \lambda$. This means that $[M^1 : L(\mu)] > 0$. The Sum Formula forces $[M(s_\alpha \cdot \lambda) : L(\mu)] > 0$ for some $\alpha \in \Phi_\lambda^+$. By induction, there exist $\alpha_1, \ldots, \alpha_r \in \Phi^+$ such that

$$\mu = (s_{\alpha_1} \cdots s_{\alpha_r})s_\alpha \cdot \lambda \uparrow (s_{\alpha_2} \cdots s_{\alpha_r})s_\alpha \cdot \lambda \uparrow \cdots \uparrow s_{\alpha_r}s_\alpha \cdot \lambda \uparrow s_\alpha \cdot \lambda.$$

Coupled with the given $s_\alpha \cdot \lambda < \lambda$, we get the desired upward chain from μ to λ. $\qquad\square$

Example. Among the examples worked out directly by Verma [**251**], the case when $\mathfrak{g} = \mathfrak{sl}(4, \mathbb{C})$ is instructive. Only in rank 3 does one begin to encounter weights λ such that $w \cdot \lambda < \lambda$ but $w \cdot \lambda$ fails to be strongly linked to λ. To construct an example, number the simple roots and fundamental weights in the usual way and abbreviate $c_1 \varpi_1 + c_2 \varpi_2 + c_3 \varpi_3$ by (c_1, c_2, c_3). Start with $\lambda = (1, -2, -1) \in \Lambda$. Set $w = s_2 s_3 s_2 s_1 s_2$ (a reduced expression in W). Here $\lambda + \rho = \alpha_1$, while $w(\lambda + \rho) = -\alpha_3$. It follows that $\lambda - w \cdot \lambda = (\lambda + \rho) - w(\lambda + \rho) = \alpha_1 + \alpha_3$, proving that $w \cdot \lambda < \lambda$. On the other hand, his computations in $U(\mathfrak{n}^-)$ (and its division ring of quotients) show that there is no possible embedding of $M(w \cdot \lambda)$ into $M(\lambda)$.

This example can be explained in terms of the Bruhat ordering and the BGG Theorem (which Verma conjectured on the basis of this and similar cases). The weight in $\Lambda^+ - \rho$ linked to λ is $\mu = (0, -1, 0)$. Here $\lambda = s_2 s_3 \cdot \mu$, whereas $w \cdot \lambda = s_2 s_3 s_2 s_1 s_2 s_2 s_3 \cdot \mu = s_3 s_2 s_1 \cdot \mu$. Inspection of the Bruhat ordering for $W = S_4$ shows that $s_3 s_2 s_1$ is unrelated to $s_2 s_3$.

5.6. Key Lemma

Next we turn to the proof of Jantzen's theorem (5.3). Before constructing his filtration, we isolate in an elementary lemma some of the technical ideas needed (following [**147**, 5.1]).

Recall some familiar facts about free modules over principal ideal domains. Let A be a principal ideal domain and M a free A-module of finite rank r. We use letters such as e and f to denote elements of M. Now suppose M has an A-valued symmetric bilinear form (e, f), "nondegenerate" in the sense that $(e, f) = 0$ for all $f \in M$ implies $e = 0$. This means that the matrix of the form relative to a basis of M has nonzero determinant

D. While D depends on the choice of the basis, it is determined up to a unit in A: changing the basis is implemented by a matrix in $\mathrm{GL}(r, A)$ whose determinant is a unit in A.

The dual module $M^* := \mathrm{Hom}_A(M, A)$ is again free of rank r. It has a submodule $M^\vee$ consisting of all functions $e^\vee : M \to A$ defined by $e^\vee(f) = (e, f)$ with $e \in M$. Nondegeneracy of the form implies that $\varphi : M \to M^\vee$ given by $\varphi(e) = e^\vee$ is an isomorphism. Thus $M^\vee$ also has rank r.

Part of the standard structure theory for modules over principal ideal domains shows that M^* has a basis $e_1^*, \ldots, e_r^*$ such that $\{d_i e_i^*\}$ is a basis of $M^\vee$ for some nonzero $d_i \in A$. This basis of M^* is dual to a unique basis $\{e_i\}$ of M: thus $e_i^*(e_j) = \delta_{ij}$. On the other hand, there is another basis $\{f_i\}$ of M mapped by φ to the elements $f_i^\vee = d_i e_i^*$. Comparing these two bases of M, we get:

$$(1) \qquad\qquad (e_i, f_j) = f_j^\vee(e_i) = d_j e_j^*(e_i) = d_j \delta_{ij}.$$

In particular, $D = \prod_i d_i$ (up to units).

To state the key lemma, we also need to fix a prime element $p \in A$, with associated valuation $v_p(a) = n$, where $p^n \mid a$ but $p^{n+1} \nmid a$. Now $\overline{M} := M/pM$ is a vector space over the field $\overline{A} := A/pA$. For each $n \in \mathbb{Z}^+$, define

$$M(n) := \{e \in M \mid (e, M) \subset p^n A\}.$$

These submodules of M form a descending chain, starting with $M = M(0)$. Their images under reduction modulo p are subspaces $\overline{M(n)}$ of $\overline{M}$.

Lemma. *Let A be a principal ideal domain, with prime element p. Suppose M is a free A-module of rank r, with a nondegenerate symmetric bilinear form (e, f) having nonzero determinant D relative to some basis of M.*

(a) *Defining $M(n)$ and $\overline{M(n)}$ as above, we have*

$$v_p(D) = \sum_{n>0} \dim_{\overline{A}} \overline{M(n)}.$$

(b) *For each n, the modified bilinear form $(e, f)_n := p^{-n}(e, f)$ on $M(n)$ induces a nondegenerate form on $\overline{M(n)}/\overline{M(n+1)}$.*

Proof. (a) The proof will show that the indicated sum is finite, with $\overline{M(n)} = \overline{0}$ for sufficiently large n.

We just have to combine the above remarks using the notation introduced there. If $f \in M$ is written as $\sum_j a_j f_j$ with $a_j \in A$, then a quick calculation using (1) above shows that $(e_i, f) = a_i d_i$ for each i. In turn, for each fixed $n > 0$,

$$
\begin{aligned}
f \in M(n) \quad &\Leftrightarrow \quad v_p((e_i, f)) \geq n \text{ for all } i \\
&\Leftrightarrow \quad v_p(a_i d_i) \geq n \text{ for all } i \\
&\Leftrightarrow \quad v_p(a_i) + v_p(d_i) \geq n \text{ for all } i \\
&\Leftrightarrow \quad v_p(a_i) \geq n - n_i,
\end{aligned}
$$

where $n_i := v_p(d_i)$. With this in hand, we see that $M(n)$ is spanned by those f_i for which $n \leq n_i$ together with those $p^{n-n_i} f_i$ for which $n > n_i$. Thus the $\overline{f_i}$ with $n \leq n_i$ form a basis of $\overline{M(n)}$, forcing $\dim \overline{M(n)} = \#\{i \mid n \leq n_i\}$. In particular $\overline{M(n)} = \overline{0}$ for sufficiently large n.

Finally, we use $D = \prod_i d_i$ to compute

$$
\sum_{n>0} \dim \overline{M(n)} = \sum_{n>0} \#\{i \mid n \leq n_i\} = \sum_{i=1}^{r} n_i = v_p(D).
$$

(b) Clearly the form $(e, f)_n$ takes values in A. It has to be checked first that it actually induces a form on $\overline{M(n)}$: in other words, any $e \in M(n)$ lying in pM satisfies $(e, M(n))_n \in pA$. This follows from

$$
(e, M(n))_n \subset p^{-n}(pM, M(n)) \subset p^{-n+1}(M, M(n)) \subset p^{n-n+1} A = pA.
$$

Now the induced form $(\overline{e}, \overline{f})_n$ makes sense. What is its radical? Clearly each $f \in M(n+1)$ satisfies

$$
(f, M(n))_n = p^{-n}(f, M(n)) \subset p^{-n}(f, M) \subset p^{-n+n+1} A = pA,
$$

forcing $\overline{f}$ to lie in the radical. On the other hand, recalling (1) above, we have $(\overline{e_i}, \overline{f_j})_n = \overline{0}$ if $i \neq j$ but otherwise $= \overline{d_i p^{-n}}$ whenever $n \leq n_i$. In particular, the $\overline{f_i}$ lie in the radical of the form just when $n < n_i$. Since the $\overline{f_i}$ with $n \leq n_i$ form a basis of $\overline{M(n)}$, we conclude that the form is nondegenerate on the quotient $\overline{M(n)}/\overline{M(n+1)}$. $\square$

5.7. Proof of Jantzen's Theorem

Now we can complete the proof of Theorem 5.3, following [**147**, 5.3], modulo a determinant computation to be carried out in 5.8–5.9. For this we work with a fixed but arbitrary highest weight $\lambda \in \mathfrak{h}^*$. To construct a filtration of $M(\lambda)$ with the stated properties requires a somewhat roundabout approach, since there is no obvious way to specify directly the submodules $M(\lambda)^i$ and then extract the Sum Formula.

Jantzen's strategy is to extend the base field to $K := \mathbb{C}(T)$, a field of rational functions in one indeterminate T. It is not difficult to check that the previous theory developed for $\mathfrak{g}$ applies equally well to the extended Lie algebra $\mathfrak{g}_K := K \otimes_{\mathbb{C}} \mathfrak{g}$ over K (which is again a splitting field for $\mathfrak{g}$). The essential ingredients we need are the construction of Verma modules, the properties of their contravariant forms (3.14–3.15), and the criterion for a Verma module to be simple (Theorem 4.8).

Let A be the principal ideal domain $\mathbb{C}[T]$ having K as field of fractions. Then the theory over K adapts easily to the base ring A, with $\mathfrak{g}_A := A \otimes_{\mathbb{C}} \mathfrak{g}$. By using Lemma 5.6 to construct filtrations in the separate weight spaces of a Verma module and then "reducing modulo T", we shall get the needed data over the residue field $\mathbb{C}$ to assemble the Jantzen filtration and Sum Formula there.

Given $\lambda \in \mathfrak{h}^*$, set $\lambda_T := \lambda + T\rho \in \mathfrak{h}_K^*$. Obviously $\langle \lambda_T + \rho, \alpha^\vee \rangle \notin \mathbb{Z}$ for all $\alpha \in \Phi$. In particular, λ_T is *antidominant*. Thanks to Theorem 4.8, $M(\lambda_T)$ is *simple*. Equivalently, the (unique up to scalars) contravariant form on $M(\lambda_T)$ is nondegenerate (3.14).

From the natural A-form $U(\mathfrak{g}_A) \cong A \otimes_{\mathbb{C}} U(\mathfrak{g})$ in $U(\mathfrak{g}_K) \cong K \otimes_{\mathbb{C}} U(\mathfrak{g})$, we get an A-form $M(\lambda_T)_A$ in which each weight space is a free A-module of finite rank. Moreover, the contravariant form on $M(\lambda_T)$ restricts to an A-valued nondegenerate symmetric bilinear form on each of these weight spaces. A typical weight occurring is $\lambda_T - \nu$ with $\nu \in \Gamma$, so for each ν we can apply Lemma 5.6 to this A-form $M_{\lambda_T - \nu}$ of $M(\lambda_T)_{\lambda_T - \nu}$. Then set

$$M(\lambda_T)_A^i := \sum_{\nu \in \Gamma} M_{\lambda_T - \nu}(i).$$

Recalling how the filtrations of weight spaces are defined, we see at once that the $M(\lambda_T)_A^i$ are $U(\mathfrak{g}_A)$-submodules and form a decreasing filtration of $M(\lambda_T)_A$. Moreover, the original contravariance property adapts at once to these A-modules.

With this set-up in place over A, setting $T = 0$ recovers $M(\lambda) \cong M(\lambda_T)_A / T M(\lambda_T)_A$ and with it a decreasing filtration $M(\lambda)^i$. Thanks to the lemma, the quotients in this filtration acquire nondegenerate contravariant forms. As the proof of the lemma shows, the filtration of each individual weight space in $M(\lambda)$ eventually ends at 0. Since only finitely many weights are linked to λ, it follows that for large enough i we have $M(\lambda)^i = 0$.

Having proved part (a) of the theorem, we immediately get part (b): the quotient $M(\lambda)/M(\lambda)^1$ is a highest weight module with a nondegenerate contravariant form, hence simple (3.14).

It remains to verify the Sum Formula (c):

$$\sum_{i>0} \operatorname{ch} M(\lambda)^i = \sum \operatorname{ch} M(s_\alpha \cdot \lambda), \text{ with the sum taken over } \alpha \in \Phi_\lambda^+.$$

Applying part (a) of Lemma 5.6, we can express the sum on the left in terms of the determinants of the contravariant forms on the filtration quotients for individual weight spaces in the A-form. For this we need to invoke an explicit formula worked out independently by Shapovalov [231] and Jantzen [144, 146]. Its proof involves technical steps unrelated to Jantzen's theorem; these will be explained in the following sections. (The computation there is done under more general hypotheses.) For our weight λ_T in the A-form, the formula gives the determinant of the universal contravariant form (3.15) specialized to the $\lambda_T - \nu$ weight space of $M(\lambda_T)$:

$$(1) \qquad D_\nu(\lambda_T) = \prod_{\alpha>0} \prod_{r>0} \left(\langle \lambda_T + \rho, \alpha^\vee \rangle - r \right)^{\mathcal{P}(\nu - r\alpha)}.$$

Here $\mathcal{P}$ is the Kostant partition function. The determinant itself depends on the choice of a basis in $U(\mathfrak{n}^-)$ and is therefore unique only up to a nonzero scalar multiple. But all we need is the value of v_T, which is easily computed and independent of the scalar. Notice that

$$\langle \lambda_T + \rho, \alpha^\vee \rangle - r = \langle \lambda + \rho, \alpha^\vee \rangle - r + T \langle \rho, \alpha^\vee \rangle.$$

Evaluating v_T for this term gives 0 unless $\langle \lambda + \rho, \alpha^\vee \rangle = r$ (and thus $\alpha \in \Phi_\lambda^+$), in which case the value is 1. Thus the contribution to the character sum for fixed ν and α is $\mathcal{P}\left(\nu - \langle \lambda + \rho, \alpha^\vee \rangle \alpha \right) e(\lambda - \nu)$. To complete the proof, we calculate as follows, using formal sums; we interchange summation over $\alpha \in \Phi_\lambda^+$ and $\nu \in \Gamma$ as needed:

$$\sum_{i>0} \mathrm{ch}\, M(\lambda)^i = \sum_\nu \sum_\alpha \mathcal{P}\left(\nu - \langle \lambda + \rho, \alpha^\vee \rangle \alpha \right) e(\lambda - \nu)$$

$$= \sum_\alpha \sum_\nu \mathcal{P}(\nu)\, e\left(\lambda - \langle \lambda + \rho, \alpha^\vee \rangle \alpha - \nu \right).$$

In the second equation we change variables while staying in Γ. Thanks to $s_\alpha \cdot \lambda = \lambda - \langle \lambda + \rho, \alpha^\vee \rangle \alpha$, the sum on the right for fixed α is just $\mathrm{ch}\, M(s_\alpha \cdot \lambda)$. This completes the proof of the Sum Formula, modulo (1). $\qquad\square$

Remark. We defined λ_T to be $\lambda + T\rho$, but of course ρ disappears when T is specialized to 0. The evaluation of v_T only requires the fact that $\langle \rho, \alpha^\vee \rangle \neq 0$. One might ask what happens if another weight is substituted for ρ. Jantzen comments briefly on this in [147, p. 149]; but only the parallel geometric approach by Beilinson–Bernstein [18] has led to a definitive answer: see 8.13 below.

5.8. Determinant Formula

Recall from 3.15 the universal bilinear form C on $U(\mathfrak{g})$ with values in $U(\mathfrak{h})$. If $\nu \in \Gamma$, write C_ν for its restriction to the finite dimensional space $U(\mathfrak{n}^-)_{-\nu}$. Relative to a chosen ordered basis for this space, the resulting **Shapovalov matrix** S_ν is symmetric and has determinant D_ν (determined only up to a nonzero scalar multiple). By specializing to a simple highest weight module such as $M(-\rho)$, one sees easily that $D_\nu \neq 0$ (Exercise 3.15); but this also follows from the main theorem. The proof, given in the following section, depends essentially on Theorem 4.12.

Theorem. *Fix $\nu \in \Gamma$. Then the determinant of the Shapovalov matrix S_ν is nonzero and is given, up to a nonzero scalar factor, by*

$$D_\nu = \prod_{\alpha > 0} \prod_{r > 0} \left(h_\alpha + \langle \rho, \alpha^\vee \rangle - r \right)^{\mathcal{P}(\nu - r\alpha)}.$$

Thus D_ν is factored completely into linear factors involving the various h_α with $\alpha > 0$. The degree of the indicated product as a polynomial in the h_α is clearly

$$(*) \qquad \sum_\alpha \sum_r \mathcal{P}(\nu - r\alpha).$$

Here is a simple illustration:

Exercise. Use the calculations in Exercise 4.12, with $\mathfrak{g} = \mathfrak{sl}(3, \mathbb{C})$ and $\nu = \alpha + \beta$, to write down the matrix of C_ν relative to the ordered basis $\{y_\alpha y_\beta, y_\gamma\}$ of $U(\mathfrak{n}^-)_{-\nu}$ (recalling that $h_\gamma = h_\alpha + h_\beta$):

$$\begin{pmatrix} h_\alpha h_\beta + h_\beta & -h_\beta \\ -h_\beta & h_\alpha + h_\beta \end{pmatrix}.$$

Check that $\det S_\nu = h_\alpha h_\beta (h_\alpha + h_\beta + 1)$, in agreement with the theorem.

Remark. In the proof of the theorem it is convenient to look at the degree of D_ν when viewed as a polynomial in m variables h_α ($\alpha > 0$). This in turn determines the degree in terms of $h_1, \ldots, h_\ell$: Recall that h_α for arbitrary $\alpha > 0$ corresponds to the coroot $\alpha^\vee$, i.e., $\beta(h_\alpha) = \langle \beta, \alpha^\vee \rangle$ for all $\beta \in \Phi$ (or for any $\lambda \in \mathfrak{h}^*$). Thus rewriting h_α as a linear combination of $h_1, \ldots, h_\ell$ is the same as writing $\alpha^\vee$ in terms of the $\alpha_i^\vee$ for $i = 1, \ldots, \ell$. Fortunately we do not need to make this explicit for the degree calculation.

5.9. Details of Shapovalov's Proof

Fix a standard basis $y_1, \ldots, y_m, h_1, \ldots, h_\ell, x_1, \ldots, x_m$ of $\mathfrak{g}$, taking $\alpha_1, \ldots, \alpha_m$ as an enumeration of Φ^+ with $\alpha_1, \ldots, \alpha_\ell$ simple and $h_1, \ldots, h_\ell$ the corresponding basis of $\mathfrak{h}$. The basic strategy involves two steps: (1) Investigate the degree of D_ν as a polynomial in the h_α with $\alpha > 0$ to see that 5.8(*) is an upper bound. This much is elementary. (2) Show that each of the (relatively prime!) linear polynomials $h_\alpha + \langle \rho, \alpha^\vee \rangle - r$ actually divides D_ν with multiplicity $\mathcal{P}(\nu - r\alpha)$. This will be done by invoking the properties of Shapovalov elements (4.12); these in turn depend on Theorem 4.6, giving embeddings $M(s_\alpha \cdot \lambda) \hookrightarrow M(\lambda)$ for sufficiently many specializations to individual weights λ.

(1) To compute explicitly the degree of D_ν we work with an (arbitrarily ordered) PBW basis of $U(\mathfrak{n}^-)_{-\nu}$ having $N := \mathcal{P}(\nu)$ elements. Each m-tuple $\omega = (r_1, \ldots, r_m)$ of nonnegative integers parametrizes a basis element $y_\omega := y_1^{r_1} \cdots y_m^{r_m}$, where $\sum r_i \alpha_i = \nu$. Write $d(\omega) := \sum_i r_i$ for the degree of y_ω as a polynomial in $y_1, \ldots, y_m$. A typical entry of the $N \times N$ matrix S_ν of C_ν is then $c(\omega, \omega') := C_\nu(y_\omega, y_{\omega'})$. Now $c(\omega, \omega')$ is just the projection to $U(\mathfrak{h})$ of the product

$$\tau(y_\omega) y_{\omega'} = x_m^{r_m} \cdots x_1^{r_1} y_1^{s_1} \cdots y_m^{s_m},$$

where $\omega = (r_1, \ldots, r_m)$ and $\omega' = (s_1, \ldots, s_m)$. So the use of commutator relations in $U(\mathfrak{g})$ to rewrite it in standard PBW order would be a nontrivial task. But we only need to isolate the degree of the pure $U(\mathfrak{h})$-summand, viewed as a polynomial in the h_α ($\alpha > 0$). This requires some elementary inductive arguments using the commutator relations in $U(\mathfrak{g})$. (Sample computations as in Exercise 4.12 may help the reader to follow the steps.)

Lemma. *Let ω be arbitrary.*

(a) *For any fixed k, we have $x_k y_\omega \equiv \sum_{i=1}^n p_i(\omega) y_{\omega_i}$ modulo the left ideal $I := U(\mathfrak{g})\mathfrak{n}$, where $p_i(\omega)$ is a polynomial in the h_α of degree ≤ 1.*

(b) $\deg c(\omega, \omega') \leq \min\{d(\omega), d(\omega')\}$ *for all ω'.*

Proof. (a) Use induction on $d(\omega)$. When this is 0, there is nothing to prove since $x_k \in I$. If $d(\omega) > 0$, let r_j be the first nonzero entry in ω and define ω' to have the same entries r_i as ω when $i \neq j$ but $r_j - 1$ in the jth position. Thus $d(\omega') < d(\omega)$. Now for the given index k compute

$$x_k y_\omega = x_k y_j y_{\omega'} = y_j x_k y_{\omega'} + [x_k y_j] y_{\omega'}.$$

Apply the induction assumption to the first summand to get (modulo I) a sum of terms of the form $y_j p(\omega') y_{\omega''}$. Bringing y_j past such a polynomial just replaces it by another polynomial of degree ≤ 1, after which further commutation in $U(\mathfrak{n}^-)$ reorganizes the term as a combination of standard PBW monomials (modulo I).

It remains to analyze the second summand, assuming $[x_k y_j] \neq 0$. There are three easy cases. If $k = j$, we get $h_{\alpha_k} y_{\omega'}$, which is of the required form. If $\alpha_k - \alpha_j < 0$, we get a product of terms in $U(\mathfrak{n}^-)$. If $\alpha_k - \alpha_j > 0$, we instead have to use the induction assumption for ω'.

(b) Since S_ν is symmetric, it is enough to prove that $\deg c(\omega, \omega') \leq d(\omega)$. Use induction on $d(\omega)$. When this is 0, clearly $c(\omega, \omega') = 1$, so its degree is also 0.

Assuming $d(\omega) > 0$, let r_i be the first nonzero entry in ω and let ω° be an m-tuple agreeing with ω except in the ith coordinate, which is $r_i - 1$; thus $d(\omega^\circ) < d(\omega)$. Next rewrite $\tau(y_\omega) y_{\omega'}$ as $\tau(y_{\omega^\circ}) x_i y_{\omega'}$. Apply part (a) to

$x_i y_{\omega'}$ to get (modulo I) a linear combination of various $y_{\omega''}$ multiplied on the left by polynomials of degrees at most 1 in the h_α. Since τ fixes $\mathfrak{h}$ pointwise and I goes to 0 under projection to $U(\mathfrak{h})$, the induction assumption can be applied to ω°. $\qquad\square$

To estimate the degree of D_ν, we now have to look at the degrees of the $N!$ products of matrix entries $c(\omega, \omega')$, one from each row and each column, which enter into the standard determinant formula. In view of the lemma, $\deg D_\nu$ cannot exceed $\sum_\omega d(\omega)$. For a fixed $\alpha = \alpha_i$, the integer $r = r_i$ occurs as the ith component in ω for precisely $\mathcal{P}(\nu - r\alpha) - \mathcal{P}(\nu - (r+1)\alpha)$ different ω. So the power of h_α occurring in the highest degree term of D_ν is at most

$$\sum_{r>0} r\big(\mathcal{P}(\nu - r\alpha) - \mathcal{P}(\nu - (r+1)\alpha)\big) = \sum_{r>0} \mathcal{P}(\nu - r\alpha).$$

Summed over $\alpha > 0$, this gives the predicted degree 5.8($*$).

(The argument below will show that the estimate just made is exact. Indeed, Shapovalov states a more precise version of (b) above, which the reader might try as an exercise: strict inequality holds unless $\omega = \omega'$.)

(2) It remains to prove that (for fixed $\alpha > 0$ and $r > 0$) each factor $h_\alpha + \rho(h_\alpha) - r$ divides D_ν at least $\mathcal{P}(\nu - r\alpha)$ times. Then the product of all these relatively prime factors as α and r vary has degree 5.8($*$) and must therefore differ from D_ν only by a scalar thanks to the estimate in (1).

Here is where the advantage of working with Shapovalov elements $\theta_{\alpha,r}$ in $U(\mathfrak{b}^-)$ becomes apparent. The idea is to replace the matrix S_ν by a matrix with entries in $U(\mathfrak{b}^-)_{-\nu}$ having the same determinant, then apply Theorem 4.12. We work with $\theta_{\alpha,r}$ for fixed $\alpha > 0$ (in place of γ there) and $r > 0$. Write $\overline{\theta}_{\alpha,r}$ for the highest degree term $\prod_{i=1}^\ell y_i^{ra_i}$ in the definition of $\theta_{\alpha,r}$ (where $\alpha = \sum_{i=1}^\ell a_i \alpha_i$).

Starting with the space $U(\mathfrak{n}^-)_{-\nu+r\alpha}$, multiply it on the right by $\overline{\theta}_{\alpha,r}$ to get a subspace $V_{\alpha,r}$ of $U(\mathfrak{n}^-)_{-\nu}$. Since right multiplication in $U(\mathfrak{g})$ is always injective, $\dim V_{\alpha,r} = \dim U(\mathfrak{n}^-)_{-\nu+r\alpha} = \mathcal{P}(\nu - r\alpha)$. Note that $V_{\alpha,r}$ involves all PBW monomials of highest degree. Now let T be the span of a set of monomials for which $U(\mathfrak{n}^-)_{-\nu} = T \oplus V_{\alpha,r}$.

Next carry out a parallel construction in $U(\mathfrak{b}^-)$ by using right multiplication by $\theta_{\alpha,r}$ in place of $\overline{\theta}_{\alpha,r}$. Let $V'_{\alpha,r} = U(\mathfrak{n}^-)_{-\nu+r\alpha}\,\theta_{\alpha,r}$. Then set $V := T \oplus V'_{\alpha,r} \subset U(\mathfrak{b}^-)_{-\nu}$. Clearly $\dim V = \dim U(\mathfrak{n}^-)_{-\nu} = \mathcal{P}(\nu)$.

From the construction of V we deduce first that the natural linear map $V \otimes U(\mathfrak{h}) \to U(\mathfrak{b}^-)_{-\nu}$ is an isomorphism. To compare V with $U(\mathfrak{n}^-)_{-\nu}$, use an ordering of PBW monomials compatible with degrees and combine bases of T and $V'_{\alpha,r}$. The description of $\theta_{\alpha,r}$ in Theorem 4.12(b) shows that there exists a *unipotent* matrix U over $U(\mathfrak{h})$ so that the restriction C_V of the form C to V has a matrix S_V related to S_ν by $S_V = U^t S_\nu U$. So their determinants are the same (up to a scalar).

Look at a column of S_V indexed by a basis element from $V'_{\alpha,r}$. Here an entry has the form $C_V(y_\omega, y\theta_{\alpha,r}) = $ the projection to $U(\mathfrak{h})$ of $\tau(y_\omega)y\theta_{\alpha,r}$. From Theorem 4.12(a) we deduce that this projection is always divisible by $h_\alpha + \rho(h_\alpha) - r$. There are precisely $\dim U(\mathfrak{n}^-)_{-\nu+r\alpha} = \mathcal{P}(\nu - r\alpha)$ of these columns. Thus D_ν is divisible by $(h_\alpha + \rho(h_\alpha) - r)^{\mathcal{P}(\nu - r\alpha)}$ as required. $\qquad\square$

Notes

The original proof of the BGG Theorem (5.1) occurs in Bernstein–Gelfand–Gelfand [**25, 26**] and is explained carefully by Dixmier [**84**, Chap. 7]. Note that the BGG convention about the Bruhat ordering is the reverse of ours: 1 is their unique maximal element. Our convention is more natural when an antidominant rather than dominant weight is used as a starting point. In [**123**], van den Hombergh provides a somewhat streamlined proof of the theorem in the special case of *integral* weights.

(5.3–5.7) We follow mainly Jantzen [**147**, Chap. 5].

The determinant formula in (5.8) was discovered independently by Shapovalov [**231**] and Jantzen [**144**, II.1] while they were doing dissertation research with different goals. Jantzen was mainly interested in representations of algebraic groups in prime characteristic p. His formulations start over $\mathbb{Z}$ using a Chevalley basis for $\mathfrak{g}$ and resulting Kostant $\mathbb{Z}$-form of $U(\mathfrak{g})$, followed by reduction modulo p. (An added remark [**144**, p. 124] calls attention to Shapovalov's just published work.) Later Jantzen [**146**] generalized the determinant calculation to the parabolic case. We follow mainly [**231**] here, but with some steps drawn from [**144**].

Ostapenko [**225**] studies the inverse of the Shapovalov matrix, showing that the poles involved are all simple. This is recovered in the setting of Virasoro algebras by Brown [**48**].

Chapter 6

Extensions and Resolutions

Most of the features of category $\mathcal{O}$ explored in the previous chapters originated before 1980, though some of the proofs have since been improved. The central problem at that time was to determine the formal characters (or composition factor multiplicities) of Verma modules. The solution of this problem in 1979–80 will be discussed in Chapter 8, but first we pause to consider some broader homological aspects of category $\mathcal{O}$ which illuminate the classical results on finite dimensional modules and also point to further problems involving higher Ext groups, projective resolutions, etc.

In 6.1–6.8 we develop a "BGG resolution" of a finite dimensional module $L(\lambda)$; its terms are direct sums of Verma modules. This realizes the Weyl–Kostant character formula as an Euler characteristic, while recovering in a transparent way Bott's classical theorem on $\mathfrak{n}$-cohomology. Note that step (E) in the proof of Theorem 6.2 is simplified by a reference to Chapter 7, where translation functors will be developed in a self-contained way. But we also explain how to do step (E) directly.

Following a broader look at homological dimension in $\mathcal{O}$ (6.9), the remainder of the chapter treats a range of issues involving higher Ext functors (6.10). These include vanishing criteria involving mixtures of simple modules, Verma modules, and dual Verma modules (6.11–6.12); an Ext criterion for a module to have a standard filtration (6.13); expressions for formal characters in terms of Ext (6.14); and a comparison with Lie algebra cohomology (6.15).

A word about the notational conventions in this chapter: To emphasize the treatment of finite dimensional modules in the first half, we usually work with a weight $\lambda \in \Lambda^+$ (which is regular for the dot-action of W) and its linkage class. This is sometimes awkward, since $w' \cdot \lambda < w \cdot \lambda$ translates into $w < w'$ in the Bruhat ordering. As a result, we modify the set-up locally in 6.7–6.8.

6.1. BGG Resolution of a Finite Dimensional Module

For any module $M \in \mathcal{O}$, it makes sense to look for a projective resolution and then to discuss projective dimension, etc. Resolutions involving the intermediate Verma modules might however be more accessible, as we discuss here in a special case.

Fix $\lambda \in \Lambda^+$. Our objective is to find a resolution of the finite dimensional module $L(\lambda)$ whose Euler character "realizes" the alternating sum formula $\operatorname{ch} L(\lambda) = \sum_{w \in W} (-1)^{\ell(w)} \operatorname{ch} M(w \cdot \lambda)$. In other words, we seek a finite exact sequence of the following form:

$$(*) \quad 0 \to M(w_\circ \cdot \lambda) = C_m \xrightarrow{\delta_m} C_{m-1} \to \cdots \xrightarrow{\delta_2} C_1 \xrightarrow{\delta_1} C_0 = M(\lambda) \xrightarrow{\varepsilon} L(\lambda) \to 0.$$

Here $m = \ell(w_\circ) = |\Phi^+|$ (which is also the dimension of $\mathfrak{n}, \mathfrak{n}^-$, and $\mathfrak{g}/\mathfrak{b}$), while $\varepsilon : M(\lambda) \to L(\lambda)$ is the canonical epimorphism. The kth term is defined by

$$C_k := \bigoplus_{w \in W^{(k)}} M(w \cdot \lambda), \quad \text{where } W^{(k)} := \{w \in W \mid \ell(w) = k\}.$$

We call any resolution of type $(*)$ a **BGG resolution** of $L(\lambda)$. Our first priority is to find such a resolution; a further natural question is to what extent it is uniquely determined.

Exercise. In any BGG resolution of $L(\lambda)$, the map δ_k must be nonzero on each summand $M(w \cdot \lambda)$ with $\ell(w) = k$.

As we observed at the end of 2.6, it is easy to get started in the direction of constructing $(*)$ by taking advantage of the description there of the unique maximal submodule $N(\lambda) \subset M(\lambda)$ as the sum of embedded Verma modules $M(s_\alpha \cdot \lambda)$ with α simple (equivalently, $\ell(s_\alpha) = 1$). Since the Hom spaces between Verma modules have dimension at most 1 (4.2), the choice of further mappings entering into a resolution of type $(*)$ is limited. But there is no easy way to make choices which yield an exact sequence. So we start with a less direct construction.

6.2. Weak BGG Resolution

Here we outline (following [**26**]) a construction of a "weak" resolution whose terms are not yet obviously direct sums of the required Verma modules. This involves a number of independent steps, for which details will be given in the following two sections. The goal is easily formulated:

Theorem. *Let $\lambda \in \Lambda^+$. Then there is an exact sequence*

$$0 \to M(w_\circ \cdot \lambda) = D_m^\lambda \to D_{m-1}^\lambda \to \cdots \to D_1^\lambda \to D_0^\lambda = M(\lambda) \to L(\lambda) \to 0,$$

where D_k^λ has a standard filtration involving exactly once each of the Verma modules $M(w \cdot \lambda)$ with $\ell(w) = k$.

Rather than work immediately with an arbitrary $\lambda \in \Lambda^+$, it is easier to focus initially on the special case $\lambda = 0$. The construction can then be readily extended. Here are the steps in the proof of the theorem:

(A) Start with the m-dimensional vector space $\mathfrak{g}/\mathfrak{b}$ (which is a $\mathfrak{b}$-module via the adjoint action) and the associated $\mathfrak{b}$-modules $\bigwedge^k(\mathfrak{g}/\mathfrak{b})$, with $0 \leq k \leq m$. The quotient $\mathfrak{g}/\mathfrak{b}$ is isomorphic as $\mathfrak{h}$-module to $\mathfrak{n}^-$, with a basis consisting of the cosets of a standard basis $y_1, \ldots, y_m$ of $\mathfrak{n}^-$. So its weights are the negative roots. In turn, the weights of $\mathfrak{h}$ on $\bigwedge^k(\mathfrak{g}/\mathfrak{b})$ are the sums of k distinct negative roots; a basis here consists of the cosets of all $y_{i_1} \wedge \cdots \wedge y_{i_k}$ with $1 \leq i_1 < \cdots < i_k \leq m$. (Among these weights will be the $w \cdot 0 = w\rho - \rho$, with $\ell(w) = k$, as discussed below in 6.4.)

(B) Now the idea is to generalize the original construction of a Verma module by forming the induced modules $D_k := U(\mathfrak{g}) \otimes_{U(\mathfrak{b})} \bigwedge^k(\mathfrak{g}/\mathfrak{b})$. Each of these has a *standard filtration*, as indicated in Example 3.6. Since $\bigwedge^0(\mathfrak{g}/\mathfrak{b})$ is the trivial $\mathfrak{b}$-module $\mathbb{C}$, we get $D_0 = M(0)$. At the other extreme, $\bigwedge^m(\mathfrak{g}/\mathfrak{b})$ is also 1-dimensional, but the $\mathfrak{b}$-action involves the weight $-\sum_{\alpha>0} \alpha = -2\rho = w_\circ \cdot 0$; so we have $D_m = M(w_\circ \cdot 0)$. The intermediate D_k are much more complicated.

(C) Next introduce homomorphisms of $U(\mathfrak{g})$-modules $\partial_k : D_k \to D_{k-1}$ for $k = 1, \ldots, m$, together with the natural map $\varepsilon : D_0 = M(0) \to L(0)$. The maps ∂_k are relative versions of the maps used in the *standard resolution* to compute the cohomology of a Lie algebra and will be recalled in 6.3 below. A direct (but slightly tricky) computation shows that one gets a complex: $\partial_k \partial_{k+1} = 0$ for $k = 1, \ldots, m-1$, while $\varepsilon \partial_1 = 0$. A less direct argument then verifies the exactness of the sequence.

(D) The resolution of $L(0)$ constructed in this way still has terms D_k which are usually much too large. The exactness is preserved by cutting down to the principal block component of each term (written D_k^0 here in place of the usual D_k^χ with $\chi = \chi_0$). In particular, D_k^0 must involve in any standard filtration at least one copy of each $M(w\cdot 0)$. It still has to be shown that no more than one copy can occur: see 6.4 below.

(E) Having arrived at a suitable resolution of the trivial module $L(0)$, we can modify the terms to get the required resolution of $L(\lambda)$. In the language of Chapter 7 this amounts to applying the translation functor T_0^λ to the sequence, which remains exact since the functor is exact (7.1). Obviously $T_0^\lambda L(0) \cong L(\lambda)$, while Theorem 7.6 shows that $T_0^\lambda M(w \cdot 0) \cong M(w \cdot \lambda)$.

Remark. A more direct argument can be given for (E) in this special case, avoiding the heavier details required to prove Theorem 7.6. Start by tensoring the resolution of $L(0)$ with $L(\lambda)$. The exactness of this functor ensures that the resulting sequence is a resolution of $L(\lambda)$. Next extract the block component for $\chi = \chi_\lambda$, again producing a resolution of $L(\lambda)$. As in the case $\lambda = 0$, it has to be shown that the resulting terms $D_k^\lambda := \left(D_k^0 \otimes L(\lambda)\right)^\chi$ with $\chi = \chi_\lambda$ are not "too large".

This requires a closer look at the Verma modules involved in a standard filtration of each $D_k^0 \otimes L(\lambda)$ (see 3.6). Their highest weights are the $w\cdot 0 + \mu$ for $\ell(w) = k$ and μ a weight of $L(\lambda)$ (counted with multiplicity). For example, $\mu = w\lambda$ gives $w \cdot 0 + w\lambda = w \cdot \lambda$. For fixed w, we claim that only this choice of μ gives a Verma module of highest weight linked to $w \cdot \lambda$ (equivalently, linked to λ). Say $w' \cdot (w\cdot 0 + \mu) = \lambda$. This simplifies to $w'w\rho - \rho + w'\mu = \lambda$, or $w'w\rho + w'\mu = \rho + \lambda$. Since both λ and ρ lie in Λ^+, the respective weights on the left occur in $L(\rho)$ and $L(\lambda)$ and are therefore $\leq \rho, \lambda$ respectively. This forces equality throughout, so $w' = w^{-1}$ and $\mu = w\lambda$.

6.3. Exactness of the Sequence

The sequence of modules D_k in 6.2 is a relative version of the standard resolution of the trivial module in Lie algebra cohomology, which is treated for example in Cartan–Eilenberg [**59**, XIII, §7], Hilton–Stammbach [**122**, VII.4], Knapp [**188**, IV.3], Weibel [**259**, 7.7]. We assume here that the reader is already acquainted with this theory.

The standard resolution involves free $U(\mathfrak{g})$-modules $U(\mathfrak{g}) \otimes_{\mathbb{C}} \bigwedge^k(\mathfrak{g})$. In our setting the modules D_k are free only over $U(\mathfrak{n}^-)$, but the map $\partial_k : D_k \to D_{k-1}$ resembles the standard one: for $u \in U(\mathfrak{g})$ and arbitrary representatives $z_1, \ldots, z_k \in \mathfrak{g}$ of cosets $\xi_1, \ldots, \xi_k$ in $\mathfrak{g}/\mathfrak{b}$, define

$$\partial_k(u \otimes \xi_1 \wedge \cdots \wedge \xi_k) := \sum_{i=1}^{k} (-1)^{i+1}(uz_i \otimes \xi_1 \wedge \cdots \wedge \widehat{\xi_i} \wedge \cdots \wedge \xi_k)$$

$$+ \sum_{1 \leq i < j \leq k} (-1)^{i+j}(u \otimes \overline{[z_i z_j]} \wedge \xi_1 \wedge \cdots \wedge \widehat{\xi_i} \wedge \cdots \wedge \widehat{\xi_j} \wedge \cdots \wedge \xi_k).$$

Here the hat indicates as usual that the term is omitted, while $\overline{[z_i z_j]}$ denotes the coset in $\mathfrak{g}/\mathfrak{b}$ containing $[z_i z_j]$.

Some routine bookkeeping is required to check that this formula is independent of the choice of coset representatives $z_1, \ldots, z_k$. Since D_k involves a tensor product over $U(\mathfrak{b})$, when a fixed z_i is replaced by $z_i + b$ with $b \in \mathfrak{g}$, one can rewrite the extra term involving ub in the first sum as

$$(-1)^{i+1}\big(u \otimes b \cdot (\xi_1 \wedge \cdots \wedge \widehat{\xi_i} \wedge \cdots \wedge \xi_k)\big).$$

Then use the fact that b acts on the wedge product as a derivation, while $b \cdot \xi_j$ is by definition the coset $\overline{[bz_j]}$ for each $j \neq i$. Comparison with the second sum in the definition of ∂_k then allows pairwise cancellation of all extra terms involving b.

Next one has to verify that the maps ∂_k and $\varepsilon : D_0 \to L(0)$ (where $D_0 = M(0)$) produce at least a *complex*. In principle this just requires a direct computation, but working out all the case distinctions for the inequalities among indices gets complicated. The proof can be organized more efficiently, as shown in the cited literature. But at any rate no really new ideas are needed for the relative version.

Why is the complex *exact*? As usual, this requires a less direct approach. Here again the essential ideas are found in the classical work, but modified slightly. As in the usual proof of the PBW Theorem, the strategy is to construct a filtration and then pass to the associated graded structures. In $U = U(\mathfrak{g})$ the usual ascending filtration by degree comes from the natural filtration of the tensor algebra of $\mathfrak{g}$, so $U^{(l)}$ is the span of the PBW basis elements (for any basis of $\mathfrak{g}$) whose degree is $\leq l$. The associated graded algebra is just the symmetric algebra $S(\mathfrak{g})$.

The details of this argument are provided by BGG [**26**, §9], who point out that everything goes through for a pair consisting of an arbitrary Lie algebra and subalgebra. They also point out (in their Proposition 9.2) that the picture is simpler if the subalgebra in question has a vector space complement which is also a subalgebra. This is true in our case, with $\mathfrak{g} = \mathfrak{n}^- \oplus \mathfrak{b}$. Then one gets a natural identification of the complex defined above for $\mathfrak{g}/\mathfrak{b}$ with the standard resolution for $\mathfrak{n}^-$: when $u, z_i \in U(\mathfrak{n}^-)$, map $u \otimes z_1 \wedge \cdots \wedge z_k$ in the kth term of the standard complex to $u \otimes \overline{z_1} \wedge \cdots \wedge \overline{z_k}$ and apply the PBW Theorem. Exactness is just a question about images and kernels of the linear maps involved, so no new arguments are actually needed for the relative version.

6.4. Weights of the Exterior Powers

As stated earlier, it is easy to specify all weights occurring in $\bigwedge^k(\mathfrak{g}/\mathfrak{b})$: these are the sums $-(\alpha_1 + \cdots + \alpha_k)$ with $\alpha_1, \ldots, \alpha_k$ *distinct* positive roots. Among these are the special ones $w \cdot 0$ for which $w \in W^{(k)}$. Indeed, each w of length k is uniquely characterized by the set of k positive roots which it sends to negative roots (0.3); the sum of the negative roots sent by w^{-1} to Φ^+ is $w\rho - \rho = w \cdot 0$. For example, the unique element of length m is $w_\circ$, which sends Φ^+ to Φ^-; here $w_\circ \cdot 0 = w_\circ \rho - \rho = -2\rho$, which is the sum over Φ^-.

Example. When $\mathfrak{g} = \mathfrak{sl}(3, \mathbb{C})$, call the simple roots α and β. The weights in $\bigwedge^k$ are listed in the middle column, while the weights $w \cdot 0$ with $w \in W^{(k)}$ appear in the last column.

0	0	0
1	$-\alpha, -\beta, -\alpha - \beta$	$-\alpha, -\beta$
2	$-2\alpha - \beta, -\alpha - 2\beta, -\alpha - \beta$	$-2\alpha - \beta, -\alpha - 2\beta$
3	$-2\alpha - 2\beta$	$-2\alpha - 2\beta$

Returning to the general case, it remains to check that weights of the form $w \cdot 0$ occur with multiplicity 1 in $\bigwedge(\mathfrak{g}/\mathfrak{b})$. For this we have to use the standard facts about the length function on W summarized in 0.3. When $w \in W$, write $\Pi_w := \Phi^+ \cap w\Phi^-$. Thus $\ell(w) = |\Pi_w|$. For any $\Pi \subset \Phi^+$, denote by $\overline{\Pi}$ the sum of the roots in Π. In particular, $\overline{\Pi}_w = -w \cdot 0 = \rho - w\rho$. Now we just need to prove:

Lemma. *Using the above notation, suppose that $\Pi \subset \Phi^+$ satisfies $\overline{\Pi} = \rho - w\rho = \overline{\Pi}_w$ for some $w \in W$. Then $\Pi = \Pi_w$.*

Proof. Use induction on $\ell(w)$, the case $w = 1$ being obvious since $\Pi_w = \emptyset$ and $\overline{\Pi}_w = 0$. When $\ell(w) = k > 0$, there exists $\alpha \in \Delta$ with $\ell(s_\alpha w) = k - 1$. In particular, $w^{-1}\alpha < 0$, so by definition $\alpha \in \Pi_w$. (This leads us to expect that $\alpha \in \Pi$, to be shown below.) Setting $w' := s_\alpha w$, we get $(w')^{-1}\alpha > 0$ and thus $\alpha \notin \Pi_{w'}$.

For use in the final step of the proof, we observe that

$$(1) \qquad\qquad \Pi_w = s_\alpha \Pi_{w'} \cup \{\alpha\}.$$

This depends mainly on the fact that s_α permutes the positive roots other than α:

$$s_\alpha \Pi_{w'} = s_\alpha \Phi^+ \cap (s_\alpha w')\Phi^- = \left((\Phi^+ \setminus \{\alpha\}) \cup \{-\alpha\}\right) \cap w\Phi^-.$$

Since $\alpha \in \Pi_w \subset w\Phi^-$, we can discard α and get

$$s_\alpha \Pi_{w'} = s_\alpha(\Phi^+ \setminus \{\alpha\}) \cap w\Phi^- = (\Phi^+ \setminus \{\alpha\}) \cap w\Phi^-,$$

from which (1) follows.

By assumption, we are given $\Pi \subset \Phi^+$ with $\overline{\Pi} = \overline{\Pi}_w = \rho - w\rho$. Thus

$$(2) \qquad\qquad \overline{s_\alpha \Pi} = s_\alpha \overline{\Pi} = (\rho - \alpha) - s_\alpha w\rho = (\rho - w'\rho) - \alpha.$$

Now we claim that $\alpha \in \Pi$. Otherwise $s_\alpha \Pi \subset \Phi^+$ (since s_α permutes the positive roots other than α), so in turn $s_\alpha \Pi \cup \{\alpha\} \subset \Phi^+$. Using (2), this implies that

$$\overline{s_\alpha \Pi \cup \{\alpha\}} = \rho - w'\rho = \overline{\Pi}_{w'}.$$

Induction would force $s_\alpha \Pi \cup \{\alpha\} = \Pi_{w'}$, leading to the contradiction $\alpha \in \Pi_{w'}$.

With these preliminaries out of the way, we can deal with the induction step. Set $\Pi' := s_\alpha(\Pi \setminus \{\alpha\})$. From (2) we get

$$\overline{\Pi'} = \rho - w'\rho = \overline{\Pi}_{w'},$$

which by induction forces $\Pi' = \Pi_{w'}$. This in turn implies $\Pi \setminus \{\alpha\} = s_\alpha \Pi' = s_\alpha \Pi_{w'}$, or $\Pi = s_\alpha \Pi_{w'} \cup \{\alpha\}$, which by (1) is Π_w as required. $\qquad\square$

6.5. Extensions of Verma Modules

Having produced a resolution of $L(\lambda)$ involving terms D_k^λ which have standard filtrations, we next claim that each D_k^λ is actually a direct sum of Verma modules. This amounts to proving that certain Ext groups vanish. (Recall from 3.1 the basic facts about these groups.)

In the case of Verma modules and extensions in $\mathcal{O}$, splitting of extensions can be understood in terms of the strong linkage relation on their highest weights (which resembles 5.1($*$) for the Hom functor). In our situation this translates also into the Bruhat ordering of W.

Theorem. *Let $\lambda \in \mathfrak{h}^*$.*

   (a) *If $\mathrm{Ext}_\mathcal{O}\bigl(M(\mu), M(\lambda)\bigr) \neq 0$ for $\mu \in \mathfrak{h}^*$, then $\mu \uparrow \lambda$ but $\mu \neq \lambda$.*

   (b) *Let $\lambda \in \Lambda^+$ and $w, w' \in W$. If $\mathrm{Ext}_\mathcal{O}\bigl(M(w' \cdot \lambda), M(w \cdot \lambda)\bigr) \neq 0$, then $w < w'$ in the Bruhat ordering. In particular, $\ell(w) < \ell(w')$.*

Proof. (a) Proposition 3.1(a) ensures that $\mu \neq \lambda$. By assumption, there is a nonsplit short exact sequence

$$(1) \qquad\qquad 0 \to M(\lambda) \to M \to M(\mu) \to 0.$$

Since $P(\mu)$ is a projective cover of $M(\mu)$ (3.9), the canonical map $P(\mu) \to M(\mu)$ lifts to a map $\varphi : P(\mu) \to M$. The image of φ must intersect the submodule $M(\lambda)$ nontrivially: otherwise the sequence (1) splits. Now recall from 3.10 that $P(\mu)$ has a standard filtration $0 = P_0 \subset P_1 \subset \cdots \subset P_n = P(\mu)$ with $P_i/P_{i-1} \cong M(\mu_i)$ for some μ_i when $i > 0$ (and $\mu_n = \mu$). Thanks to BGG Reciprocity (3.11) and the BGG Theorem 5.1, $\mu \uparrow \mu_i$ for all $i \leq n$.

Since the image of φ intersects $M(\lambda)$ nontrivially, there is a least i for which $\varphi(P_i) \cap M(\lambda) \neq 0$. This implies that $M(\lambda)$ has a nonzero submodule which is a homomorphic image of $M(\mu_i)$; therefore $[M(\lambda) : L(\mu_i)] \neq 0$.

Applying the BGG Theorem again, we get $\mu_i \uparrow \lambda$. This combines with $\mu \uparrow \mu_i$ to complete the proof of (a).

(b) Under these assumptions, part (a) shows that $w' \cdot \lambda \uparrow w \cdot \lambda$ (and the weights are unequal). Here all the linked weights in question are regular and integral, so the strong linkage relation translates into the Bruhat ordering of W as in 5.2. Thus $w < w'$. $\qquad\square$

The special case allows us to conclude (following Rocha [**226**]) that the "weak" BGG resolution is actually a resolution of type 6.1($*$). Indeed, an easy induction on the number of Verma subquotients shows that each D_k^λ must split into a direct sum of Verma modules: nonsplit extensions as in the theorem require $w < w'$ and thus $\ell(w) < \ell(w')$.

Corollary. *If $\lambda \in \Lambda^+$, the resolution of $L(\lambda)$ in* 6.2 *is a BGG resolution.* $\qquad\square$

Remark. One might ask which infinite dimensional simple modules $L(\lambda)$ have analogous resolutions. For example, in the extreme case when λ is *antidominant*, we know that $L(\lambda) = M(\lambda)$ (Theorem 4.8); so $0 \to M(\lambda) \to L(\lambda) \to 0$ already realizes the character formula. But in general there are serious obstacles, starting with the fact that the maximal submodule of $M(\lambda)$ need not be a sum of embedded Verma submodules (unless all composition factor multiplicities are 1: Exercise 5.1). This is seen in the example at the end of BGG [**25**] in case $\mathfrak{g} = \mathfrak{sl}(4, \mathbb{C})$ and $\lambda = -\varpi_1 - \varpi_3$ in the usual numbering of fundamental weights. Notice that in this case the integral weight λ is dominant (for the dot-action of W) but not regular.

Gabber–Joseph [**105**, 2.6] proved the existence of BGG resolutions under fairly general hypotheses. To state their result using our notation, fix a *regular antidominant* weight $\lambda \in \mathfrak{h}^*$, with associated root system $\Phi_{[\lambda]}$, simple system $\Delta_{[\lambda]}$, and Weyl group $W_{[\lambda]}$. Let $I \subset \Delta_{[\lambda]}$ and let t be the length of the longest element w_I of W_I (relative to $\Delta_{[\lambda]}$). Then the simple module $L(w_I \cdot \lambda)$ has a resolution of the form

$$0 \to C_0 \to C_1 \to \cdots \to C_t \to L(w_I \cdot \lambda) \to 0,$$

where C_k is the direct sum of all $M(w \cdot \lambda)$ with $w \in W_I$ of length k. In particular, the maximal submodule of $M(w_I \cdot \lambda)$ is a sum of Verma modules. When λ is integral, this recovers the finite dimensional case $w_\circ \cdot \lambda$ as well as the antidominant case. (See also Jantzen's discussion of the Gabber–Joseph result in [**148**, 4.18, 4A.6].)

In spite of these positive results, there seems to be no reasonable necessary and sufficient condition on λ ensuring that $L(\lambda)$ $(\lambda \in \mathfrak{h}^*)$ has a finite resolution by direct sums of Verma modules.

6.6. Application: Bott's Theorem

At the end of his influential 1957 paper on the cohomology of vector bundles on homogeneous spaces such as flag varieties of semisimple Lie groups, Bott [**43**, §15] obtained what he described as a "curious corollary": an explicit formula in terms of W for the dimensions of certain Lie algebra cohomology groups. (He also remarked that this formula can be shown "without much trouble" to imply Weyl's character formula.) Although he worked in the setting of compact Lie groups and their complexifications, the essential point of the corollary is to describe the cohomology of a maximal nilpotent subalgebra of $\mathfrak{g}$ (conjugate under the adjoint group to $\mathfrak{n}$ or $\mathfrak{n}^-$) with coefficients in a finite dimensional simple module $L(\lambda)$. (Kostant [**196**] later developed these ideas further in the setting of Lie algebra cohomology.)

Bott's formula was recovered by BGG as an easy consequence of their weak resolution of $L(\lambda)$ in [**26**, §9], by using some standard homological algebra for the Ext and Tor functors. We follow Jantzen's version of the argument [**153**, 1.7]. To exploit the BGG resolution, it is convenient to state Bott's formula as follows.

Theorem. *If $\lambda \in \Lambda^+$, then $\dim H^k\big(\mathfrak{n}^-, L(\lambda)\big) = |W^{(k)}|$, where as before $W^{(k)}$ denotes the set of elements in W having length k.*

Proof. By definition, $H^k\big(\mathfrak{n}^-, L(\lambda)\big) = \mathrm{Ext}^k_{\mathfrak{n}^-}\big(\mathbb{C}, L(\lambda)\big)$, where $\mathbb{C}$ has trivial $\mathfrak{n}^-$-module action. Note that this vector space has an $\mathfrak{h}$-module structure, since $\mathfrak{h}$ normalizes $\mathfrak{n}^-$. Dualizing, we obtain an isomorphism of $\mathfrak{h}$-modules

$$\mathrm{Ext}^k_{\mathfrak{n}^-}\big(\mathbb{C}, L(\lambda)\big) \cong \mathrm{Ext}^k_{\mathfrak{n}^-}\big(L(\lambda)^*, \mathbb{C}\big).$$

Here $L(\lambda)^* \cong L(\lambda^*)$, where $\lambda^* = -w_\circ\lambda$. To compute the right side, we observe that the BGG resolution provides a convenient resolution of $L(\lambda^*)$ by free $U(\mathfrak{n}^-)$-modules: the direct sums of Verma modules $M(w \cdot \lambda^*)$. Now $\mathrm{Ext}^k_{\mathfrak{n}^-}(L(\lambda^*), \mathbb{C})$ is the cohomology of the complex $\mathrm{Hom}_{\mathfrak{n}^-}(M^\bullet, \mathbb{C})$ of $\mathfrak{h}$-modules for which $M^k := \bigoplus_{w \in W^{(k)}} M(w \cdot \lambda^*)$. For any $\mathfrak{n}^-$-module M we can naturally identify $\mathrm{Hom}_{\mathfrak{n}^-}(M, \mathbb{C})$ with $(M/\mathfrak{n}^- M)^*$. In particular, when $M = M(\mu)$ we have $M/\mathfrak{n}^- M \cong \mathbb{C}_\mu$ as $\mathfrak{h}$-modules, while the dual becomes $\mathbb{C}_{-\mu}$. Applied to the direct sum M^k, this shows that the kth term in the complex $\mathrm{Hom}_{\mathfrak{n}^-}(M^\bullet, \mathbb{C})$ is isomorphic to $\bigoplus_{w \in W^{(k)}} \mathbb{C}_{-w \cdot \lambda^*}$. Because the $w \cdot \lambda^*$ are pairwise distinct, all maps in the complex are zero. This computes $\mathrm{Ext}^k_{\mathfrak{n}^-}(L(\lambda^*), \mathbb{C})$ and thus proves Bott's dimension formula (with additional information about the $\mathfrak{h}$-module structure of the cohomology).

The end result of this computation can be restated more neatly in terms of the cohomology of $\mathfrak{n}$, by doing a little bookkeeping. If we change the system of positive roots so that $\mathfrak{n}$ corresponds to negative roots, we have to replace λ by the new highest weight $w_\circ\lambda$ of $L(\lambda)$ and also replace ρ by $-\rho$ in order to apply the argument above. Now we get an $\mathfrak{h}$-module isomorphism between $H^k\big(\mathfrak{n}, L(\lambda)\big)$ and $\bigoplus_{w\in W^{(k)}} \mathbb{C}_{-w((w_\circ\lambda)^*-\rho)-\rho}$. Since $(w_\circ\lambda)^* = -\lambda$, this expression simplifies: $H^k\big(\mathfrak{n}, L(\lambda)\big) \cong \bigoplus_{w\in W^{(k)}} \mathbb{C}_{w\cdot\lambda}$. $\qquad\square$

Exercise. Show that the freeness of each term in the BGG resolution as a $U(\mathfrak{n}^-)$-module does not require the splitting result in 6.5 above, since only the action of $\mathfrak{n}^-$ is in question.

Remark. There are other alternating sum character formulas in representation theory besides Weyl's character formula. For example, the classical work of Frobenius on the characters of finite symmetric groups expresses an irreducible character as an alternating sum of characters of certain easily constructed induced characters (from subgroups which are direct products of smaller symmetric groups). Zelevinsky [**261**] exploited BGG resolutions for the reductive Lie algebra $\mathfrak{gl}(n,\mathbb{C})$ (whose representation theory is very close to that of $\mathfrak{sl}(n,\mathbb{C})$) to realize this formula. Here Schur–Weyl duality provides a natural transition between the two settings. (Independent work along this line was done by Akin [**2**] from a somewhat different viewpoint.) Zelevinsky also treats other alternating sum formulas arising in Lie theory.

6.7. Squares

Although we have shown the existence of at least one BGG resolution of the finite dimensional module $L(\lambda)$, we have avoided an explicit description of the maps involved. But more precise information is needed for deciding to what extent such a resolution might be *unique*. In this and the next section we explore this question. (It was of more critical interest in [**26**]: there it was not yet known that the "weak" resolution in 6.2 above is actually of the "strong" type 6.1(∗), so a more direct construction was needed.)

Suppose we are given a BGG resolution with $C_k = \bigoplus_{\ell(w)=k} M(w\cdot\lambda)$ and maps $\delta_k : C_k \to C_{k-1}$ along with $\varepsilon : M(\lambda) \to L(\lambda)$. The map $\delta_1 : C_1 \to C_0 = M(\lambda)$ was already described in 2.6: here each Verma module $M(s_\alpha\cdot\lambda)$ with $\alpha\in\Delta$ embeds naturally in $M(\lambda)$. Beyond this the details are less obvious.

To analyze maps between Verma modules, we employ the Bruhat ordering in the spirit of 5.1. As noted in 0.4, we may limit the use of the arrow $w \to w'$ here to cases when $\ell(w') = \ell(w)+1$. Now the Verma module maps are easiest to visualize as arrows when we use the *antidominant* weight $\lambda^\circ := w_\circ\cdot\lambda$ as the starting point. Accordingly, we depart from the BGG convention in [**26**] by rewriting the BGG resolution in an equivalent form:

$$0 \to M(\lambda^\circ) = C_0^\circ \xrightarrow{\varepsilon_0} C_1^\circ \to \cdots \xrightarrow{\varepsilon_{m-2}} C_{m-1}^\circ \xrightarrow{\varepsilon_{m-1}} C_m^\circ = M(\lambda) \xrightarrow{\varepsilon} L(\lambda) \to 0.$$

For $k = 0, \ldots, m-1$, the kth term is defined by

$$C_k^\circ := \bigoplus_{w \in W^{(k)}} M(w \cdot \lambda^\circ).$$

Consider the pairs (w, w') with $\ell(w') = \ell(w) + 1$. In general we know that $\dim \operatorname{Hom}_{\mathcal{O}}(M(w \cdot \lambda^\circ), M(w' \cdot \lambda^\circ)) \leq 1$; the dimension is 1 precisely when $w < w'$ (5.1–5.2). In our case this means $w \to w'$.

The map ε_k is uniquely determined by its restriction to the summands $M(w \cdot \lambda^\circ)$ of C_k°, mapping $M(w \cdot \lambda^\circ)$ into various summands $M(w' \cdot \lambda^\circ)$ of C_{k+1}°. Fix an embedding of each $M(w \cdot \lambda^\circ)$ into $M(\lambda)$, which also fixes embeddings $M(w \cdot \lambda^\circ) \hookrightarrow M(w' \cdot \lambda^\circ)$ whenever $w < w'$. (This amounts to specifying a maximal vector of weight $w \cdot \lambda^\circ$ in $M(w \cdot \lambda^\circ)$, expressed as $u \cdot v^+$ for some $u \in U(\mathfrak{n}^-)$ when v^+ is a fixed maximal vector of weight λ in $M(\lambda)$.)

Now maps between Verma modules are specified by scalars. In particular, the map $M(w \cdot \lambda^\circ) \to M(w' \cdot \lambda^\circ)$ induced by ε_k is determined by a scalar $e(w, w')$ (possibly zero!). If no map of this type exists, just define $e(w, w') = 0$. So the possible maps ε_k correspond to collections of scalars $e(w, w')$. These scalars are not arbitrary, since $\varepsilon_{k+1}\varepsilon_k = 0$. To analyze this we focus on special configurations in the Bruhat ordering. If elements $w_i \in W$ are related as shown in the figure below, we say they form a **square**.

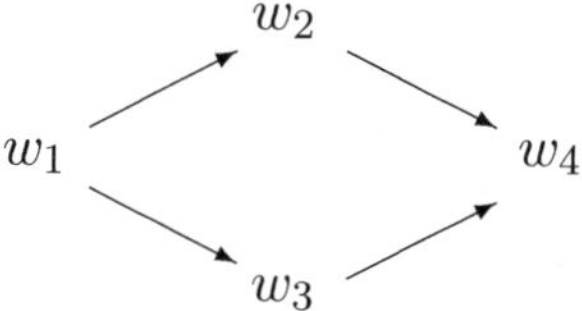

Now we invoke a special feature of the Bruhat ordering stated in Proposition 0.4(d): whenever $\ell(w_4) = \ell(w_1) + 2$ and some w satisfies $w_1 < w < w_4$, there are precisely two elements w_2, w_3 between w_1 and w_4 (thus completing a square as above). This was proved inductively in [**26**, Lemma 10.3] (see [**84**, 7.7.6]) for Weyl groups using the lemma above. (The reader might try to do this as an exercise.) Later Verma showed that it resulted for arbitrary Coxeter groups from the nature of the Möbius function. Another general proof comes from Kazhdan–Lusztig theory. See [**129**, 8.5] for details.

This property of the ordering, combined with the condition $\varepsilon_{k+1}\varepsilon_k = 0$, translates immediately into the equation

$$(1) \qquad e(w_2, w_4)\, e(w_1, w_2) + e(w_3, w_4)\, e(w_1, w_3) = 0.$$

Exercise. Suppose that all scalars $e(w_i, w_j)$ in (1) lie in $\mathbb{R}$. Prove that the factors in one term have like signs ($+$ or $-$), while those in the other term have opposite signs.

For use in the following section, we prove a key lemma showing how squares may be formed by combining two configurations:

Lemma. *Let $\alpha \in \Delta$ and $\beta > 0$ with $\beta \neq \alpha$. Given the diagram on the left, the diagram on the right exists, and conversely.*

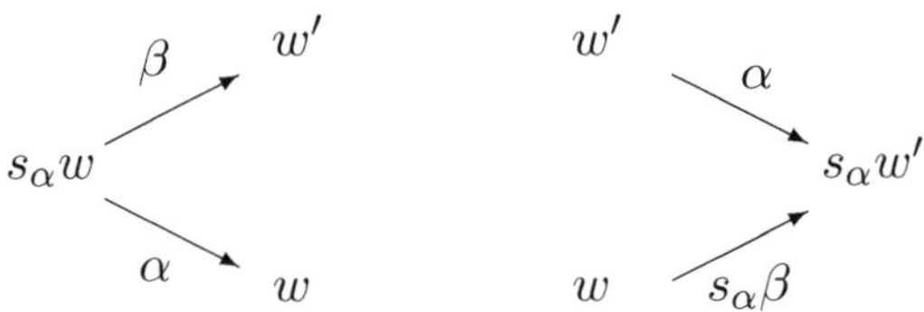

Proof. On the left we are given $s_\beta s_\alpha w = w'$, or $s_\alpha w = s_\beta w'$, while $\ell(w') = \ell(w) = \ell(s_\alpha w) + 1$. We have to show that the diagram on the right is correct. (It is left as an exercise for the reader to argue in the opposite direction.)

Set $\gamma := s_\alpha \beta$; this translates into $s_\gamma = s_\alpha s_\beta s_\alpha$. So in the diagram on the right, $s_\gamma w = s_\alpha w'$. Since $\ell(s_\alpha w') = \ell(w') \pm 1$ and $\ell(w) = \ell(w')$, it suffices to show that $\ell(s_\alpha w') > \ell(w')$ (the length difference being 1). This requires some standard facts from 0.3.

By assumption $\ell(s_\alpha w) < \ell(w)$, or equivalently, $\ell(s_\beta w') < \ell(w')$. From 0.3 we get $(w')^{-1}\beta < 0$. Since $\beta = s_\alpha \gamma$, we have $(s_\alpha w')^{-1}\gamma = (w')^{-1} s_\alpha \gamma = (w')^{-1}\beta < 0$. This forces $\ell(s_\gamma s_\alpha w') < \ell(s_\alpha w')$. Substituting $s_\gamma = s_\alpha s_\beta s_\alpha$, we get $\ell(s_\gamma s_\alpha w') = \ell(s_\alpha s_\beta w') = \ell(w) = \ell(w')$ (using $s_\beta w' = s_\alpha w$). Thus $\ell(w') < \ell(s_\alpha w')$, as required. $\qquad\square$

6.8. Maps in a BGG Resolution

Now we are in a position to show that in any BGG resolution, all scalars $e(w, w')$ which could be nonzero because $w < w'$ are in fact nonzero.

Theorem. *Given a BGG resolution with maps $\varepsilon_k : C_k^\circ \to C_{k+1}^\circ$, all scalars $e(w, w')$ are nonzero when $w \to w'$ with $\ell(w) = k$ and $\ell(w') = k + 1$.*

Proof. First we observe (as indicated earlier in Exercise 6.1) that the restriction of ε_k to each summand $M(w \cdot \lambda^\circ)$ of C_k° must be nonzero. By exactness, this is clear already for ε_0; so assume $k > 0$. If $M(w \cdot \lambda^\circ)$ lies in $\mathrm{Ker}\, \varepsilon_k$, exactness forces it to be in the image of ε_{k-1}. This is impossible: all maps of Verma module summands of C_{k-1}° into $M(w \cdot \lambda^\circ)$ have *proper* submodules as images, so the sum of images is also a proper submodule.

To complete the proof, use downward induction on k. The case $k = m$ is already clear, so let $k < m$. We need to apply Lemma 6.7 twice.

(1) Since $\varepsilon_k \neq 0$ on at least one summand $M(w \cdot \lambda^\circ)$ of C_k°, there exists $\beta > 0$ such that $w \xrightarrow{\beta} w'$ with $\ell(w') = k + 1$ and $e(w, w') \neq 0$. On the other hand, since $k < m$, there must exist $\alpha \in \Delta$ with $w \xrightarrow{\alpha} s_\alpha w$, so $\ell(s_\alpha w) > \ell(w)$. In case $\beta = \alpha$, we can skip to step (2) below. Otherwise we get the diagram on the left side of the figure below, which by the lemma implies the diagram on the right. Now the induction hypothesis ensures that both $e(w', s_\alpha w')$

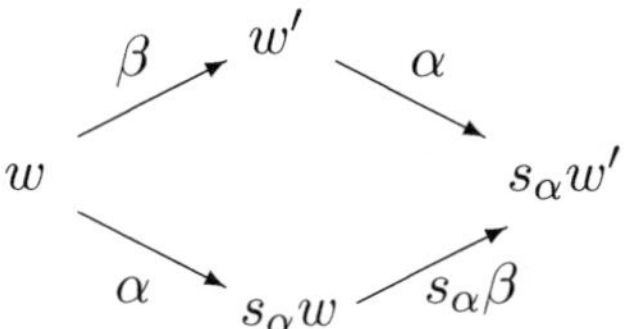

and $e(s_\alpha w, s_\alpha w')$ are nonzero, while $e(w, w')$ was chosen to be nonzero. So 6.7(1) forces $e(w, s_\alpha w) \neq 0$.

(2) At this point we know that $e(w, s_\alpha w) \neq 0$. Suppose $w \xrightarrow{\gamma} w''$ with $w'' \neq s_\alpha w$ and $\ell(w'') = k + 1$. A second application of Lemma 6.7 produces another square:

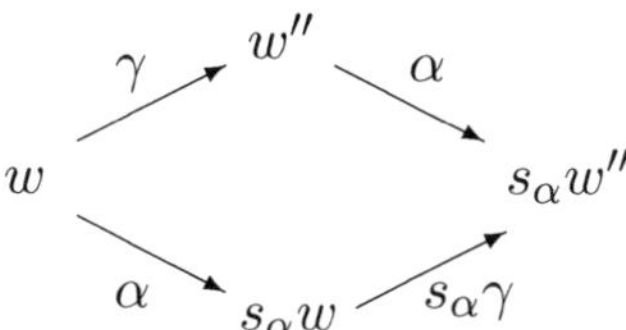

Now the proof can be completed as in step (1). $\qquad\qquad\square$

Going back to the earlier notation for a BGG resolution, note that the proof above only requires that the sequence 6.1(∗) be a *complex*, together with the fact (based on exactness) that δ_k is nonzero on each Verma summand of C_k.

Remark. As indicated above, BGG developed some of these ideas in preparation for an explicit construction of a resolution of type 6.1(∗). Rocha [**226**, §§10–11] instead uses the theorem above to streamline their construction and to explain the sense in which a BGG resolution is unique. For this she works in a suitable $\mathbb{R}$-split form of $\mathfrak{g}$, without any significant loss of generality. In particular, the scalars $e(w, w')$ above are real, as well as nonzero when $w < w'$.

When $e(w, w') \neq 0$, define $e'(w, w') := e(w, w')/|e(w, w')| = \pm 1$; otherwise set $e'(w, w') = 0$. As pointed out in Exercise 6.7, it is clear that for any square in the Bruhat ordering, the factors in one term in 6.7(1) have like signs while the factors in the other term have opposite signs. This carries over immediately to the modified scalars $e'(w, w')$, so 6.7(1) remains valid for these scalars. This recovers a key lemma [**26**, Lemma 10.4]: it is possible to assign the scalars ± 1 to all $w \to w'$ in the Bruhat ordering in such a way that this equation holds for each square.

Now use the scalars $e'(w, w')$ with $\ell(w) = k$ (some may be 0) to define new homomorphisms $\varepsilon'_k : C^\circ_k \to C^\circ_{k+1}$. The analogue of 6.7(1) shows that these maps together with $\varepsilon : M(\lambda) \to L(\lambda)$ form at least a *complex*. By complicated arguments BGG proved that this complex is the sought-for resolution of $L(\lambda)$. But Rocha can prove the exactness more simply. Using downward induction on k (with the step $k = m$ being obvious), she shows the existence of an automorphism σ_k of C°_k so that $\varepsilon'_k \sigma_k = \sigma_{k+1} \varepsilon_k$ for all k. This comparison with the known BGG resolution shows that the new complex is exact and also clarifies the uniqueness question.

6.9. Homological Dimension

In the study of $\mathcal{O}$ or its blocks as abelian categories with finiteness conditions, it is natural to look at projective (or injective) resolutions. By a *projective resolution* of $M \in \mathcal{O}$ of length n we mean an exact sequence

$$0 \to P_n \to P_{n-1} \to \cdots \to P_1 \to P_0 \to M \to 0,$$

where all P_i are projective and $P_n \neq 0$. The **projective dimension** $\operatorname{pd} M$ of M is defined to be the smallest possible length of such a resolution, or ∞ if there is no finite projective resolution; thus $\operatorname{pd} M = 0$ just when M is projective. In turn, define the **global** or **homological dimension** $\operatorname{gl.dim} \mathcal{O} \leq \infty$ to be $\sup\{\operatorname{pd} M\}$ taken over all $M \in \mathcal{O}$; define $\operatorname{gl.dim} \mathcal{O}_\lambda$ similarly. It is nontrivial to compute such dimensions, but our study of highest weight modules allows us to say something definite at this point.

We shall need some standard facts, which are developed in texts on homological algebra such as Weibel [**259**] or in the notes of Bass [**14**, I.6].

Lemma. *Let* $0 \to A \to B \to C \to 0$ *be a short exact sequence in a module category or other abelian category admitting projective resolutions. Then:*

    (a) $\operatorname{pd} A \leq \max\{\operatorname{pd} B, \operatorname{pd} C - 1\}$.

    (b) $\operatorname{pd} B \leq \max\{\operatorname{pd} A, \operatorname{pd} C\}$.

    (c) $\operatorname{pd} C \leq \max\{\operatorname{pd} A + 1, \operatorname{pd} B\}$.

With the lemma in hand, several inequalities can be obtained right away, following Bernstein–Gelfand–Gelfand [**27**, §7]. We focus especially on the principal block $\mathcal{O}_0$ (1.13), whose simple modules are of the form $L(w \cdot 0)$ with $w \in W$. Set $m := \ell(w_\circ)$; so $\ell(w) \leq m$ for all $w \in W$.

Proposition. *In the principal block $\mathcal{O}_0$, projective dimensions satisfy for all $w \in W$:*

(a) $\operatorname{pd} M(w \cdot 0) \le \ell(w)$.

(b) $\operatorname{pd} L(w \cdot 0) \le 2m - \ell(w)$.

As a result, $\operatorname{gl.dim} \mathcal{O}_0 \le 2m$.

Proof. (a) Use induction on $\ell(w)$. Since $0 \in \Lambda^+$, we know that $M(0)$ is projective (Proposition 3.8); thus $\operatorname{pd} M(0) = 0 = \ell(1)$. For the induction step, recall from the discussion in 3.9 that $P(w \cdot 0)$ is the projective cover of $M(w \cdot 0)$ as well as $L(w \cdot 0)$. Consider the short exact sequence

$$0 \to N \to P(w \cdot 0) \to M(w \cdot 0) \to 0.$$

Here N has a standard filtration (3.7). Thanks to Theorem 3.10, the highest weights of all Verma modules $M(w' \cdot 0)$ involved in this filtration satisfy $w' \cdot 0 > w \cdot 0$. Thus $\ell(w') < \ell(w)$. Now part (c) of the lemma, applied to the above sequence, combines with iteration of part (b) and induction:

$$\operatorname{pd} M(w \cdot 0) \le \operatorname{pd} N + 1 \le \max\{\operatorname{pd} M(w \cdot 0) \mid \ell(w') < \ell(w)\} + 1 \le \ell(w).$$

(b) Now use downward induction on $\ell(w)$, starting with $w = w_\circ$. Here Theorem 4.4 shows that $L(w_\circ \cdot 0) = M(w_\circ \cdot 0)$. By part (a), $\operatorname{pd} L(w_\circ \cdot 0) \le m = 2m - \ell(w_\circ)$. For the induction step, consider the short exact sequence

$$0 \to N(w \cdot 0) \to M(w \cdot 0) \to L(w \cdot 0) \to 0.$$

All composition factors $L(w' \cdot 0)$ of $N(w \cdot 0)$ have weights below $w \cdot 0$, forcing $\ell(w') > \ell(w)$. Since $\operatorname{pd} M(w \cdot 0) \le \ell(w) \le 2m - \ell(w)$, part (c) of the lemma and induction yield

$$\operatorname{pd} L(w \cdot 0) \le \max\{\operatorname{pd} L(w' \cdot 0) \mid \ell(w') > \ell(w)\} + 1 \le 2m - \ell(w).$$

To apply these estimates to the global dimension of $\mathcal{O}_0$, use part (b) of the lemma inductively to see that $\operatorname{pd} M$ for any $M \in \mathcal{O}_0$ is bounded by the projective dimensions of its composition factors. Thanks to part (b) of the proposition, $2m$ is always an upper bound for these. $\square$

Notice that the proof of the proposition requires only a few basic facts. By using more theory, one can see that equality actually holds in (a) and (b); then $\operatorname{gl.dim} \mathcal{O}_0 = 2m$. The idea is to build on the proposition, starting with $\operatorname{pd} M(0) = 0$ and again using induction on $\ell(w)$. To bound $\operatorname{pd} M(w \cdot 0)$ below by $\ell(w)$, use Corollary 5.2 of the BGG Theorem together with BGG Reciprocity and the fact that adjacent elements in the Bruhat ordering differ in length by 1. Then part (a) of the lemma and the induction hypothesis take over. A similar method yields the reverse inequality in part (b) of the proposition. We leave the details as an exercise for the reader.

Theorem. *For the principal block $\mathcal{O}_0$, projective and global dimensions are given by:*

(a) $\operatorname{pd} M(w \cdot 0) = \ell(w)$.

(b) $\operatorname{pd} L(w \cdot 0) = 2m - \ell(w)$.

(c) $\operatorname{gl.dim} \mathcal{O}_0 = 2m$.

Here the weight 0 can be replaced by any $\lambda \in \Lambda^+$, which is again dot-regular. Part (c) was stated by BGG in remarks at the end of [**27**]. Their idea is to show that for all $\lambda \in \Lambda^+$, the equality $\operatorname{pd} L(\lambda) = 2m$ holds: in fact, $\dim \operatorname{Ext}_{\mathcal{O}}^{2m}\big(L(\lambda), L(\lambda)\big) = 1$. This results from an isomorphism between the Ext algebra $\operatorname{Ext}_{\mathcal{O}}^{\bullet}\big(L(\lambda), L(\lambda)\big)$ and the cohomology algebra of the flag variety G/B associated with a semisimple group G and Borel subgroup B whose respective Lie algebras are $\mathfrak{g}$ and $\mathfrak{b}$.

Recently Mazorchuk [**214**] has recovered the above theorem and obtained related results within a unified framework, relying on homological methods. In the process he formulates dimension conjectures about some other classes of modules in $\mathcal{O}$.

6.10. Higher Ext Groups

In 3.1 we recalled the way in which right derived functors such as $\operatorname{Ext} = \operatorname{Ext}^1$ arise from the Hom functor. Although Ext^n usually lacks the explicit interpretation of Ext, the study of these functors often helps to organize better the interaction of objects in a module category. As in the case of Ext, long exact sequences derived from short exact sequences play a basic role.

It is a natural problem to determine Ext^n for various pairs of standard modules in $\mathcal{O}$ such as $L(\lambda), M(\lambda), M(\lambda)^\vee$. In some cases this turns out to be quite difficult, but in other cases we can make effective use of the theory developed so far. Recall first what is already known about Verma modules and their duals, starting with the functor $\operatorname{Hom} = \operatorname{Ext}^0$.

- Any nonzero homomorphism $M(\mu) \to M(\lambda)$ is injective, while $\dim \operatorname{Hom}_{\mathcal{O}}\big(M(\mu), M(\lambda)\big) \leq 1$ (4.2).

- The precise criterion for existence of a nonzero homomorphism (embedding) $M(\mu) \to M(\lambda)$ is that $\mu \uparrow \lambda$ (Theorem 5.1).

- For all $\lambda, \mu \in \mathfrak{h}^*$, $\dim \operatorname{Hom}_{\mathcal{O}}\big(M(\mu), M(\lambda)^\vee\big) = \delta_{\lambda\mu}$ (Theorem 3.3(c)).

- For all $\lambda, \mu \in \mathfrak{h}^*$, $\operatorname{Ext}_{\mathcal{O}}\big(M(\mu), M(\lambda)^\vee\big) = 0$ (Theorem 3.3(d)).

- For all $\lambda, \mu \in \mathfrak{h}^*$, $\operatorname{Ext}_{\mathcal{O}}\big(M(\mu), M(\lambda)\big) \neq 0$ only if $\mu \uparrow \lambda$ but $\mu \neq \lambda$ (Theorem 6.5).

6.11. Vanishing Criteria for Ext^n

It is easy, using some inductive arguments together with long exact sequences, to obtain vanishing criteria for Ext^n in the spirit of Theorem 6.5. This requires some careful bookkeeping, but no really new ideas.

Theorem. *Let $\lambda \in \Lambda^+$ and $w, w' \in W$.*

    (a) *Unless $w' \cdot \lambda \uparrow w \cdot \lambda$ with $w' \neq w$, we have for all $n > 0$:*

$$\mathrm{Ext}^n_\mathcal{O}\left(M(w' \cdot \lambda), M(w \cdot \lambda)\right) = 0 = \mathrm{Ext}^n_\mathcal{O}\left(M(w' \cdot \lambda), L(w \cdot \lambda)\right).$$

    (b) *If $w' \cdot \lambda \leq w \cdot \lambda$, then for all $n > \ell(w') - \ell(w)$:*

$$\mathrm{Ext}^n_\mathcal{O}\left(M(w' \cdot \lambda), M(w \cdot \lambda)\right) = 0 = \mathrm{Ext}^n_\mathcal{O}\left(M(w' \cdot \lambda), L(w \cdot \lambda)\right).$$

Proof. In the Bruhat ordering of W, the statement $w' \cdot \lambda \uparrow w \cdot \lambda$ is equivalent to $w \leq w'$.

(a) Assuming that $w \not< w'$, the first vanishing statement is true for $n = 1$ by Theorem 6.5, so we proceed by induction on n. Consider the short exact sequence

$$0 \to N \to P(w' \cdot \lambda) \to M(w' \cdot \lambda) \to 0.$$

Thanks to Theorem 3.10, N has a standard filtration involving quotients $M(x \cdot \lambda)$ with $w' \cdot \lambda < x \cdot \lambda$; here $x < w'$, so by assumption $w \not< x$. Part of the resulting long exact sequence is

$$\cdots \to \mathrm{Ext}^n_\mathcal{O}\left(N, M(w \cdot \lambda)\right) \to \mathrm{Ext}^{n+1}_\mathcal{O}\left(M(w' \cdot \lambda), M(w \cdot \lambda)\right)$$

$$\to \mathrm{Ext}^{n+1}_\mathcal{O}\left(P(w' \cdot \lambda), M(w \cdot \lambda)\right) \to \cdots .$$

The last term is 0 since $P(w' \cdot \lambda)$ is projective, while the first term vanishes by the induction hypothesis coupled with a subsidiary induction on the standard filtration length of N.

For the second vanishing statement, use downward induction on $\ell(w)$, starting with the case $w = w_\circ$: here $L(w_\circ \cdot \lambda) = M(w_\circ \cdot \lambda)$, permitting us to apply the first vanishing statement for all $n > 0$. Now assume the result for all $w \not< w'$. Start with the short exact sequence

$$0 \to N(w \cdot \lambda) \to M(w \cdot \lambda) \to L(w \cdot \lambda) \to 0.$$

All composition factors $L(x \cdot \lambda)$ of $N(w \cdot \lambda)$ satisfy $x \cdot \lambda < w \cdot \lambda$, or $w < x$; in particular, $x \not< w'$. Since $\ell(x) > \ell(w)$, we know by induction that the vanishing statement holds for $L(x \cdot \lambda)$. So a subsidiary induction on the length of $N(w \cdot \lambda)$ shows that $\mathrm{Ext}^k_\mathcal{O}\left(M(w' \cdot \lambda), N(w \cdot \lambda)\right) = 0$ for all $k > 0$. Finally consider part of the long exact sequence:

$$\cdots \to \mathrm{Ext}^n_\mathcal{O}\left(M(w' \cdot \lambda), M(w \cdot \lambda)\right) \to \mathrm{Ext}^n_\mathcal{O}\left(M(w' \cdot \lambda), L(w \cdot \lambda)\right)$$

$$\to \mathrm{Ext}^{n+1}_\mathcal{O}\left(M(w' \cdot \lambda), N(w \cdot \lambda)\right) \to \cdots .$$

The first term is 0 by the first vanishing statement, while the third term has just been seen to be 0, so the middle term is 0 as desired.

(b) Here the starting point is $\ell(w) \leq \ell(w')$. The idea is to imitate the strategy in part (a), but using induction on $\ell(w') - \ell(w)$. If this is 0, then $w' = w$ and part (a) applies. We leave to the reader the task of filling in the remaining details. $\qquad\square$

6.12. Computation of $\mathrm{Ext}_{\mathcal{O}}^{n}(M(\mu), M(\lambda)^{\vee})$

Here we extend Theorem 3.3(d) to all Ext^{n}, while broadening the formulation to cover all modules in $\mathcal{O}$ having a standard filtration. In the proof essential use is made of the fact that a projective module has a standard filtration (Theorem 3.10).

Theorem. *For all $n > 0$ and all $\lambda, \mu \in \mathfrak{h}^{*}$, we have $\mathrm{Ext}_{\mathcal{O}}^{n}(M(\mu), M(\lambda)^{\vee}) = 0$. More generally, this is true when $M(\mu)$ is replaced by any module having a standard filtration.*

Proof. The idea is to use induction on n. For the induction step we actually need the more general formulation.

(1) In case $n = 1$, Theorem 3.3(d) shows that $\mathrm{Ext}_{\mathcal{O}}(M(\mu), M(\lambda)^{\vee}) = 0$.

To extend the vanishing to an arbitrary M having a standard filtration, use induction on the filtration length of M; we have just treated the length 1 case. Because M has a standard filtration, there is a short exact sequence

$$0 \to N \to M \to M(\mu) \to 0$$

for some $\mu \in \mathfrak{h}^{*}$, with N of smaller filtration length. Part of the resulting long exact sequence is

$$\cdots \to \mathrm{Ext}(M(\mu), M(\lambda)^{\vee}) \to \mathrm{Ext}_{\mathcal{O}}(M, M(\lambda)^{\vee}) \to \mathrm{Ext}_{\mathcal{O}}(N, M(\lambda)^{\vee}) \to \cdots .$$

We have seen that the first term vanishes, while induction shows the same for the third term. This forces the middle term to vanish.

(2) For the induction step, assume the result for n. Given $M(\mu)$, we get a short exact sequence

$$0 \to N \to P(\mu) \to M(\mu) \to 0,$$

where N has a standard filtration because $P(\mu)$ does. (See Remark 3.10.) Part of the resulting long exact sequence is

$$\to \mathrm{Ext}_{\mathcal{O}}^{n}(N, M(\lambda)^{\vee}) \to \mathrm{Ext}_{\mathcal{O}}^{n+1}(M(\mu), M(\lambda)^{\vee}) \to \mathrm{Ext}_{\mathcal{O}}^{n+1}(P(\mu), M(\lambda)^{\vee}) \to .$$

By induction, the first term vanishes, while the third term vanishes because $P(\mu)$ is projective; so the middle term vanishes.

Finally, replace $M(\mu)$ by any module M having a standard filtration and argue inductively as in the case $n = 1$ to complete the proof. $\qquad\square$

Exercise. If M, N have standard filtrations, prove that $\mathrm{Ext}_{\mathcal{O}}^{n}(M, N^{\vee}) = 0$ for all $n > 0$. [Imitate the proof of the theorem, but dualize the short exact sequence in each step.]

6.13. Ext Criterion for Standard Filtrations

While the generalization from $M(\mu)$ to M serves mainly a technical function in the proof of Theorem 6.12, there is in fact a precise converse to the theorem which characterizes modules with a standard filtration. These include projective modules (a fact used in the proof below, via Theorem 6.12), as well as the "tilting" modules introduced later in Chapter 11.

Theorem. *For any $M \in \mathcal{O}$, the following three conditions are equivalent.*

(a) *M has a standard filtration.*

(b) *$\mathrm{Ext}_{\mathcal{O}}^{n}(M, M(\lambda)^{\vee}) = 0$ for all $n > 0$ and all $\lambda \in \mathfrak{h}^{*}$.*

(c) *$\mathrm{Ext}_{\mathcal{O}}^{1}(M, M(\lambda)^{\vee}) = 0$ for all $\lambda \in \mathfrak{h}^{*}$.*

Proof. It was already shown in Theorem 6.12 that (a) implies (b), while it is obvious that (b) implies (c). To show that (c) implies (a), use induction on the length of M. The length one case (with M simple) is covered by the first three steps below, as the reader can check. We may assume $M \neq 0$, since by definition the 0 module has a standard filtration.

(1) Let λ be minimal so that $\mathrm{Hom}_{\mathcal{O}}\left(M, L(\lambda)\right) \neq 0$; thus there is a surjective map $M \to L(\lambda)$. The first problem is to see that this map factors through the natural map $M(\lambda) \to L(\lambda)$, yielding a (necessarily surjective) map $M \to M(\lambda)$.

(2) By the choice of λ, we have $\mathrm{Hom}_{\mathcal{O}}\left(M, L(\mu)\right) = 0$ for all $\mu < \lambda$. We claim that this forces $\mathrm{Ext}_{\mathcal{O}}^{1}\left(M, L(\mu)\right) = 0$ for all $\mu < \lambda$. Using $L(\mu)^{\vee} \cong L(\mu)$, we have a short exact sequence $0 \to L(\mu) \to M(\mu)^{\vee} \to M(\mu)^{\vee}/L(\mu) \to 0$. Part of the resulting long exact sequence is

$$\to \mathrm{Hom}_{\mathcal{O}}\left(M, M(\mu)^{\vee}/L(\mu)\right) \to \mathrm{Ext}_{\mathcal{O}}^{1}\left(M, L(\mu)\right) \to \mathrm{Ext}_{\mathcal{O}}^{1}\left(M, M(\mu)^{\vee}\right) \to .$$

The term on the right is 0 by (c), while the term on the left is 0 because $\mu < \lambda$ forces $\mathrm{Hom}_{\mathcal{O}}\left(M, L(\nu)\right) = 0$ for any simple submodule $L(\nu)$ of the quotient module. So the middle term vanishes, proving our claim.

From this it follows by an easy induction on the length of M' that $\mathrm{Ext}_{\mathcal{O}}^{1}(M, M') = 0$ whenever all composition factors $L(\nu)$ of M' satisfy $\nu < \lambda$.

(3) Now we can prove that $M \to L(\lambda)$ lifts to $M \to M(\lambda)$ by analyzing:

$$\mathrm{Hom}_{\mathcal{O}}\left(M, N(\lambda)\right) \to \mathrm{Hom}_{\mathcal{O}}\left(M, M(\lambda)\right) \to \mathrm{Hom}_{\mathcal{O}}\left(M, L(\lambda)\right)$$

$$\to \mathrm{Ext}_{\mathcal{O}}^{1}\left(M, N(\lambda)\right).$$

The first term is 0, since all weights in $N(\lambda)$ are $< \lambda$, while the last term is 0 by step (2). The resulting isomorphism of the middle terms permits the desired lifting.

(4) At this point we have constructed a short exact sequence

$$0 \to N \to M \to M(\lambda) \to 0.$$

For our induction step, we have to check that N satisfies the vanishing condition in (c). Consider part of the long exact sequence when $\mu \in \mathfrak{h}^*$ is arbitrary:

$$\to \mathrm{Ext}^1_{\mathcal{O}}(M, M(\mu)^{\vee}) \to \mathrm{Ext}^1_{\mathcal{O}}(N, M(\mu)^{\vee}) \to \mathrm{Ext}^2_{\mathcal{O}}(M(\lambda), M(\mu)^{\vee}) \to .$$

By assumption, the first term vanishes, while by Theorem 6.12 the third term is 0. This forces the middle term to be 0.

(5) Now the induction hypothesis, applied to N, shows that N has a standard filtration. Then the definition shows that M also does. $\square$

6.14. Characters in Terms of $\mathrm{Ext}^{\bullet}_{\mathcal{O}}$

In his study of extensions in category $\mathcal{O}$ and their relationship with Lie algebra cohomology, Delorme [**75**, Thm. 3(i)] works out an expression for the formal character of an arbitrary $N \in \mathcal{O}$ in terms of higher Ext groups. This uses some elementary homological algebra together with BGG Reciprocity (3.11) and will play a role in Chapter 8 below. First we introduce a notation for certain Euler characteristics:

$$\chi\big(\mathrm{Ext}^{\bullet}_{\mathcal{O}}(M(\mu), N)\big) := \sum_{n \geq 0} (-1)^n \dim \mathrm{Ext}^n_{\mathcal{O}}(M(\mu), N).$$

The sum on the right is actually finite, since modules in $\mathcal{O}$ have finite projective (or injective) dimension by 6.9.

Proposition. *For all $N \in \mathcal{O}$ we have*

$$(*) \qquad \mathrm{ch}\, N = \sum_{\mu \in \mathfrak{h}^*} \chi\big(\mathrm{Ext}^{\bullet}_{\mathcal{O}}(M(\mu), N)\big)\, \mathrm{ch}\, M(\mu).$$

Proof. Clearly only finitely many μ contribute nonzero terms to the right side. Note too that both sides are additive in N: if $(*)$ holds for N_1 and N_2, then it holds for $N_1 \oplus N_2$.

The strategy is to verify $(*)$ first for *injectives* in $\mathcal{O}$, for which it is enough to consider the various indecomposable modules $Q(\lambda) = P(\lambda)^{\vee}$. On one hand, BGG Reciprocity gives

$$\mathrm{ch}\, Q(\lambda) = \mathrm{ch}\, P(\lambda) = \sum_{\mu} [M(\mu) : L(\lambda)]\, \mathrm{ch}\, M(\mu).$$

On the other hand, $\mathrm{Ext}_{\mathcal{O}}^{n}\big(M(\mu), Q(\lambda)\big) = 0$ for all $n > 0$ since $Q(\lambda)$ is injective, whereas by definition

$$\mathrm{Ext}_{\mathcal{O}}^{0}\big(M(\mu), Q(\lambda)\big) = \mathrm{Hom}_{\mathcal{O}}\big(M(\mu), Q(\lambda)\big) \cong \mathrm{Hom}_{\mathcal{O}}(P(\lambda), M(\mu)^{\vee}).$$

Thanks to Theorem 3.9(c), the dimension of the right side is equal to $[M(\mu)^{\vee} : L(\lambda)] = [M(\mu) : L(\lambda)]$. Since the Euler characteristic collapses in this case, comparison with $\mathrm{ch}\, Q(\lambda)$ above yields $(*)$.

If N is arbitrary, it has a (finite) injective resolution

$$0 \to N \to I_0 \to I_1 \to \cdots \to I_n \to 0.$$

Exactness then gives $\mathrm{ch}\, N = \sum_i (-1)^i \, \mathrm{ch}\, I_i$. On the other hand, for each fixed μ the Euler characteristic on the right side of $(*)$ can be computed using the injective resolution of N. This yields

$$\chi\big(\mathrm{Ext}_{\mathcal{O}}^{\bullet}(M(\mu), N)\big) = \sum_i (-1)^i \chi\big(\mathrm{Ext}_{\mathcal{O}}^{\bullet}(M(\mu), I_i)\big).$$

Finally, applying $(*)$ to each I_i (and interchanging summations over i and μ) proves $(*)$ for N. $\qquad\qquad\square$

6.15. Comparison of $\mathrm{Ext}_{\mathcal{O}}^{\bullet}$ and Lie Algebra Cohomology

To round out this chapter we outline briefly some connections between the Ext^{n} functors for $\mathcal{O}$ and the more traditional Lie algebra cohomology, following Delorme [**75, 76**]. (He works over the complex numbers.) While this comparison does not immediately lead to new computations, it makes the functors $\mathrm{Ext}_{\mathcal{O}}^{n}$ look less artificial and also permits interesting translations of older cohomology results such as Bott's Theorem (6.6).

The main idea is to view modules in $\mathcal{O}$ as "$(\mathfrak{g}, \mathfrak{h})$-modules" in the spirit of Harish-Chandra modules coming from Lie group representations. For pairs such as $(\mathfrak{g}, \mathfrak{h})$ there is a well-developed theory of relative Lie algebra cohomology (see for example Borel–Wallach [**39**, Chap. I]), which in this case is related to $\mathfrak{n}$-cohomology. Delorme's first two theorems can be stated as follows, using our notation.

Theorem. *Let* $M, N \in \mathcal{O}$. *Then*

(a) *For all* $n \geq 0$, $\mathrm{Ext}_{\mathcal{O}}^{n}(M, N) \cong \mathrm{Ext}_{(\mathfrak{g}, \mathfrak{h})}^{n}(M, N)$, *which in turn is isomorphic to* $H^{n}\big(\mathfrak{g}, \mathfrak{h}, \mathrm{Hom}(M, N)\big)$.

(b) *If* $M = M(\mu)$, *then* $\mathrm{Ext}_{\mathcal{O}}^{n}\big(M(\mu), N\big) \cong H^{n}(\mathfrak{n}, N)_{\mu}$, *a weight space for the natural action of* $\mathfrak{h}$ *on* $\mathfrak{n}$-*cohomology.*

In view of Bott's Theorem, (b) yields an interesting interpretation of $\mathrm{Ext}_{\mathcal{O}}^{k}\big(M(w \cdot \lambda), L(\lambda)\big)$ when $\lambda \in \Lambda^{+}$ and $w \in W$. More generally, (b) contributes to the formulation of Vogan's version of the Kazhdan–Lusztig Conjecture (8.10).

Using Proposition 6.14, together with Rocha's vanishing criterion (6.5) for $\mathrm{Ext}_{\mathcal{O}}$ (which can be avoided here with more work), Delorme arrives at vanishing criteria for $\mathrm{Ext}^n_{\mathcal{O}}$ which are more general than those in Theorem 6.11 above. These involve a kind of length function $\ell(\mu, \lambda)$ expressing the distance between $M(\mu)$ and $M(\lambda)$ in certain "strongly" standard filtrations:

$$\mathrm{Ext}^n_{\mathcal{O}}\big(M(\mu), M(\lambda)\big) = 0 \text{ for all } n > \ell(\mu, \lambda),$$

$$\mathrm{Ext}^n_{\mathcal{O}}\big(M(\mu), L(\lambda)\big) = 0 \text{ for all } n > \ell(\mu, \lambda).$$

Combined with the earlier comparisons, this recovers and generalizes some vanishing criteria for $\mathfrak{n}$-cohomology with coefficients in $L(\lambda)$ or $M(\lambda)$ found earlier by W. Schmid.

Notes

The BGG resolution goes back to the fundamental work of Bernstein–Gelfand–Gelfand [**26**, Part II], culminating in their Theorem 10.1'. Afterwards the treatment was streamlined considerably by Rocha [**226**], who developed the approach in 6.5 to show that their "weak" resolution is actually of the "strong" type they construct separately. She worked in the more general setting of parabolic category $\mathcal{O}$, as did Lepowsky [**202**] in an earlier variant of the resolution using parabolic Verma modules.

The idea of a BGG resolution resonates in a number of directions, such as the work of Gabber and Joseph [**105**] related to primitive ideals in $U(\mathfrak{g})$. There are also generalizations to the setting of Kac–Moody algebras (13.5–13.6 below) or quantum groups.

(6.9) Apart from the original source [**27**], unpublished work by König [**189**] and a recent paper by Mazorchuk [**214**] go further into these questions.

In [**137**, §4] Irving places results such as those in 6.11 and 6.14 in the axiomatic setting of "BGG algebras".

(6.12) The results here are "well-known"; we have drawn on the lecture notes of Gaitsgory [**107**].

(6.14–6.15) These sections are based on a 1978 preprint [**75**] by Delorme, later published in a summary version [**76**] with the author identified only as "M. Delorm".

Chapter 7

Translation Functors

Beyond the techniques involved in proving the theorems of Verma and BGG discussed in Chapters 4 and 5, progress in the algebraic study of Verma modules has been mainly due to the use of translation functors. These were introduced in category $\mathcal{O}$ by Jantzen [**147**].

Foundations are laid in 7.1–7.4, after which the effect of the functors on Verma modules is examined in 7.6. This already shows how to obtain category equivalences between certain blocks (7.8).

It is a more subtle matter to apply translation functors to simple modules (7.9), but then the effect on projectives becomes transparent (7.11). Deeper ideas begin to emerge when we translate from a wall of a Weyl chamber into the chamber or compose translation functors to create "wall-crossing" functors (7.12–7.15). By combining these methods, we arrive at a general proof that self-dual projectives are those corresponding to antidominant weights (7.16); the case of integral weights was treated more directly in Theorem 4.10.

Most of the ideas in this chapter were developed by Jantzen to aid in the study of composition factor multiplicities in Verma modules. Though his strong partial results in this direction were overtaken by the more definitive Kazhdan–Lusztig Conjecture [**170**] (soon a theorem), translation functors have continued to be a vital tool in the study of $\mathcal{O}$ and analogous categories.

In order to bypass the extra technical details needed to treat nonintegral weights, the reader may want to focus at first just on the integral case.

7.1. Translation Functors

The study of category $\mathcal{O}$ reduces to some extent to the study of the subcategories $\mathcal{O}_\chi$, with further refinements involving the dot-orbits of groups $W_{[\lambda]}$ in cosets of $\mathfrak{h}^*$ modulo Λ_r. But it is natural to ask how the different $\mathcal{O}_\chi$ are related to each other as χ varies over the central characters. In particular, which of these categories are equivalent to others? Here we introduce "translation functors" on $\mathcal{O}$, exact functors which take modules in one $\mathcal{O}_\chi$ to another. The idea has been developed extensively by Jantzen in the algebraic setting (including the closely analogous study of algebraic groups) but also arises independently in the work of Zuckerman [262] on Lie group representations.

More generally, as we shall see in later chapters, the structure of category $\mathcal{O}$ is typically studied in terms of a variety of endofunctors: one can learn much from seeing how they act on standard objects and how they interact with other functors. We have already accumulated some elementary examples of exact functors on $\mathcal{O}$: duality, tensoring with a fixed finite dimensional module, projecting $\mathcal{O}$ onto one of the subcategories $\mathcal{O}_\chi$. By composing such functors we get new ones whose effect on modules may be much less obvious.

Translation functors are constructed as composites of projections to various $\mathcal{O}_\chi$ and tensoring with finite dimensional modules. Fix $\lambda, \mu \in \mathfrak{h}^*$, writing pr_λ and pr_μ for the natural projections of $\mathcal{O}$ onto $\mathcal{O}_{\chi_\lambda}$ and $\mathcal{O}_{\chi_\mu}$. If L is finite dimensional and $M \in \mathcal{O}$, then $M \mapsto \mathrm{pr}_\mu\left(L \otimes (\mathrm{pr}_\lambda M)\right)$, followed by inclusion into $\mathcal{O}$, defines an exact functor on $\mathcal{O}$. Its restriction to $\mathcal{O}_{\chi_\lambda}$ (without the inclusion) relates just the two subcategories in question. But when L is arbitrary, examples quickly show that nothing very useful can be expected from the construction. Instead, require that $\mu - \lambda \in \Lambda$; in this case we say that λ and μ are **compatible**, which is automatic if $\lambda, \mu \in \Lambda$. (This implies that $\Phi_{[\lambda]} = \Phi_{[\mu]}$ and $W_{[\lambda]} = W_{[\mu]}$.) Now the W-orbit of $\nu := \mu - \lambda$ contains a unique weight $\overline{\nu}$ in Λ^+; set $L := L(\overline{\nu})$ and write T_λ^μ for the resulting functor on $\mathcal{O}$ (or on $\mathcal{O}_{\chi_\lambda}$). We call T_λ^μ a **translation functor**.

Here are a couple of initial observations which do not yet require restrictions on the weights involved:

Proposition. *If $\lambda, \mu \in \mathfrak{h}^*$, the exact functor T_λ^μ commutes with the duality functor and takes projective modules to projective modules.*

Proof. To see that $T_\lambda^\mu(M^\vee) \cong (T_\lambda^\mu M)^\vee$, use Theorem 3.2(b),(d) with the exercise in that section. The second assertion follows from the corresponding properties of tensoring with a finite dimensional module (Proposition 3.8(b)) and projecting to a direct summand. $\qquad\square$

The rationale for requiring λ and μ to be compatible will soon become more obvious. For now just observe that in the special case $M = M(\lambda)$, a standard filtration of $L(\overline{\nu}) \otimes M(\lambda)$ (as in 3.6–3.7) involves unique occurrences of the Verma modules $M(\lambda + w\overline{\nu})$, $w \in W$; one of these highest weights is μ. Thus $M(\mu)$ must be a subquotient of $T_\lambda^\mu M(\lambda)$. (This module also has a standard filtration, thanks to Proposition 3.7(b).)

7.2. Adjoint Functor Property

How are the exact functors T_λ^μ and T_μ^λ related to each other? This is easy to see, by invoking a basic adjointness property:

Lemma. *Let $L, M, N \in \mathcal{O}$ with $\dim L < \infty$. Then there is a natural isomorphism of vector spaces* $\mathrm{Hom}_\mathcal{O}(L \otimes M, N) \cong \mathrm{Hom}_\mathcal{O}(M, L^* \otimes N)$.

Proof. Omitting the action of $U(\mathfrak{g})$, the vector space isomorphism is a standard fact from linear algebra (making no assumption about the dimensions of M and N). Compatibility with the $U(\mathfrak{g})$-module structures is then routine to check. $\qquad\square$

Proposition. *Let $\lambda, \mu \in \mathfrak{h}^*$ be compatible. Then T_λ^μ is left and right adjoint to T_μ^λ:*

(a) *For all $M, N \in \mathcal{O}$,*
$$\mathrm{Hom}_\mathcal{O}(T_\lambda^\mu M, N) \cong \mathrm{Hom}_\mathcal{O}(M, T_\mu^\lambda N).$$

(b) *For all $M, N \in \mathcal{O}$,*
$$\mathrm{Ext}_\mathcal{O}^n(T_\lambda^\mu M, N) \cong \mathrm{Ext}_\mathcal{O}^n(M, T_\mu^\lambda N).$$

Similar isomorphisms hold with λ, μ interchanged.

Proof. By symmetry it is enough to check the displayed isomorphisms.

(a) With $\overline{\nu}$ the W-conjugate in Λ^+ of $\nu = \mu - \lambda$ as above, recall that $L(\overline{\nu})^* \cong L(-w_\circ\overline{\nu})$ (1.6). If $\overline{\nu} = w\nu$, then $-w_\circ\overline{\nu} = -w_\circ w\nu = w_\circ w(-\nu) \in W(-\nu) \cap \Lambda^+$. Thus $-w_\circ\overline{\nu}$ is the dominant weight conjugate to $\lambda - \mu$. Combine this observation with the lemma.

Then (b) follows from general properties of derived functors (and is easy to verify directly). $\qquad\square$

Like other exact functors, T_λ^μ preserves short exact sequences and therefore induces an endomorphism of the Grothendieck group $K(\mathcal{O})$ (or equally well, of the subgroup of $\mathcal{X}$ generated by formal characters of modules in $\mathcal{O}$). Since the $[M(\lambda)]$ or the $[L(\lambda)]$ form a basis for $K(\mathcal{O})$, it is enough for many purposes to know how translation functors act on Verma modules and simple modules. This will be explored below. But without further restrictions on the choice of λ and μ beyond the natural requirement that $\mu - \lambda \in \Lambda$, the effect of T_λ^μ on standard modules is hard to characterize.

Exercise. Let $\mathfrak{g} = \mathfrak{sl}(2,\mathbb{C})$. Show that T_λ^μ need not take Verma modules to Verma modules. [For example, let $\lambda = 1$ and $\mu = -3$.]

7.3. Weyl Group Geometry

To deal with T_λ^μ when λ and μ are *integral* weights, we can take advantage of the special geometry associated with the action of W on the euclidean space E. This is developed by Bourbaki [**45**, Ch. V] in the general setting of reflection groups, and in the case of finite reflection groups by Humphreys [**129**, 1.15]. For the application to translation functors, see Jantzen [**147**, Kap. 2] as well as [**152**, II.7].

Recall from 0.6 that the **Weyl chambers** in E are the connected components in the complement of the union of all hyperplanes orthogonal to roots; in particular, they are open sets (open cones if Φ is irreducible). The chambers are permuted in simply transitive fashion by W, while the closure of any fixed chamber is a fundamental domain for the action. The geometry is unaffected by the shift of origin from 0 to $-\rho$, but now a typical hyperplane becomes $H_\alpha := \{\lambda \in E \mid \langle \lambda + \rho, \alpha^\vee \rangle = 0\}$.

Under the simply transitive action of W, the chambers are in natural bijection with simple systems in Φ: if C corresponds to Δ, then C lies on the positive side of all hyperplanes H_α with $\alpha \in \Delta$ (hence all H_β with $\beta > 0$). Since a simple reflection s_α permutes the positive roots other than α (0.3), $s_\alpha \cdot C$ lies on the positive side of all hyperplanes H_β with $\beta > 0$ but $\beta \neq \alpha$.

In the study of translation functors we need a refined decomposition of E into **facets**. A facet F is a *nonempty* subset of E determined by a partition of Φ^+ into disjoint subsets $\Phi_F^0, \Phi_F^+, \Phi_F^-$:

$$\lambda \in F \Leftrightarrow \begin{cases} \langle \lambda + \rho, \alpha^\vee \rangle = 0 \text{ when } \alpha \in \Phi_F^0, \\ \langle \lambda + \rho, \alpha^\vee \rangle > 0 \text{ when } \alpha \in \Phi_F^+, \\ \langle \lambda + \rho, \alpha^\vee \rangle < 0 \text{ when } \alpha \in \Phi_F^-. \end{cases}$$

Clearly the closure $\overline{F}$ is obtained by replacing $>$ by $\geq$ and $<$ by $\leq$.

Remarks. (1) Given a facet F, it is easy to check that the roots α for which $\pm\alpha \in \Phi_F^0$ form a root system, whose Weyl group may be characterized as the subgroup of W fixing F pointwise (such an isotropy group is generated by the reflections it contains: see 0.3).

(2) Observe that if $\lambda \in F$ and $w \cdot \lambda < \lambda$ for some $w \in W$, then the same inequality holds for all points of F. Moreover, $w \cdot \mu \leq \mu$ for all $\mu \in \overline{F}$.

Next define the **upper closure** $\widehat{F}$ of the facet F by the conditions

$$\lambda \in \widehat{F} \Leftrightarrow \begin{cases} \langle \lambda + \rho, \alpha^\vee \rangle = 0 & \text{when } \alpha \in \Phi_F^0, \\ \langle \lambda + \rho, \alpha^\vee \rangle > 0 & \text{when } \alpha \in \Phi_F^+, \\ \langle \lambda + \rho, \alpha^\vee \rangle \leq 0 & \text{when } \alpha \in \Phi_F^-. \end{cases}$$

The **lower closure** is then defined similarly.

Exercise. Let F be a facet in E.

(a) Prove that $\overline{F}$ is the union of the facets F' for which $\Phi_{F'}^+ \subset \Phi_F^+$ and $\Phi_{F'}^- \subset \Phi_F^-$.

(b) Prove that $\widehat{F}$ is the union of the facets $F' \subset \overline{F}$ satisfying for every $\alpha > 0$ the condition:

$$\langle \lambda + \rho, \alpha^\vee \rangle \leq 0 \text{ for all } \lambda \in F' \Leftrightarrow \langle \lambda + \rho, \alpha^\vee \rangle \leq 0 \text{ for all } \lambda \in F.$$

Now a facet C is called a **chamber** if $\Phi_C^0 = \emptyset$. The closure $\overline{C}$ is a fundamental domain for the dot-action of W. Those facets F of $\overline{C}$ for which $|\Phi_F^0| = 1$ are called **walls**; each of these lies in either the upper or the lower closure of C. When Δ is fixed, there is a (shifted) **dominant chamber** whose upper closure is itself:

$$C_\circ := \{\lambda \in E \mid \langle \lambda + \rho, \alpha^\vee \rangle > 0 \text{ for all } \alpha \in \Delta\}.$$

All other chambers are of the form $C = wC_\circ$ for $w \in W$. It is easy to check that $\Phi_C^+ = \Phi^+ \cap w\Phi^+$ and $\Phi_C^- = \Phi^+ \cap w\Phi^-$.

A moment's thought shows that a given (nonempty!) facet F lies in the upper closure of the chamber C defined by $\Phi_C^- = \Phi_F^0 \cup \Phi_F^-$ and $\Phi_C^+ = \Phi_F^+$. This evidently characterizes C uniquely, proving part (a) of the following proposition. Part (b) follows straightforwardly from the above description of an upper closure:

Proposition. *Upper closures of facets have the following properties:*

(a) *Each facet lies in the upper closure of a unique chamber.*

(b) *If the facet F lies in the upper closure of the chamber C, then $\widehat{F} \subset \widehat{C}$.* $\square$

How does this geometry enter into the study of translation functors? In the important special case when λ and μ lie in Λ (in which case their difference also does), the work of Jantzen gives the most definitive results when λ and μ both lie in the closure of a single facet F or when μ lies in the upper closure of the facet containing λ. Moreover, the chamber geometry will sometimes permit the use of inductive arguments, if we define the "distance" between chambers C and C' to be the number of root hyperplanes separating C' from C.

7.4. Nonintegral Weights

Having refined the formulation of translation functors in the case of integral weights by taking account of the facets in which weights lie, we now consider the general case: $\lambda, \mu \in \mathfrak{h}^*$ with $\mu - \lambda \in \Lambda$. Recalling the discussion in 3.4 of the group $W_{[\lambda]}$ and the parametrization of its dot-orbits by antidominant weights, we can use the subspace $E(\lambda)$ of E spanned by $\Phi_{[\lambda]}$ to capture all of the "integral" behavior of the weight λ. To this end we associate with an arbitrary $\lambda \in \mathfrak{h}^*$ the *integral* weight $\lambda^\natural$ (relative to $\Phi_{[\lambda]}$) in $E(\lambda)$ characterized uniquely by the requirement that

$$\langle \lambda^\natural + \rho, \alpha^\vee \rangle = \langle \lambda + \rho, \alpha^\vee \rangle \text{ for all } \alpha \in \Phi_{[\lambda]}.$$

Equivalently, $\langle \lambda^\natural, \alpha^\vee \rangle = \langle \lambda, \alpha^\vee \rangle$ for $\alpha \in \Phi_{[\lambda]}$.

Denote by W_λ° the stabilizer of λ in $W_{[\lambda]}$, relative to the dot-action. It depends on λ and not just its coset modulo Λ. (In [147] Jantzen uses the notation $\mathcal{R}\lambda$ rather than $\lambda^\natural$. Since he uses the notation W_λ in place of our $W_{[\lambda]}$, he writes W_λ° for the isotropy group.)

It is easy to check a few elementary facts:

Proposition. *Let $\lambda \in \mathfrak{h}^*$, with stabilizer W_λ° in $W_{[\lambda]}$.*

    (a) *For all $w \in W_{[\lambda]}$, we have $\lambda - w\cdot\lambda = \lambda^\natural - w\cdot\lambda^\natural$, while $(w\cdot\lambda)^\natural = w\cdot\lambda^\natural$.*

    (b) *The stabilizer of $\lambda^\natural$ in $W_{[\lambda]}$ is W_λ°.*

    (c) *Suppose $\Phi_{[\lambda]} = \Phi_{[\mu]}$. Then $(\lambda + \mu)^\natural = \lambda^\natural + \mu^\natural$.*

Proof. (a) Since the s_α with $\alpha \in \Phi_{[\lambda]}$ generate $W_{[\lambda]}$, it is enough to check these assertions when $w = s_\alpha$. The first follows immediately from the way $\lambda^\natural$ is defined. The second follows from the equality valid for all $\beta \in \Phi_{[\lambda]}$:

$$\langle s_\alpha(\lambda^\natural + \rho), \beta^\vee \rangle = \langle \lambda^\natural + \rho, s_\alpha\beta^\vee \rangle = \langle \lambda + \rho, s_\alpha\beta^\vee \rangle = \langle s_\alpha(\lambda + \rho), \beta^\vee \rangle.$$

(b) Part (a) shows that W_λ° stabilizes $\lambda^\natural$. In the other direction, we know from the theory of reflection groups (0.3) that the stabilizer of $\lambda^\natural$ in $W_{[\lambda]}$ is generated by those s_α for which $\langle \lambda^\natural + \rho, \alpha^\vee \rangle = 0$, or equivalently, $\langle \lambda + \rho, \alpha^\vee \rangle = 0$. These reflections also stabilize λ, whence (b).

(c) It follows from the assumption that $\Phi_{[\lambda+\mu]} = \Phi_{[\lambda]}$. In turn, for all $\alpha \in \Phi_{[\lambda]} = \Phi_{[\mu]}$ we compute

$$\langle (\lambda+\mu)^\natural + \rho, \alpha^\vee \rangle = \langle \lambda+\mu+\rho, \alpha^\vee \rangle = \langle \lambda^\natural+\rho, \alpha^\vee \rangle + \langle \mu^\natural, \alpha^\vee \rangle = \langle \lambda^\natural+\mu^\natural+\rho, \alpha^\vee \rangle.$$

$\square$

Remarks. (1) As noted in the proof of (b), W_λ° is generated by the reflections s_α which fix $\lambda^\natural$ (or λ). More precisely, if $\lambda^\natural$ lies in the closure of a chamber C in $E(\lambda)$, the theory of reflection groups (7.3) shows that W_λ°

is generated by the reflections in the walls which contain $\lambda^\natural$. For example, when λ is either dominant or antidominant, these reflections belong to simple roots in $\Delta_{[\lambda]}$.

(2) Similarly, the isotropy group of λ in W is generated by those s_α for which $\langle \lambda + \rho, \alpha^\vee \rangle = 0$. Since these are the reflections fixing $\lambda^\natural$, we see that W_λ° is also the subgroup of W fixing λ.

Exercise. Show that W_λ° is the group which fixes pointwise the facet F in $E(\lambda)$ to which $\lambda^\natural$ belongs; in turn, W_λ° is the group fixing $\overline{F}$ pointwise.

How does the transition from λ to $\lambda^\natural$ affect the study of translation functors? Given $\lambda, \mu \in \mathfrak{h}^*$ with λ and μ compatible, the study of T_λ^μ can often be reduced to the study of how integral weights such as $\lambda^\natural$ and $\mu^\natural$ are located in the geometry of $E(\lambda) = E(\mu)$ relative to $W_{[\lambda]} = W_{[\mu]}$. Here Jantzen's refinements in terms of the facets of E are replaced by similar refinements involving the facets of $E(\lambda)$. In particular, part (a) of the proposition shows that the partial order of the linkage class of λ is mirrored by that of $\lambda^\natural$.

7.5. Key Lemma

Now we can begin to explore the effect of translation functors on both Verma modules and simple modules in $\mathcal{O}$. This leads to category equivalences between certain blocks (as well as an alternative proof of the BGG Theorem [**147**, 2.20]). The ideas here are due to Jantzen, as indicated in the Notes.

We start with the easier case of Verma modules, where Theorem 3.6 can be applied directly. The following lemma captures, without extraneous notation, the essential information about what happens when a Verma module is tensored with a finite dimensional simple module and then projected to an appropriate block.

Lemma. *Let $\lambda, \mu \in \mathfrak{h}^*$ be compatible, with $\nu := \mu - \lambda \in \Lambda$ and $\overline{\nu}$ the unique W-conjugate of ν lying in Λ^+. Assume that $\lambda^\natural$ lies in a facet F of $E(\lambda)$, while $\mu^\natural$ lies in the closure $\overline{F}$. Then for all weights $\nu' \neq \nu$ of $L(\overline{\nu})$, the weight $\lambda + \nu'$ is not linked by $W_{[\lambda]} = W_{[\mu]}$ to $\lambda + \nu = \mu$.*

Proof. First we observe that the proof reduces to a corresponding statement about linkage of *integral* weights in $E(\lambda)$. Suppose there exists $w \in W_{[\lambda]}$ such that $w \cdot (\lambda + \nu') = \lambda + \nu$. Here $w \cdot (\lambda + \nu') = w \cdot \lambda + w\nu'$, so we have $\lambda - w \cdot \lambda = w\nu' - \nu$. Thanks to Proposition 7.4(a), the left side is the same as $\lambda^\natural - w \cdot \lambda^\natural$. Rewriting, we get $w \cdot (\lambda^\natural + \nu') = \lambda^\natural + \nu$. So it is enough to show that this forces $\nu' = \nu$.

Suppose on the contrary $\nu' \neq \nu$ with $\lambda^\natural + \nu'$ linked to $\lambda^\natural + \nu$ and eventually reach a contradiction. The geometry comes into play by placing $\lambda^\natural + \nu$, $\lambda^\natural + \nu'$ in the closures of two chambers C, C' and using induction on the *distance* $d(C, C')$ between them: this is a geometric version of the difference between lengths in W and is defined to be the number of root hyperplanes H_α separating C from C' (which is 0 just when $C = C'$). Since we are working with a hypothetical configuration of weights which turns out to be impossible, it is unfortunately not easy to gain insight into the proof by drawing a sketch. But the individual steps are straightforward.

Suppose that $\nu' \neq \nu$ is a weight of $L(\bar{\nu})$ with $\lambda^\natural + \nu'$ linked to $\lambda^\natural + \nu$. Say the given facet F is contained in the closure $\overline{C}$ of some chamber C and choose ν' to make $d(C, C')$ as small as possible with $\lambda^\natural + \nu' \in \overline{C'}$. Using induction on this distance, we will eventually arrive at a contradiction.

(1) The case $d(C, C') = 0$ (which means $C = C'$) is impossible, since $\overline{C}$ is a fundamental domain for the dot-action of W while $\lambda^\natural + \nu' \neq \lambda^\natural + \nu$.

(2) Now $d(C, C') > 0$. Since a chamber is uniquely determined by the hyperplanes which define its walls, there exists $\alpha > 0$ such that $H_\alpha \cap C'$ contains a wall of C' and H_α separates C' from C. Say C' lies on the positive side and C on the negative side; so for all $\xi \in \overline{F}$ we have $\langle \xi + \rho, \alpha^\vee \rangle \leq 0$.

(3) Set $C'' := s_\alpha \cdot C'$. Then $d(C, C'') < d(C, C')$. To see this, assume without loss of generality that C' corresponds to the simple system $\Delta_{[\lambda]}$ (with $\alpha \in \Delta_{[\lambda]}$); thus C is separated from C'' by the same hyperplanes (except H_α) which separate C from C'.

(4) Since $\lambda^\natural + \nu' \in \overline{C'}$, we have $\langle \lambda^\natural + \nu' + \rho, \alpha^\vee \rangle \geq 0$ (thanks to (2)).

(5) Since $\lambda^\natural \in F \subset \overline{C}$, we have $\langle \lambda^\natural + \rho, \alpha^\vee \rangle \leq 0$ (also thanks to (2)).

(6) From (4) we get $s_\alpha \cdot (\lambda^\natural + \nu') = \lambda^\natural - \langle \lambda^\natural + \rho, \alpha^\vee \rangle \alpha + s_\alpha \nu' \leq \lambda^\natural + \nu'$.

(7) Combining (5) and (6), we get $s_\alpha \nu' \leq s_\alpha \nu' - \langle \lambda^\natural + \rho, \alpha^\vee \rangle \alpha \leq \nu'$.

(8) Setting $\nu'' := s_\alpha \nu' - \langle \lambda^\natural + \rho, \alpha^\vee \rangle \alpha$, we deduce from (6) and (7) that $s_\alpha \cdot (\lambda^\natural + \nu') = \lambda^\natural + \nu'' \in s_\alpha \cdot \overline{C'} = \overline{C''}$.

(9) Since ν' and $s_\alpha \nu'$ are weights of $L(\bar{\nu})$, while $s_\alpha \nu' \leq \nu'' \leq \nu'$ by (7), it follows from 1.6 that the intermediate weight ν'' in this α-string is also a weight of $L(\bar{\nu})$.

(10) The minimality assumed for ν', coupled with (8) and (9), implies that $\nu'' = \nu$.

(11) In turn, the inequalities $s_\alpha \nu' \leq \nu \leq \nu'$ in (9) force $s_\alpha \nu' = \nu$: here $\nu \neq \nu'$ by assumption, whereas not both $\nu + \alpha$ and $\nu - \alpha$ can be weights of $L(\bar{\nu})$ because ν is an extremal weight (see 1.6).

(12) But $\nu = \nu''$, which by definition is $s_\alpha \nu' - \langle \lambda^\natural + \rho, \alpha^\vee \rangle \alpha$. So $s_\alpha \nu' = \nu$ in (11) forces $\langle \lambda^\natural + \rho, \alpha^\vee \rangle = 0$.

(13) Since $\lambda^\natural \in F$, all $\xi \in \overline{F}$ must also satisfy $\langle \xi + \rho, \alpha^\vee \rangle = 0$, in particular $\xi = \lambda^\natural + \nu$. This implies that $\langle \nu, \alpha^\vee \rangle = 0$.

(14) Combining $s_\alpha \nu' = \nu$ from (11) with $\langle \nu, \alpha^\vee \rangle = 0$ from (13), we reach the contradiction $\nu = \nu'$. $\qquad\square$

Exercise. How does the proof simplify if both $\lambda^\natural$ and $\mu^\natural$ are assumed to lie in C?

7.6. Translation Functors and Verma Modules

Now it is easy to see how Verma modules behave under translation from a facet into its closure.

Theorem. *Let $\lambda, \mu \in \mathfrak{h}^*$ be antidominant and compatible. Assume that $\lambda^\natural$ lies in a facet F of $E(\lambda)$ relative to the dot-action of $W_{[\lambda]}$, while $\mu^\natural$ lies in $\overline{F}$. Then $T_\lambda^\mu M(w \cdot \lambda) \cong M(w \cdot \mu)$ for all $w \in W_{[\lambda]}$. Similarly, $T_\lambda^\mu M(w \cdot \lambda)^\vee \cong M(w \cdot \mu)^\vee$ for all $w \in W_{[\lambda]}$.*

Proof. The idea is to apply Lemma 7.5 to $w \cdot \lambda$ and $w \cdot \mu$ in place of λ and μ, letting $\nu := w \cdot \mu - w \cdot \lambda$. The assumption that $\lambda^\natural \in F$ and $\mu^\natural \in \overline{F}$ transfers immediately to this setting, with F replaced by the facet $w \cdot F$.

Now T_λ^μ involves tensoring $M(w \cdot \lambda)$ with the finite dimensional module $L(\overline{\nu})$. Theorem 3.6 ensures that the resulting module has a filtration whose quotients are the Verma modules $M(w \cdot \lambda + \nu')$, with ν' ranging over the weights of $L(\overline{\nu})$ counted according to multiplicity; in particular, $M(w \cdot \mu) = M(w \cdot \lambda + \nu)$ occurs precisely once. After passing to the linkage class of $w \cdot \mu$, the only question is whether other Verma modules can also be involved. But this is ruled out immediately by the lemma.

The last statement of the theorem follows from the fact that translation functors commute with taking duals (Proposition 7.1). $\qquad\square$

It is not immediately obvious what T_λ^μ does to $M(w \cdot \lambda)$ if the assumptions on λ and ν in the theorem are reversed: $\mu^\natural \in F$ while $\lambda^\natural \in \overline{F}$. This will be explored in 7.12 below; it turns out to be intimately involved with the deeper study of composition factors of Verma modules in Chapter 8. But it is easy to extend the theorem from Verma modules to other modules with a standard filtration:

Corollary. *Let λ, μ be as in the theorem. If $M \in \mathcal{O}_\lambda$ has a standard filtration, then $T_\lambda^\mu M$ has a standard filtration in $\mathcal{O}_\mu$.*

Proof. Use induction on the filtration length of M. The theorem disposes of the case when M is a Verma module. In general, we have a short exact sequence $0 \to N \to M \to M(w \cdot \lambda) \to 0$. Since T_λ^μ is an exact functor, the theorem yields $0 \to T_\lambda^\mu N \to T_\lambda^\mu M \to M(w \cdot \mu) \to 0$. By induction, $T_\lambda^\mu N$ has a standard filtration in $\mathcal{O}_\mu$; so $T_\lambda^\mu M$ also does. $\qquad\square$

7.7. Translation Functors and Simple Modules

With Theorem 7.6 in hand, it is natural to ask under the same hypotheses whether T_λ^μ permits a comparison of composition factor multiplicities in Verma modules. In particular, what is the effect of T_λ^μ on a simple module $L(w \cdot \lambda)$ (assuming for simplicity that $\lambda \in \Lambda$)? It is easy to get started in this direction, but not so easy to formulate a definitive theorem.

Proposition. *Let* $\lambda, \mu \in \mathfrak{h}^*$ *be antidominant and compatible. Assume that* $\lambda^\natural$ *lies in a facet* F *of* $E(\lambda)$ *relative to the dot-action of* $W_{[\lambda]}$, *while* $\mu^\natural$ *lies in* $\overline{F}$. *If* $w \in W_{[\lambda]}$, *then* $T_\lambda^\mu L(w \cdot \lambda)$ *is either* 0 *or else isomorphic to* $L(w \cdot \mu)$.

Proof. Here the exactness of T_λ^μ can be invoked. Applied to the surjective map $M(w \cdot \lambda) \to L(w \cdot \lambda)$, it produces (using Theorem 7.6) a surjective map $M(w \cdot \mu) \to T_\lambda^\mu L(w \cdot \lambda)$. So $M := T_\lambda^\mu L(w \cdot \lambda)$ is a highest weight module of weight $w \cdot \mu$ or else 0.

Suppose $M \neq 0$. Another application of the theorem shows that T_λ^μ, applied to the injective map $L(w \cdot \lambda) \to M(w \cdot \lambda)^\vee$, produces an injective map $M \to M(w \cdot \mu)^\vee$. Then Theorem 3.3(c) forces $M \cong L(w \cdot \mu)$. $\qquad\square$

Exercise. In the proposition, assume that $T_\lambda^\mu L(w \cdot \lambda) \cong L(w \cdot \mu)$. For arbitrary $M \in \mathcal{O}$, prove that $[M : L(w \cdot \lambda)] = [T_\lambda^\mu M : L(w \cdot \mu)]$.

It remains to decide when translation gives a nonzero result. Here is a cautionary example which points the way to a general theorem.

Example. Let $\mathfrak{g} = \mathfrak{sl}(2, \mathbb{C})$ and consider just integral weights (identified with integers). In case $\lambda \geq 0$, it lies inside the facet $F := \mathbb{Z}^+$. Start with the exact sequence $0 \to L(-\lambda - 2) \to M(\lambda) \to L(\lambda) \to 0$. We know that $L(-\lambda - 2) = M(-\lambda - 2)$. Now set $\mu = -1$ (that is, $-\rho$), so $M(\mu) = L(\mu)$. Here μ lies in $\overline{F}$ as required in the proposition. By Theorem 7.6, applying T_λ^μ produces the exact sequence $0 \to M(\mu) \to M(\mu) \to T_\lambda^\mu L(\lambda) \to 0$. This forces $T_\lambda^\mu L(\lambda) = 0$.

On the other hand, $T_\lambda^\mu L(-\lambda - 2) \cong L(\mu)$. So the behavior of simple modules under translation into the closure of a facet is more complicated than the behavior of Verma modules. Note however that μ fails to lie in the upper closure of F whereas it does lie in the upper closure of the opposite facet (7.3).

7.8. Application: Category Equivalences

With the help of Theorem 7.6 and Proposition 7.7, we can exploit the exactness properties of translation functors (7.2) to define an equivalence between certain blocks of $\mathcal{O}$. Recall from 4.9 that the blocks $\mathcal{O}_\lambda$ are parametrized by antidominant weights $\lambda \in \mathfrak{h}^*$: the objects in $\mathcal{O}_\lambda$ are the modules whose composition factors all have highest weights in $W_{[\lambda]} \cdot \lambda$.

As a step toward proving a category equivalence, consider first the associated Grothendieck groups (which are isomorphic to the groups spanned by formal characters): see 1.11.

Proposition. *Let* $\lambda, \mu \in \mathfrak{h}^*$ *be antidominant and compatible. Assume that* $\lambda^\natural, \mu^\natural$ *lie in the same facet of* $E(\lambda) = E(\mu)$ *relative to* $W_{[\lambda]} = W_{[\mu]}$. *Then* T_λ^μ *induces an isomorphism between the Grothendieck groups* $K(\mathcal{O}_\lambda)$ *and* $K(\mathcal{O}_\mu)$ *of the associated blocks, sending* $[M(w \cdot \lambda)]$ *to* $[M(w \cdot \mu)]$ *and* $[L(w \cdot \lambda)]$ *to* $[L(w \cdot \mu)]$ *for all* $w \in W_{[\lambda]}$.

Proof. Recall that the symbols $[M(w \cdot \lambda)]$ form a $\mathbb{Z}$-basis of $K(\mathcal{O}_\lambda)$, as do the symbols $[L(w \cdot \lambda)]$; there are similar bases for $K(\mathcal{O}_\mu)$. Thanks to Theorem 7.6, $T_\lambda^\mu M(w \cdot \lambda) \cong M(w \cdot \mu)$ and inversely using T_μ^λ. When an arbitrary basis element $[M]$ of $\mathcal{O}_\lambda$ is written uniquely as a $\mathbb{Z}$-linear combination of the $[M(w \cdot \lambda)]$, it follows at once that the map induced by $T_\mu^\lambda T_\lambda^\mu$ sends $[M]$ to $[M]$ and similarly for the maps in reverse order on $K(\mathcal{O}_\mu)$. This gives an isomorphism of Grothendieck groups.

Similarly, Proposition 7.7 implies that $T_\mu^\lambda T_\lambda^\mu$ sends $[L(w \cdot \lambda)]$ to either $[L(w \cdot \lambda)]$ or 0. But the latter is impossible because of the above isomorphism of Grothendieck groups. In turn, T_λ^μ must send $[L(w \cdot \lambda)]$ to $[L(w \cdot \mu)]$; indeed, Proposition 7.7 forces $T_\lambda^\mu L(w \cdot \lambda) \cong L(w \cdot \mu)$. $\square$

In view of the isomorphism (1.15) between $K(\mathcal{O})$ and the group $\mathcal{X}_0$ of associated formal characters induced by $M \mapsto \operatorname{ch} M$, the proposition implies that composition factor multiplicities in $\mathcal{O}_\lambda$ match those in $\mathcal{O}_\mu$:

$$[M : L(w \cdot \lambda)] = [T_\lambda^\mu M : L(w \cdot \mu)] \text{ for all } w \in W_{[\lambda]}.$$

In the general setting of artinian abelian categories, Gaitsgory observes in his lecture notes [**107**, Lemma 4.27] that isomorphism on the level of Grothendieck groups is actually enough to imply category equivalence. Here we show this directly in our situation:

Theorem. *Let* $\lambda, \mu \in \mathfrak{h}^*$ *be antidominant and compatible. In case* $\lambda^\natural$ *and* $\mu^\natural$ *lie in the same facet for the action of* $W_{[\lambda]} = W_{[\mu]}$ *on* $E(\lambda) = E(\mu)$, *the functors* T_λ^μ *and* T_μ^λ *define inverse equivalences of categories between* $\mathcal{O}_\lambda$ *and* $\mathcal{O}_\mu$. *In particular,* T_λ^μ *sends simple modules to simple modules.*

Proof. It has to be shown that $T_\mu^\lambda T_\lambda^\mu$ is naturally isomorphic to the identity functor on $\mathcal{O}_\lambda$ (and similarly in the other order). For this we use induction on the length of an arbitrary module M in $\mathcal{O}_\lambda$ to show that $T_\mu^\lambda T_\lambda^\mu M \cong M$. Coupled with the functorial properties of T_λ^μ and T_μ^λ, this will prove the theorem.

By the adjointness property (7.1), we have for all $M, N \in \mathcal{O}$

$$\mathrm{Hom}_{\mathcal{O}}(T_\mu^\lambda M, N) \cong \mathrm{Hom}_{\mathcal{O}}(M, T_\lambda^\mu N).$$

In particular, for all $M \in \mathcal{O}_\lambda$ we get

$$\mathrm{Hom}_{\mathcal{O}}(T_\mu^\lambda T_\lambda^\mu M, M) \cong \mathrm{Hom}_{\mathcal{O}}(T_\lambda^\mu M, T_\lambda^\mu M).$$

Let φ_M be the unique homomorphism on the left corresponding to the identity map on the right. By the naturality of the isomorphisms involved, each short exact sequence $0 \to N \to M \to L \to 0$ in $\mathcal{O}_\lambda$ with L simple induces a commutative diagram

$$
\begin{array}{ccccccccc}
0 & \longrightarrow & T_\mu^\lambda T_\lambda^\mu N & \longrightarrow & T_\mu^\lambda T_\lambda^\mu M & \longrightarrow & T_\mu^\lambda T_\lambda^\mu L & \longrightarrow & 0 \\
& & \varphi_N \downarrow & & \varphi_M \downarrow & & \varphi_L \downarrow & & \\
0 & \longrightarrow & N & \longrightarrow & M & \longrightarrow & L & \longrightarrow & 0
\end{array}
$$

Now proceed by induction on length. In the case of length 1, the proposition shows that φ_L is an isomorphism. By induction, φ_N is an isomorphism, so φ_M is also an isomorphism. $\qquad\square$

Remark. Under the hypotheses of the theorem, we know from Theorem 7.6 that $T_\lambda^\mu M(w \cdot \lambda) \cong M(w \cdot \mu)$ for all $w \in W_{[\lambda]}$. Each of the Verma modules has a Jantzen filtration (5.3), leading Jantzen [**147**, 5.17] to ask whether T_λ^μ induces isomorphisms $M(w \cdot \lambda)^i \cong M(w \cdot \mu)^i$. (One can ask more generally what happens if $\mu^\natural$ is only required to lie in the closure of the facet of $\lambda^\natural$.)

Exercise. Given compatible antidominant weights $\lambda, \mu, \nu \in \mathfrak{h}^*$, assume that $\mu^\natural$ lies in the closure of the facet containing $\lambda^\natural$ while $\nu^\natural$ lies in the closure of the facet containing $\mu^\natural$. Prove that $\mathrm{ch}(T_\mu^\nu T_\lambda^\mu M) = \mathrm{ch}(T_\lambda^\nu M)$ for all $M \in \mathcal{O}_\lambda$.

7.9. Translation to Upper Closures

The next task is to determine what happens to simple modules under translation into the closure of a facet. Recall from 7.3 how to define the upper closure $\widehat{F}$ of a facet F for the dot-action of a reflection group (which will be $W_{[\lambda]}$ in our situation) in terms of a partition of the positive roots.

Theorem. *Let* $\lambda, \mu \in \mathfrak{h}^*$ *be antidominant and compatible. Assume that* $\lambda^\natural \in F$ *for some facet* $F \subset E(\lambda)$, *while* $\mu^\natural \in \overline{F}$.

    (a) *For all* $w \in W_{[\lambda]}$, *if* $w \cdot \mu^\natural$ *lies in the upper closure of* $w \cdot F$, *then* $T_\lambda^\mu L(w \cdot \lambda) \cong L(w \cdot \mu)$.

    (b) *If* $w \in W_{[\lambda]}$ *but* $w \cdot \mu^\natural \notin \widehat{w \cdot F}$, *then* $T_\lambda^\mu L(w \cdot \lambda) = 0$.

Proof. Proposition 7.7 shows that $T_\lambda^\mu L(w \cdot \lambda)$ is either $L(w \cdot \mu)$ or else 0, while Theorem 7.6 shows that $T_\lambda^\mu M(w \cdot \lambda) \cong M(w \cdot \mu)$. The exact functor T_λ^μ must therefore take some composition factor $L(w'w \cdot \lambda)$ to $L(w \cdot \mu)$; here

$w'w{\cdot}\lambda \le w{\cdot}\lambda$. Exactness of T_λ^μ also ensures that no other composition factor can be translated to $L(w \cdot \mu)$. On the other hand, $T_\lambda^\mu L(w'w \cdot \lambda) \cong L(w'w \cdot \mu)$ because it is nonzero. This forces $w'w \cdot \mu = w \cdot \mu$, and similarly for $\mu^\natural$. So w' lies in the subgroup W' of $W_{[\lambda]}$ generated by the reflections s_α for which $\alpha \in w\Phi_F^0$.

(a) Suppose $w \cdot \mu^\natural$ lies in $\widehat{w \cdot F}$. Then each reflection s_α with $\alpha \in w\Phi_F^0$ satisfies $s_\alpha w \cdot \lambda^\natural \ge w \cdot \lambda^\natural$. By Remark (1) in 7.3, the group W' is itself the Weyl group of a root system attached to F. Applying Lemma 0.6 to this group, one gets $w'w{\cdot}\lambda^\natural \ge w{\cdot}\lambda^\natural$ and similarly for λ in place of $\lambda^\natural$. Combined with the choice of w' above, this implies equality. Thus $L(w \cdot \lambda) = L(w'w \cdot \lambda)$ is the unique composition factor of $M(w \cdot \lambda)$ taken to $L(w \cdot \mu)$ by T_λ^μ.

(b) If $w{\cdot}\mu^\natural$ fails to lie in the upper closure of $w{\cdot}F$, then by definition it lies in some hyperplane for $s_\alpha \in W_{[\lambda]}$ bounding $w{\cdot}F$ below. Thus $s_\alpha w{\cdot}\lambda < w{\cdot}\lambda$ but $s_\alpha w \cdot \mu = w \cdot \mu$. By Verma's Theorem, there is a (proper) embedding of $M(s_\alpha w \cdot \lambda)$ into $M(w \cdot \lambda)$. But T_λ^μ maps both of these Verma modules to $M(w \cdot \mu)$. In other words, given the short exact sequence

$$0 \to M(s_\alpha w \cdot \lambda) \to M(w \cdot \lambda) \to Q \to 0,$$

we get $T_\lambda^\mu Q = 0$. In turn, the quotient $T_\lambda^\mu L(w \cdot \lambda)$ of $T_\lambda^\mu Q$ is 0. $\qquad\square$

Remark. The proof in (a) above is an adaptation of one given by Jantzen for algebraic groups in [**152**, II.7.15]; it is more direct than his earlier proof in [**147**, 2.11–2.13]. But either proof avoids use of the BGG Theorem, which allows him to provide a new proof of this theorem using translation functors (see [**147**, 2.20]). Here is an alternative approach to (a) which does use the BGG Theorem.

Suppose $w \cdot \mu^\natural$ lies in $\widehat{w \cdot F}$. In this case we claim that $w'w \cdot \lambda < w \cdot \lambda$ is impossible, forcing equality to hold. Thus $L(w \cdot \lambda) = L(w'w \cdot \lambda)$ is the unique composition factor of $M(w \cdot \lambda)$ taken to $L(w \cdot \mu)$ by T_λ^μ.

If the strict inequality holds, the BGG Theorem (5.1) implies the existence of reflections $s_1, \ldots, s_r \in W_{[\lambda]}$ such that

$$w'w \cdot \lambda = s_1 \cdots s_r w \cdot \lambda < \cdots < s_r w \cdot \lambda < w \cdot \lambda.$$

The same inequalities hold when λ is replaced by $\lambda^\natural$ and then apply equally well to all elements of F. Since $\mu^\natural \in \overline{F}$, we get

$$w'w \cdot \mu^\natural = s_1 \cdots s_r w \cdot \mu^\natural \le \cdots \le s_r w \cdot \mu^\natural \le w \cdot \mu^\natural.$$

But $w'w{\cdot}\mu^\natural = w{\cdot}\mu^\natural$, so equality holds everywhere in this chain. In particular, $\widehat{w \cdot F}$ meets the root hyperplane for s_r, contrary to $s_r w \cdot \lambda^\natural < w \cdot \lambda^\natural$.

Exercise. Let λ, μ satisfy the hypotheses of the above theorem. Prove that for all $w \in W_{[\lambda]}$, the projective module $T_\lambda^\mu P(w \cdot \lambda)$ is nonzero.

7.10. Character Formulas

The results on translation functors obtained so far in this chapter enable us to refine further the problem of finding composition factor multiplicities for arbitrary Verma modules. This problem has already been reduced to the study of the blocks $\mathcal{O}_\lambda$, where λ runs over the antidominant weights. Thanks to the theorems of Verma and BGG (5.1), the formal character of each $L(w \cdot \lambda)$ with $w \in W_{[\lambda]}$ can be expressed as a $\mathbb{Z}$-linear combination of the ch $M(w' \cdot \lambda)$ for which $w' \cdot \lambda \uparrow w \cdot \lambda$. When λ is *regular*, this relationship simplifies to $w' \le w$ in the Bruhat ordering of $W_{[\lambda]}$ (5.2).

In any case, it is equivalent to write $[L(w \cdot \lambda)]$ as a $\mathbb{Z}$-linear combination of the $[M(w' \cdot \lambda)]$ in the Grothendieck group $K(\mathcal{O}_\lambda)$:

$$(*) \qquad [L(w \cdot \lambda)] = \sum_{w' \in W_{[\lambda]},\, w' \cdot \lambda \uparrow w \cdot \lambda} b^\lambda_{w',w}[M(w' \cdot \lambda)],$$

where $b^\lambda_{w,w} = 1$. This format for the coefficients emphasizes the role of $W_{[\lambda]}$ while raising the question of how these integers depend on λ.

To exploit translation functors, look first at the case of *integral* weights; here the formulations are a bit more straightforward. Since translation functors are exact, they induce homomorphisms between the Grothendieck groups $K(\mathcal{O}_\lambda)$. Now in case λ and μ lie in the same facet for W, Proposition 7.8 ensures that T_λ^μ sends $[M(w \cdot \lambda)]$ to $[M(w \cdot \mu)]$ for all $w \in W$. This in turn transforms the formula $(*)$ for $[L(w \cdot \lambda)]$ into one for $[L(w \cdot \mu)]$ involving precisely the same coefficients, which are thus seen to depend just on W. So the category equivalence between $\mathcal{O}_\lambda$ and $\mathcal{O}_\mu$ obtained in Theorem 7.8 behaves well with respect to formal characters.

More generally, if $w \cdot \mu$ lies in the *upper closure* of the facet to which $w \cdot \lambda$ belongs, then Theorem 7.9 shows that $T_\lambda^\mu L(w \cdot \lambda) \cong L(w \cdot \mu)$. So again the formula for $[L(w \cdot \lambda)]$ yields the formula for $[L(w \cdot \mu)]$. But here one has to combine some coefficients, taking into account the isotropy group W_μ°. For example, if λ is regular, then after translation the coefficient of $[M(w' \cdot \mu)]$ in $(*)$ becomes

$$\sum_{w'z \le w,\, z \in W_\mu^\circ} b^\lambda_{w'z,w}.$$

Example. Let $\mathfrak{g} = \mathfrak{sl}(3, \mathbb{C})$, with $\Delta = \{\alpha, \beta\}$. If C is the antidominant Weyl chamber, suppose $\lambda \in \Lambda$ is regular and antidominant, while $\mu \in \Lambda$ is antidominant and lies in just the β-hyperplane. Setting $w = s_\beta s_\alpha$, we know from 5.4 that

$$[L(w \cdot \lambda)] = [M(w \cdot \lambda)] - [M(s_\alpha \cdot \lambda)] - [M(s_\beta \cdot \lambda)] + [M(\lambda)].$$

Now $w \cdot \mu$ lies in the α-hyperplane, in the upper closure of $w \cdot C$. Since $W_\mu = \{1, s_\beta\}$, applying T_λ^μ involves some cancellation and results simply in $[L(w \cdot \mu)] = [M(w \cdot \mu)] - [M(s_\alpha \cdot \mu)]$. (This is a typical pattern for $\mathfrak{sl}(2, \mathbb{C})$.)

The general case involves two compatible antidominant weights $\lambda, \mu \in \mathfrak{h}^*$ and the reflection subgroup $W_{[\lambda]} = W_{[\mu]}$ of W. Here the above discussion adapts readily, taking into account the relative positions of $\lambda^\natural$ and $\mu^\natural$ in the alcove geometry of $E(\lambda) = E(\mu)$.

7.11. Translation Functors and Projective Modules

As observed in Proposition 7.1, translation functors take projectives to projectives. By combining Theorem 7.9 with the adjointness property of T_λ^μ and T_μ^λ we can make a precise statement:

Theorem. *Let $\lambda, \mu \in \mathfrak{h}^*$ be antidominant and compatible. Further assume that $\mu^\natural$ lies in the closure of the facet F to which $\lambda^\natural$ belongs. If $w \cdot \mu^\natural$ lies in $\widehat{w \cdot F}$ for some $w \in W_{[\lambda]}$, then $T_\mu^\lambda P(w \cdot \mu) \cong P(w \cdot \lambda)$.*

Proof. We know that $P := T_\mu^\lambda P(w \cdot \mu)$ is projective. Theorem 3.9(b) allows us to determine the indecomposable summands $P(w' \cdot \lambda)$ of P by looking at homomorphisms:

$$\dim \operatorname{Hom}_{\mathcal{O}}\big(P, L(w' \cdot \lambda)\big) = \text{ multiplicity of } P(w' \cdot \lambda) \text{ as a summand of } P.$$

In our situation, adjointness implies

$$\operatorname{Hom}_{\mathcal{O}}\big(T_\mu^\lambda P(w \cdot \mu), L(w' \cdot \lambda)\big) \cong \operatorname{Hom}_{\mathcal{O}}\big(P(w \cdot \mu), T_\lambda^\mu L(w' \cdot \lambda)\big).$$

Thanks to Theorem 7.9, $T_\lambda^\mu L(w' \cdot \lambda)$ is isomorphic to $L(w' \cdot \mu)$ when $w' \cdot \mu^\natural$ lies in $\widehat{w' \cdot F}$, but is otherwise 0. Since $L(w \cdot \mu)$ is the unique simple quotient of $P(w \cdot \mu)$, we must get 0 unless $L(w' \cdot \mu) \cong L(w \cdot \mu)$ (in which case $w' \cdot \mu = w \cdot \mu$).

The assumption on $w \cdot \mu$, together with Theorem 7.9, ensures that $P(w \cdot \lambda)$ occurs once as a summand of P. Suppose another summand $P(w' \cdot \lambda)$ also occurs: here $w' \cdot \mu^\natural = w \cdot \mu^\natural$ lies in the upper closure of the respective facets $w' \cdot F$ and $w \cdot F$ in which $w' \cdot \lambda^\natural$ and $w \cdot \lambda^\natural$ lie. We have to show that $w' \cdot \lambda = w \cdot \lambda$.

By Proposition 7.3(a), the facet $w \cdot F$ lies in $\widehat{C}$ for a unique chamber C. In turn, Proposition 7.3(b) implies that $w \cdot \mu^\natural \in \widehat{C}$. Then C is also the unique chamber whose upper closure contains the facet of $w \cdot \mu^\natural = w' \cdot \mu^\natural$. In turn, $w' \cdot F$ must lie in $\widehat{C}$. But $\overline{C}$ is a fundamental domain for the action of $W_{[\lambda]}$, so $w \cdot \lambda^\natural = w' \cdot \lambda^\natural$. This shows that $w^{-1} w'$ lies in the stabilizer of $\lambda^\natural$ in $W_{[\lambda]}$, which is the same as the stabilizer W_λ° of λ in $W_{[\lambda]}$ (or in W): see Proposition 7.4(b). Thus $w' \cdot \lambda = w \cdot \lambda$. $\square$

7.12. Translation from a Facet Closure

Now we study more closely translation functors T_μ^λ, with $\mu^\natural$ in the closure of the facet of $\lambda^\natural$. The effect of such functors on Verma modules (or their simple quotients) is potentially quite complicated, but also interesting for the pursuit of composition factor multiplicities. Following Jantzen [**147**, 2.17], we can make an explicit statement about formal characters. Since $T_\mu^\lambda M(w \cdot \mu)$ has a standard filtration (Corollary 7.6), this specifies the filtration multiplicities. Recall that notation such as W_μ° indicates the isotropy group of μ relative to the dot-action of W. (This is the same when μ is replaced by $\mu^\natural$ and W by $W_{[\mu]}$.)

Theorem. *Let λ, μ be antidominant and compatible. Assume that $\mu^\natural$ lies in the closure of the facet of $\lambda^\natural$. Then for all $w \in W_{[\lambda]} = W_{[\mu]}$,*

$$(1) \qquad \operatorname{ch} T_\mu^\lambda M(w \cdot \mu) = \sum_{w' \in W_\mu^\circ / W_\lambda^\circ} \operatorname{ch} M(ww' \cdot \lambda).$$

In particular, all Verma modules which occur as quotients in a standard filtration of $T_\mu^\lambda M(w \cdot \mu)$ have filtration multiplicity 1.

Proof. We take advantage of the characterization of standard filtration multiplicities given by Theorem 3.7, together with the adjointness property of translation functors. Given $w \in W_{[\lambda]}$, it is convenient to denote an arbitrary weight linked to λ by $ww' \cdot \lambda$ with $w' \in W_{[\lambda]}$. Then:

$$
\begin{aligned}
\big(T_\mu^\lambda M(w \cdot \mu) : M(ww' \cdot \lambda)\big) &= \dim \operatorname{Hom}_{\mathcal{O}} \big(T_\mu^\lambda M(w \cdot \mu), M(ww' \cdot \lambda)^\vee\big) \\
&= \dim \operatorname{Hom}_{\mathcal{O}} \big(M(w \cdot \mu), T_\lambda^\mu M(ww' \cdot \lambda)^\vee\big) \\
&= \dim \operatorname{Hom}_{\mathcal{O}} \big(M(w \cdot \mu), (T_\lambda^\mu M(ww' \cdot \lambda))^\vee\big) \\
&= \dim \operatorname{Hom}_{\mathcal{O}} \big(M(w \cdot \mu), M(ww' \cdot \mu)^\vee\big).
\end{aligned}
$$

By Theorem 3.3(c), this last dimension is 0 unless $ww' \cdot \mu = w \cdot \mu$ (equivalent to $w' \in W_\mu^\circ$), in which case it is 1. Since $ww' \cdot \lambda = w \cdot \lambda$ just when $w' \in W_\lambda^\circ$, the character equality in (1) follows. $\qquad\square$

In the framework of 7.5, the theorem recovers indirectly the fact that the only weights ν' of the finite dimensional module $L(\nu)$ with highest weight W-conjugate to $\lambda - \mu$ for which $\mu + \nu' \in W_{[\mu]} \cdot \lambda$ are the extremal weights. (In [**147**] this was the approach taken to the theorem, parallel to the proof of Theorem 7.6.)

Corollary. *Let λ, μ be as in the theorem. For arbitrary $M \in \mathcal{O}$, we have* $\operatorname{ch} T_\lambda^\mu T_\mu^\lambda M = |W_\mu^\circ / W_\lambda^\circ| \operatorname{ch} M$.

Proof. First write $\operatorname{ch} M$ as a $\mathbb{Z}$-linear combination of characters of Verma modules; then apply the theorem together with Theorem 7.6. $\qquad\square$

What can be said about the module structure of $T_\mu^\lambda M(w \cdot \mu)$ in the situation of the theorem? Since we do not in general understand well the submodule structure of a single Verma module (including its composition factor multiplicities), there are obvious limits to what we can learn by applying translation functors. However, it is reasonable to use the adjoint property as in the proof above to investigate the possible simple submodules or simple quotients of $T_\mu^\lambda M(w \cdot \mu)$, using the results of 7.9. This may help to decide, for example, whether $T_\mu^\lambda M(w \cdot \mu)$ is indecomposable. Rather than attempt this in general, we turn to the most important special case.

7.13. Example

Theorem 7.12 can be used to recover Theorem 4.10 in a more general setting, as follows. Keep the notation above: λ, μ are antidominant and compatible, with $\mu^\natural$ in the closure of the facet of $\lambda^\natural$. Denote by w_λ the longest element of $W_{[\lambda]} = W_{[\mu]}$, and let w_μ° be the longest element of its (parabolic) subgroup W_μ°. It follows from the definitions that $\lambda_\circ := w_\lambda \cdot \lambda$ and $\mu_\circ := w_\lambda \cdot \mu$ are dominant. In particular, Proposition 3.8(a) shows that $M(\mu_\circ)$ is *projective*. In turn, $P := T_\mu^\lambda M(\mu_\circ)$ is projective, thanks to Proposition 7.1.

Now we can apply Theorem 7.12 with w taken to be w_λ. The character formula there applies to any standard filtration of P. To simplify the notation we assume that λ is *regular*, leaving the details in the general case to the reader. Now $W_\lambda^\circ = 1$, so the sum in the theorem is taken over the characters of all $M(w_\lambda w' \cdot \lambda)$ with $w' \in W_\mu^\circ$.

Recall from 7.4 that the reflection group W_μ° is generated by the simple reflections in $W_{[\lambda]}$ which fix μ; in particular, its intrinsic Bruhat ordering is induced by the Bruhat ordering of $W_{[\lambda]}$ (0.4). Thus $w_\mu^\circ \cdot \lambda \geq w' \cdot \lambda$ for all $w' \in W_\mu^\circ$, a relationship which is reversed when w_λ is applied to both weights. Now Remark 3.6 implies that $M(w_\lambda w_\mu^\circ \cdot \lambda)$ occurs as a quotient of P. This in turn forces $P(w_\lambda w_\mu^\circ \cdot \lambda)$ to occur as a summand of P.

We claim there are no other summands, which would necessarily be of the form $P(w_\lambda w' \cdot \lambda)$ with $w' \in W_\mu^\circ$. From Verma's Theorem 4.6 and the ordering of weights, we know that $M(w_\lambda w_\mu^\circ \cdot \lambda)$ embeds in $M(w_\lambda w' \cdot \lambda)$; in particular, $L(w_\lambda w_\mu^\circ \cdot \lambda)$ is a composition factor of the latter. By BGG Reciprocity (3.11), $M(w_\lambda w' \cdot \lambda)$ occurs in any standard filtration of $P(w_\lambda w_\mu^\circ \cdot \lambda)$. Comparison with the character formula in Theorem 7.12 proves our claim.

Proposition. *Let $\lambda, \mu \in \mathfrak{h}^*$ be antidominant and compatible, with $\mu^\natural$ in the closure of the chamber containing $\lambda^\natural$. Denote by w_λ the longest element in $W_{[\lambda]} = W_{[\mu]}$ and by w_μ° the longest element in W_μ°. Then $T_\mu^\lambda M(w_\lambda \cdot \mu) \cong P(w_\lambda w_\mu^\circ \cdot \lambda)$. Moreover, the Verma modules occurring as quotients in a* *standard filtration of $P(w_\lambda \cdot \lambda)$ are the $M(w_\lambda w \cdot \lambda)$ with $w \in W_\mu^\circ$; each has multiplicity 1.*　　$\square$

To recover the special case treated in 4.10, take λ to be integral and $\mu = -\rho$: then $T_{-\rho}^\lambda$ takes $M(-\rho) = P(-\rho)$ to $P(\lambda)$.

7.14. Translation from a Wall

As before, we assume that λ and μ are antidominant and compatible. But from now on we focus on the situation when λ is *regular*, so $W_\lambda^\circ = \{1\}$ and $\lambda^\natural$ lies in a chamber C of $E(\lambda) = E(\mu)$. We then require $\mu^\natural$ to lie in a single wall of C, the intersection of $\overline{C}$ with the root hyperplane H_α for some $\alpha > 0$. Write $s = s_\alpha$, so $W_\mu^\circ = \{1, s\}$. Thanks to Corollary 7.12, for all $w \in W_{[\lambda]}$ we have $\mathrm{ch}\, T_\lambda^\mu T_\mu^\lambda L(w \cdot \mu) = 2\,\mathrm{ch}\, L(w \cdot \mu)$. But since simple modules cannot extend themselves nontrivially (Proposition 3.1), this forces

$$(1) \qquad T_\lambda^\mu T_\mu^\lambda L(w \cdot \mu) \cong L(w \cdot \mu) \oplus L(w \cdot \mu).$$

Now consider $w \in W_{[\lambda]}$ for which $w\alpha > 0$; thus $\ell(ws) > \ell(w)$. In particular, $w \cdot \mu^\natural$ lies in the upper closure of $w \cdot C$. Thanks to Theorem 7.9, $T_\lambda^\mu L(w \cdot \lambda) \cong L(w \cdot \mu)$, whereas $T_\lambda^\mu L(ws \cdot \lambda) = 0$.

In this special situation we want to describe some of the structure and composition factor multiplicities of $T_\mu^\lambda M(w \cdot \mu)$ and its quotient $T_\mu^\lambda L(w \cdot \mu)$. If we express $\mathrm{ch}\, L(w \cdot \mu)$ as a $\mathbb{Z}$-linear combination of various $\mathrm{ch}\, M(w' \cdot \mu)$ and then apply T_μ^λ, Theorem 7.12 yields a similar (but more complicated) expression for $\mathrm{ch}\, T_\mu^\lambda L(w \cdot \mu)$ as a $\mathbb{Z}$-linear combination of characters of Verma modules. This can be used effectively in the proof of the following theorem.

Theorem. *Let λ, μ be antidominant and compatible. Assume that λ is regular, so $\lambda^\natural$ lies in a chamber C, while $\mu^\natural$ lies in a single wall of C corresponding to $\alpha > 0$. Suppose $w \in W_{[\lambda]}$ satisfies $w\alpha > 0$, so $\ell(ws) > \ell(w)$ with $s = s_\alpha$.*

(a) *There is a short exact sequence*

$$0 \to M(ws \cdot \lambda) \to T_\mu^\lambda M(w \cdot \mu) \to M(w \cdot \lambda) \to 0.$$

(b) *$\mathrm{Hd}\, T_\mu^\lambda M(w \cdot \mu) \cong L(w \cdot \lambda)$. In particular, $T_\mu^\lambda M(w \cdot \mu)$ is indecomposable and the above exact sequence nonsplit.*

(c) *The module $T_\mu^\lambda L(w \cdot \mu)$ is self-dual, with head and socle both isomorphic to $L(w \cdot \lambda)$; thus $T_\mu^\lambda L(w \cdot \mu)$ is indecomposable.*

(d) *$[T_\mu^\lambda L(w \cdot \mu) : L(w \cdot \lambda)] = 2$.*

(e) *$[T_\mu^\lambda L(w \cdot \mu) : L(ws \cdot \lambda)] = 1$.*

(f) *Let $w' \in W_{[\lambda]}$ with $w' \cdot \lambda \neq w \cdot \lambda$. If $[T_\mu^\lambda L(w \cdot \mu) : L(w' \cdot \lambda)] > 0$, then $w's \cdot \lambda < w' \cdot \lambda$. Moreover, $T_\lambda^\mu L(w' \cdot \lambda) = 0$.*

(g) *If $w' \in W_{[\lambda]}$ satisfies $w's \cdot \lambda < w' \cdot \lambda$, then*

$$\mathrm{Ext}_\mathcal{O}\left(L(w \cdot \lambda), L(w' \cdot \lambda)\right) \cong \mathrm{Hom}_\mathcal{O}\left(\mathrm{Rad}\, T_\mu^\lambda L(w \cdot \mu), L(w' \cdot \lambda)\right).$$

Proof. (a) Theorem 7.12 implies that

$$\mathrm{ch}\, T_\mu^\lambda M(w \cdot \mu) = \mathrm{ch}\, M(w \cdot \lambda) + \mathrm{ch}\, M(ws \cdot \lambda).$$

Since the translated module has a standard filtration (Corollary 7.6), it follows from Proposition 3.7(a) that $M(ws \cdot \lambda)$ occurs as a submodule.

(b) By adjointness,

$$\mathrm{Hom}_\mathcal{O}\left(T_\mu^\lambda M(w \cdot \mu), L(w' \cdot \lambda)\right) \cong \mathrm{Hom}_\mathcal{O}\left(M(w \cdot \mu), T_\lambda^\mu L(w' \cdot \lambda)\right),$$

which by Theorem 7.9 is nonzero only when $w' \cdot \mu = w \cdot \mu$ (hence $w' = w$ or $w' = ws$) and $w' \cdot \mu^\natural$ lies in the upper closure of the chamber containing $w' \cdot \lambda^\natural$. The upper closure requirement is met only by $w' = w$.

(c) By exactness of T_μ^λ, the module $T_\mu^\lambda L(w \cdot \mu)$ is a quotient of $T_\mu^\lambda M(w \cdot \mu)$. So the simple quotients of the former are also simple quotients of the latter. Thanks to (b), only $L(w \cdot \lambda)$ occurs (and has multiplicity 1). Since $L(w \cdot \mu)$ is self-dual in $\mathcal{O}$ and translation commutes with duality (Proposition 7.1), the module $T_\mu^\lambda L(w \cdot \mu)$ is also self-dual and therefore has socle $L(w \cdot \lambda)$.

(d) From (c) we get $[T_\mu^\lambda L(w \cdot \mu) : L(w \cdot \lambda)] \geq 1$, with equality holding just when $T_\mu^\lambda L(w \cdot \mu) \cong L(w \cdot \lambda)$. Why is this impossible? Applying T_λ^μ would yield $T_\lambda^\mu T_\mu^\lambda L(w \cdot \mu) \cong L(w \cdot \mu)$, which contradicts (1) above.

On the other hand, suppose $[T_\mu^\lambda L(w \cdot \mu) : L(w \cdot \lambda)] \geq 3$. Then Exercise 7.7 would force $[T_\lambda^\mu T_\mu^\lambda L(w \cdot \lambda) : L(w \cdot \mu)] \geq 3$, again contradicting (1).

(e) From (a) we see that $L(ws \cdot \lambda)$ occurs just once as a composition factor of $T_\mu^\lambda M(w \cdot \mu)$, with a maximal vector of weight $ws \cdot \lambda$ generating a submodule M isomorphic to $M(ws \cdot \lambda)$. So $[T_\mu^\lambda L(w \cdot \mu) : L(ws \cdot \lambda)] \leq 1$. If M were in the kernel of the surjection $T_\mu^\lambda M(w \cdot \mu) \to T_\mu^\lambda L(w \cdot \mu)$, there would just be a single composition factor $L(w \cdot \lambda)$ in the latter module, contrary to (d). So M has nonzero image under the surjection, forcing $T_\mu^\lambda L(w \cdot \mu)$ to have a composition factor isomorphic to $L(ws \cdot \lambda)$.

(f) In case $w' = ws$ (which satisfies the hypothesis by (e)), we have $w's \cdot \lambda = w \cdot \lambda < ws \cdot \lambda = w' \cdot \lambda$. Then Theorem 7.9(b) implies that $T_\lambda^\mu L(w' \cdot \lambda) = 0$.

In the contrary case, $w' \neq w, ws$, so $L(w' \cdot \mu)$ is not isomorphic to $L(w \cdot \mu)$. By hypothesis, $[T_\mu^\lambda L(w \cdot \mu) : L(w' \cdot \lambda)] > 0$. Thanks to Exercise 7.7, this implies $[T_\lambda^\mu T_\mu^\lambda L(w \cdot \mu) : T_\lambda^\mu L(w' \cdot \lambda)] > 0$. But $T_\lambda^\mu L(w' \cdot \lambda)$ is either 0 or isomorphic to $L(w' \cdot \mu)$. Since $L(w' \cdot \mu)$ is not isomorphic to $L(w \cdot \mu)$, the second alternative is impossible by (1) above. In turn, Theorem 7.9 forces $w's \cdot \lambda < w' \cdot \lambda$.

(g) Conversely, assume that $w's \cdot \lambda < w' \cdot \lambda$, so $T_\lambda^\mu L(w' \cdot \lambda) = 0$ by Theorem 7.9. By adjointness, this implies that both $\mathrm{Hom}_\mathcal{O}\left(T_\mu^\lambda L(w \cdot \mu), L(w' \cdot \lambda)\right)$ and $\mathrm{Ext}_\mathcal{O}\left(T_\mu^\lambda L(w \cdot \mu), L(w' \cdot \lambda)\right)$ are 0. Apply the functor $\mathrm{Hom}_\mathcal{O}\left(?, L(w' \cdot \lambda)\right)$ to the short exact sequence

$$0 \to \mathrm{Rad}\, T_\mu^\lambda L(w \cdot \mu) \to T_\mu^\lambda L(w \cdot \mu) \to L(w \cdot \lambda) \to 0.$$

The resulting long exact sequence together with the vanishing results above complete the proof. $\square$

Exercise. Keep the hypotheses of the theorem. For all $w' \in W_{[\lambda]}$, prove that $[T_\mu^\lambda L(w \cdot \mu) : L(w' \cdot \lambda)] \leq 2[M(ws \cdot \lambda) : L(w' \cdot \lambda)]$.

How much does the theorem reveal about the structure and composition factors of $M := T_\mu^\lambda L(w \cdot \mu)$? This module is self-dual, with $L(w \cdot \lambda)$ as head and socle. There is one occurrence of the composition factor $L(ws \cdot \lambda)$. Since $w \cdot \lambda < ws \cdot \lambda$, it follows from (a) that any composition factor $L(w' \cdot \lambda)$ different from these must satisfy $w' \cdot \lambda < ws \cdot \lambda$. Moreover, each composition factor $L(w' \cdot \lambda)$ with $w' \neq w$ (including $L(ws \cdot \lambda)$) satisfies $w's \cdot \lambda < w' \cdot \lambda$, forcing $T_\lambda^\mu L(w' \cdot \lambda) = 0$. This last point is already evident from (1) above: $T_\lambda^\mu M \cong L(w \cdot \mu) \oplus L(w \cdot \mu)$. Beyond this we still know very little about the potentially complicated structure of $\mathrm{Rad}\, M / \mathrm{Soc}\, M$.

In the simplest nontrivial case, M just has Loewy length 3: If $w = s$, then $M(\lambda) = L(\lambda)$ and $T_\lambda^\mu L(\lambda) \cong L(\mu) = M(\mu)$. Then $T_\mu^\lambda L(\mu)$ has only $L(\lambda)$ and $L(s \cdot \lambda)$ as possible composition factors. But in general the structure of M cannot be worked out using the tools we have developed so far.

7.15. Wall-Crossing Functors

To round out this chapter, we introduce some important functorial language whose significance will become clearer in the following chapters. In the above setting, with λ regular antidominant and $\mu^\natural$ lying only in the s-wall of the chamber containing $\lambda^\natural$, the functor $\Theta_s := T_\mu^\lambda T_\lambda^\mu$ may be thought of as "translation across a wall". Call it a **wall-crossing functor** on $\mathcal{O}$ or $\mathcal{O}_\lambda$. It is sometimes referred to as an operator of "coherent continuation". (The notation θ_s is somewhat more common in the literature; in any case, the use of a greek symbol has become well-entrenched by now.)

To indicate explicitly the dependence of Θ_s on the choice of λ, one might write Θ_s^λ. Note that Θ_s seems to depend not just on λ and s but also on the choice of μ. One can prove that any other antidominant weight satisfying the hypotheses on μ will produce an isomorphic functor. (There seems to be no direct argument for this, but Example 10.8 below indicates how to derive it from the classification of "projective functors".)

In Theorem 7.14 we considered $\Theta_s L(w \cdot \lambda) = T_\mu^\lambda L(w \cdot \mu)$, a module with simple head and simple socle both isomorphic to $L(w \cdot \lambda)$. What remains mysterious is the structure of $\operatorname{Rad} \Theta_s L(w \cdot \lambda)/\operatorname{Soc} \Theta_s L(w \cdot \lambda)$. In the Lie group context where Zuckerman's analogous translation functors appear, Vogan [**253**] conjectured that this module should be *semisimple*, in which case a recursive computation of characters would be possible. This will be explored in more detail in 8.10 in the context of the Kazhdan–Lusztig Conjecture. (The subquotient of $\Theta_s L(w \cdot \lambda)$ in question is sometimes referred to as the "Jantzen middle"; he investigated a parallel semisimplicity conjecture in the prime characteristic setting.)

From the adjoint property of translation functors we see that for all $M \in \mathcal{O}$,

$$\operatorname{Hom}_{\mathcal{O}}(T_\lambda^\mu M, T_\lambda^\mu M) \cong \operatorname{Hom}_{\mathcal{O}}(M, T_\mu^\lambda T_\lambda^\mu M).$$

In particular, the identity map on $T_\lambda^\mu M$ corresponds to an **adjunction morphism** $M \to \Theta_s M$. (See Gelfand–Manin [**110**, II.3.24].) This morphism will not in general be injective (for example when $M \neq 0$ but $T_\lambda^\mu M = 0$); but there is always a cokernel to consider:

$$M \to \Theta_s M \to \operatorname{Sh}_s M \to 0.$$

In this way we introduce a "shuffling" functor $\operatorname{Sh}_s = \operatorname{Sh}_s^\lambda$ on $\mathcal{O}$, to be explored further in Chapter 12.

Exercise. Let $\mathfrak{g} = \mathfrak{sl}(2, \mathbb{C})$. For a fixed regular antidominant weight λ, set $\mu := -\rho$ and let s be the nontrivial element of W. Work out the effect of the functors Θ_s and Sh_s on the Verma modules and simple modules in $\mathcal{O}_\lambda$, using some of the previous theory.

7.16. Self-Dual Projectives

In Chapter 4 we observed that a projective module $P(\lambda)$ cannot be self-dual unless λ is *antidominant* (Exercise 4.8). When λ is *integral* and antidominant, we showed directly in Theorem 4.10 that $P(\lambda)$ is self-dual and has a standard filtration involving each $M(w \cdot \lambda)$ exactly once. The proof amounted to showing that $P(\lambda) \cong T_{-\rho}^\lambda M(-\rho)$. Nothing quite so straightforward can be done if λ fails to be integral; but the self-duality remains true. This was first shown by Irving [**133**], whose proof using translation and wall-crossing functors is given here. (See the Notes below.)

Theorem. *Let $\lambda \in \mathfrak{h}^*$ be antidominant. Then:*

(a) $P(\lambda) \cong P(\lambda)^\vee$.

(b) *Any standard filtration of $P(\lambda)$ involves each $M(w\cdot\lambda)$ with $w \in W_{[\lambda]}$ precisely once as a quotient. Equivalently, BGG Reciprocity implies that $[M(w \cdot \lambda) : L(\lambda)] = 1$ for all $w \in W_{[\lambda]}$.*

Proof. (1) First we observe that it is enough to prove the theorem when λ is *regular*, which will allow us to apply Theorem 7.14 in that case. Assume for the moment the theorem is true for all regular antidominant λ.

If μ is an arbitrary antidominant weight, subtracting $n\rho$ from it for large enough n will obviously produce a compatible regular antidominant weight λ. Since $P(\lambda)$ is self-dual by assumption, $T_\lambda^\mu P(\lambda)$ is also projective and self-dual by Proposition 7.1. We claim that this module is nonzero. Thanks to Corollary 4.8, $\operatorname{Soc} P(\lambda)$ is a direct sum of copies of $L(\lambda)$. If $L(\lambda)$ is such a submodule, exactness of T_λ^μ ensures that $T_\lambda^\mu L(\lambda)$ is a submodule of $T_\lambda^\mu P(\lambda)$. Since $\mu^\natural$ lies in the upper closure of the chamber containing $\lambda^\natural$, Theorem 7.9 shows that $T_\lambda^\mu L(\lambda) \cong L(\mu)$ and is therefore nonzero.

Now $T_\lambda^\mu P(\lambda)$ is isomorphic to a direct sum of $n > 0$ indecomposable projectives $P(w \cdot \mu)$ with $w \in W_{[\mu]} = W_{[\lambda]}$. Each $P(w \cdot \mu)$ has head $L(w \cdot \mu)$ and contains $L(\mu)$ in its socle. Accordingly $\operatorname{Hd} T_\lambda^\mu P(\lambda)$ is the direct sum of the n modules $L(w \cdot \mu)$ while $\operatorname{Soc} T_\lambda^\mu P(\lambda)$ contains the direct sum of n copies of $L(\mu)$. But $T_\lambda^\mu P(\lambda)$ is *self-dual,* so its head and socle are isomorphic. It follows that $T_\lambda^\mu P(\lambda)$ is the direct sum of n copies of $P(\mu)$. Self-duality further implies that $T_\lambda^\mu P(\lambda)$ is the direct sum of n copies of $P(\mu)^\vee$. Since these modules are indecomposable, they must all be isomorphic to $P(\mu)$.

Part (b) of the theorem for $P(\mu)$ then follows easily from the corresponding fact for $P(\lambda)$.

(2) From now on λ is both *regular* and *antidominant*. In particular, $L(\lambda) = M(\lambda)$ is self-dual and a Verma module. It also occurs as the socle of each $M(w \cdot \lambda)$. If w_λ is the longest element in $W_{[\lambda]}$, it follows from the definition that the weight $\lambda_\circ := w_\lambda \cdot \lambda$ is dominant. So $M(\lambda_\circ) = P(\lambda_\circ)$ is projective (by Proposition 3.8(a)) as well as a Verma module.

The objective now is to pass from $M(\lambda_\circ) = P(\lambda_\circ)$ to $P(\lambda)$ using wall-crossing functors Θ_s. Fix a reduced expression $w_\lambda = s_1 \cdots s_n$ with $s_i = s_{\alpha_i}$ for some $\alpha_i \in \Delta_{[\lambda]}$ and set $\Theta := \Theta_{s_1} \circ \cdots \circ \Theta_{s_n}$. Note that the exactness of Θ and the embedding $L(\lambda) \subset M(\lambda_\circ)$ yield an embedding $\Theta L(\lambda) \subset \Theta M(\lambda_\circ)$.

(3) Since λ is regular, we can take advantage of the precise way in which a wall-crossing functor Θ_s applies to a Verma module $M(w \cdot \lambda)$ in $\mathcal{O}_\lambda$ such as $M(\lambda_\circ)$ or $M(\lambda) = L(\lambda)$. Taking into account Theorem 7.6, we see from Theorem 7.14(a) that $\Theta_s M(w \cdot \lambda)$ is a nonsplit extension of the two Verma modules $M(w \cdot \lambda)$ and $M(ws \cdot \lambda)$ when $ws > w$. Here $L(w \cdot \lambda)$ occurs just bonce as a composition factor of each Verma module.

(4) From parts (d) and (f) of Theorem 7.14, we see that $\Theta_s L(w \cdot \lambda)$ is either 0 or has just one composition factor isomorphic to $L(\lambda)$ when $w \neq 1$, while $\Theta_s L(\lambda)$ has two composition factors of this type (one in each of the Verma modules produced by wall-crossing). In particular, the only composition factors of type $L(\lambda)$ in $\Theta M(\lambda_\circ)$ are produced by applying wall-crossing functors to $L(\lambda)$. In view of step (3), it follows that

$$(*) \qquad [\Theta M(\lambda_\circ) : L(\lambda)] = [\Theta L(\lambda) : L(\lambda)].$$

(5) Iteration of wall-crossings, starting at either $M(\lambda)$ or $M(\lambda_\circ)$, shows that each resulting Verma module $M(w \cdot \lambda)$ has the composition factor $L(\lambda)$ only once (and then it is the socle). Moreover, it is easy to check that any $w \in W_{[\lambda]}$ can occur here: write $w_\lambda = ww'$ with $\ell(w_\lambda) = \ell(w) + \ell(w')$ and choose a corresponding reduced expression for w_λ.

(6) Since $\lambda_\circ$ is dominant, $M(\lambda_\circ)$ is *projective*. So $\Theta M(\lambda_\circ)$ is also projective (Proposition 7.1). By Theorem 7.14(a), each wall-crossing functor Θ_s takes a Verma module $M(w \cdot \lambda)$ to a nonsplit extension involving the two Verma modules $M(w \cdot \lambda)$ and $M(ws \cdot \lambda)$. Therefore the projective module $\Theta M(\lambda_\circ)$ acquires in n steps a standard filtration involving the final Verma module $M(\lambda) = L(\lambda)$ just once. Thanks to BGG Reciprocity, $P(\lambda)$ occurs once as a direct summand of $\Theta M(\lambda_\circ)$.

(7) On the other hand, $\Theta L(\lambda)$ is *self-dual* (because $L(\lambda)$ is). Since $L(\lambda) \subset \mathrm{Soc}\, \Theta L(\lambda)$, we get $L(\lambda) \subset \mathrm{Hd}\, \Theta L(\lambda)$. From $(*)$ it follows that the image of $\Theta L(\lambda) \subset \Theta M(\lambda_\circ)$ under the projection $\Theta M(\lambda_\circ) \to P(\lambda) \to L(\lambda)$ cannot be zero. So the image of $\Theta L(\lambda)$ in $P(\lambda)$ cannot be a proper submodule (lest it lie in the radical). Using the projective property, we get $P(\lambda) \subset \Theta L(\lambda)$.

(8) In turn, the self-duality of $\Theta L(\lambda)$ implies that the injective module $Q(\lambda) = P(\lambda)^\vee \subset \Theta L(\lambda) \subset \Theta M(\lambda_\circ)$. Being injective, $Q(\lambda)$ must be a direct summand of $\Theta M(\lambda_\circ)$; the latter being projective, $Q(\lambda)$ must also be projective (and indecomposable). In particular, its head is simple and therefore isomorphic to the simple submodule $L(\lambda)$ of its dual $P(\lambda)$. In other words, $P(\lambda)^\vee \cong P(\lambda)$, which proves (a).

(9) For part (b) of the theorem, we use the fact developed in step (5): each $M(w \cdot \lambda)$ has $L(\lambda)$ as a composition factor exactly once. This implies, by BGG Reciprocity, that all $M(w \cdot \lambda)$ with $w \in W_{[\lambda]}$ occur just once in each standard filtration of $P(\lambda)$. $\qquad\square$

Exercise. It is instructive to follow the details of the proof in the case $\mathfrak{g} = \mathfrak{sl}(3, \mathbb{C})$, with λ integral. There are two reduced expressions $w_\circ = s_1 s_2 s_1 = s_2 s_1 s_2$, each leading to different standard filtrations of $\Theta L(\lambda)$ and $\Theta L(\lambda_\circ)$. The latter module is a direct sum of $P(\lambda)$ and one other indecomposable projective. Both $\Theta L(\lambda)$ and $\Theta L(\lambda_\circ)$ have standard filtrations involving 8 Verma modules, with 6 distinct ones involved in $P(\lambda)$ in each case.

Notes

(7.1–7.4) Translation functors were extensively developed by Jantzen, in both the category $\mathcal{O}$ setting and the parallel characteristic p theory (where W is replaced by an affine Weyl group). See [**147**, Kap. 2] as well as later refinements for algebraic groups in [**152**, II.7]. (Independently, Zuckerman [**262**] worked out similar ideas for the category of Harish-Chandra modules.) Around the same time a broader view of some of the ideas involved in translation was developed by Bernstein–Gelfand [**24**] under the rubric "projective functors"; this will be discussed in Chapter 10.

The argument in Lemma 7.5 originates in Jantzen's paper [**145**, Satz 5].

For the equivalence of blocks in 7.8 we follow Jantzen [**147**, 2.15, Bem. 2)]; he refers to similar arguments used by Zuckerman [**262**] in the setting of Harish-Chandra modules. The special case involving regular integral blocks is formulated by BGG in [**27**, Thm. 4].

The arguments in 7.9 are analogues of those in Jantzen [**152**, II.7.15]. (For his earlier approach in $\mathcal{O}$, see [**147**, 2.11–2.13].)

(7.11) The proof of the theorem differs from Jantzen's [**147**, 2.24], which is somewhat longer and relies on a formal character comparison, together with BGG Reciprocity.

(7.12–7.15) These results are drawn from Jantzen [**147**, Chap. 2] and [**152**, II.7]. The terminology "wall-crossing functor" did not appear explicitly in Jantzen [**147**] but was soon in common use and appears in [**152**, II.7.21] as well as [**152**, II.C], where Lusztig's algebraic group analogue of the Kazhdan–Lusztig Conjecture is explained in detail.

The proof of Proposition 7.13 was suggested by Jantzen.

In unpublished notes [**132**, 5.3], Irving first worked out a lengthy proof of Theorem 7.16, which he later streamlined in [**133**, §3] in the context of a more general conjecture for parabolic category $\mathcal{O}$. In other unpublished notes, Joseph [**160**, 3.13] then shortened the argument further by taking advantage of the work of Bernstein–Gelfand [**24**] on projective functors.

Chapter 8

Kazhdan–Lusztig Theory

Following Verma's 1966 thesis and the realization that the composition factor multiplicities of Verma modules could not as readily be pinned down as he expected at first, a considerable amount of effort went into finding an algebraic solution of the problem: work of BGG, Deodhar, Jantzen, Joseph, and others. In the previous chapters we have seen how to refine the problem, but not how to solve it completely except in isolated cases.

The landmark 1979 paper of Kazhdan–Lusztig [170] proposed a precise conjecture about the multiplicities for the principal block of $\mathcal{O}$, in terms of the values at 1 of certain polynomials attached to the Iwahori–Hecke algebra of W. Following Deodhar's suggestion in [81], we refer to the developments inspired by [170] as "Kazhdan–Lusztig theory" (or KL theory, for short). Their approach was motivated in part by geometric questions involving the flag variety and unipotent variety associated with $\mathfrak{g}$, as we shall recall below. New ideas coming from geometry, topology, and analysis were already in the air at the time and pointed quickly to a proof of the conjecture.

After reviewing briefly in 8.1 what was known by 1979, we outline the construction of Kazhdan–Lusztig polynomials (8.2–8.3) before stating the Kazhdan–Lusztig Conjecture formally (8.4). Its relationship with singularities of Schubert varieties and intersection cohomology is then explored in 8.5–8.6, followed by a comparison with Jantzen's earlier criterion for a composition factor of a Verma module to have multiplicity 1 (8.7).

Independent proofs of the conjecture were published in 1981 by Beilinson–Bernstein [16] and Brylinski–Kashiwara [55], based on similar techniques involving $\mathcal{D}$-modules and intersection cohomology. In 8.8 we can only introduce this literature, which requires ideas far beyond the scope of the earlier algebraic study of category $\mathcal{O}$.

The KL Conjecture leads to structural information about the modules involved, not just multiplicity data, as suggested by an equivalent formulation due to Vogan in terms of Ext^n functors together with a deeper study of wall-crossing functors (8.10–8.11). It turns out that the *coefficients* of KL polynomials yield information about extensions of simple modules and the layer structure of naturally occurring Loewy filtrations (8.14–8.15). A refinement due to Beilinson–Bernstein of the geometric methods used to prove the KL Conjecture ultimately settles the Jantzen Conjecture as well (8.13).

This chapter marks a shift from "textbook" mode to "survey" mode, in which we no longer attempt to provide full details of proofs but rather guide the reader into the more advanced literature. In line with this goal, we often focus just on *integral* weights, starting with the principal block.

Notational convention: When $x \le w$ in W, the difference $\ell(w) - \ell(x)$ occurs frequently; it will be abbreviated to $\ell(x, w)$.

8.1. The Multiplicity Problem for Verma Modules

To put things in perspective, recall the discussion in 7.10, where we saw that translation functors can be used for *integral* weights to reduce the computation of formal characters of simple modules (or composition factor multiplicies of Verma modules) to a single regular block such as the principal block $\mathcal{O}_0$. Together with the BGG–Verma Theorem (5.1), this shifts attention to the role of W and its Bruhat ordering. But the substantial algebraic progress made on the problem in the 1970s fell short of reaching a complete solution. The Kazhdan–Lusztig Conjecture [**170**, Conj. 1.5] offered a conceptual (and computable) interpretation of multiplicities for $\mathcal{O}_0$, based on a new set of ideas.

On the other hand, the treatment of arbitrary weights $\lambda \in \mathfrak{h}^*$ requires further work, even though Jantzen correctly foresaw the leading role to be played by the reflection group $W_{[\lambda]}$. A basic reduction, due to Soergel [**237**, Thm. 11], will be explained in 13.13.

Before taking up Kazhdan–Lusztig theory we review briefly the progress actually made before 1979. Some low rank computations were done independently by several people. But the most definitive results are due to Jantzen in [**147**] and earlier papers, based on algebraic tools he developed: contravariant forms, Jantzen filtration and Sum Formula, translation functors, together with other methods we have not discussed related to primitive ideals of $U(\mathfrak{g})$ and associated varieties of modules. The results are of two types: (1) Computation of all multiplicities when Φ (or sometimes $\Phi_{[\lambda]}$) involves just irreducible components of rank ≤ 3 or types A_4, D_4. These can now be reproduced more routinely by computations of the type shown in Goresky's tables [**114**]. (2) More general qualitative or recursive formulations concerning the multiplicities. Here is a quick summary of both types:

- In the background are the early examples found by BGG [**25**] and Conze–Dixmier [**72**], exhibiting weights for which $[M(\lambda) : L(\mu)] > 1$; here the weights are dominant but singular. Deodhar–Lepowsky [**83**] then showed that in rank ≥ 3 there always exist similar examples involving regular dominant weights.

- To work out all multiplicities for types A_2, B_2, G_2 is already somewhat challenging: for A_2, Exercise 5.3 indicates how to fill in details for integral weights by using the Jantzen Sum Formula. Jantzen himself treated the other rank 2 cases and found in general that for blocks involving a group $W_{[\lambda]}$ of rank 2, all composition factor multiplicities of Verma modules are 1.

- Jantzen [**147**, 5.24] was able to determine all multiplicities in ranks up to 3 as well as in type A_4. He remarked also that he could use recent methods of Vogan to complete the case D_4. This and his more general results prompted him to observe that in all examples treated so far, multiplicities depend essentially just on the group $W_{[\lambda]}$ (and the root system $\Phi_{[\lambda]}$) and to ask whether this is true in general.

- The role of Weyl groups was exhibited most generally by Jantzen [**147**, 4.13–4.14] as follows, for weights which are not necessarily integral. Suppose λ, μ are regular, antidominant, compatible, while all components of $\Phi_{[\lambda]}$ are of classical type A_ℓ–D_ℓ; then λ, μ have the "same multiplicities": for all $w, w' \in W$,

$$[M(w \cdot \lambda) : L(w' \cdot \lambda)] = [M(w \cdot \mu) : L(w' \cdot \mu)].$$

- In a similar spirit he used his full arsenal of methods to find, for any $\mathfrak{g}$ and arbitrary $\lambda, \mu \in \mathfrak{h}^*$ satisfying $\mu \uparrow \lambda$, necessary and sufficient conditions for $[M(\lambda) : L(\mu)] = 1$ [**147**, 5.20]. (This will be revisited in 8.7.) The conditions here are essentially recursive in nature: for all $\nu \in \mathfrak{h}^*$ such that $\mu \uparrow \nu \uparrow \lambda$ but $\nu \neq \lambda$, we have $[M(\nu) : L(\mu)] = 1$ together with

$$|\Phi_\lambda^+| - |\Phi_\mu^+| = |\{\alpha \in \Phi_\lambda^+ \mid \mu \uparrow s_\alpha \cdot \lambda\}|.$$

From this one recovers by an easy induction: $[M(\lambda) : L(\mu)] = 1$ whenever $\mu \uparrow \lambda$ and μ is *antidominant*.

From the early work emerged a strong sense that a full determination of composition factor multiplicities in Verma modules would depend mostly (perhaps entirely) on the reflection group $W_{[\lambda]}$ and would require a *recursive* approach rather than a closed formula for each pair of weights. The correct framework is provided by Kazhdan–Lusztig theory.

8.2. Hecke Algebras and Kazhdan–Lusztig Polynomials

Here is a concise outline of the elementary algebraic development due to Kazhdan–Lusztig [170], which leads to the introduction of their special polynomials. Our notation, following Humphreys [129, Chap. 7], differs slightly from theirs. Everything here can be done uniformly for an arbitrary Coxeter group (without reference to Lie theory), but our immediate concern is just with the Weyl group W attached to $\mathfrak{g}$.

Inspired by earlier work of Iwahori on representations of finite Chevalley groups, one starts with a deformation $\mathcal{H}$ of the integral group ring of W using an indeterminate q as parameter. Setting $q = 1$ recovers $\mathbb{Z}W$. Call $\mathcal{H}$ the **Hecke algebra** (or **Iwahori–Hecke algebra**) of W. More precisely, $\mathcal{H}$ is defined by generators and relations over the ring of Laurent polynomials $A := \mathbb{Z}[q, q^{-1}]$, the generators T_w being in bijection with elements of W. For example, when s is a simple reflection, $T_s^2 = (q-1)T_s + qT_1$, with T_1 acting as the identity in the ring structure on $\mathcal{H}$.

The elements T_w turn out to be an A-basis of $\mathcal{H}$. Moreover, these are units in the ring $\mathcal{H}$:

$$(T_{w^{-1}})^{-1} = (-1)^{\ell(w)} q^{-\ell(w)} \sum_{x \leq w} (-1)^{\ell(x)} R_{x,w}(q) T_x.$$

Here we use the Bruhat ordering of W (which is seen to be unavoidable!), while $R_{x,w} \in \mathbb{Z}[q]$ is a recursively computable monic polynomial of degree $\ell(x, w) = \ell(w) - \ell(x)$, with $R_{w,w}(q) = 1$. For example, the relation quoted above leads to

$$T_s^{-1} = q^{-1}T_s - (1 - q^{-1})T_1.$$

The next ingredient is an involution of $\mathcal{H}$, denoted by bar, which interchanges q with q^{-1} in the ring A and sends T_w to $(T_{w^{-1}})^{-1}$. The objective is to locate a new basis of $\mathcal{H}$ indexed by W consisting of elements C_w which satisfy $\overline{C_w} = C_w$. This requires working with a square root of q, as seen in the formula

$$C_s := q^{-\frac{1}{2}}(T_s - qT_1).$$

So the base ring A is replaced by the ring of Laurent polynomials in $q^{\frac{1}{2}}$ and $q^{-\frac{1}{2}}$. Now we can state the theorem.

Theorem (Kazhdan–Lusztig). *There are unique elements C_w ($w \in W$) fixed by the involution on $\mathcal{H}$ and satisfying the conditions*

$$C_w = (-1)^{\ell(w)} q^{\ell(w)/2} \sum_{x \leq w} (-1)^{\ell(x)} q^{-\ell(x)} \overline{P_{x,w}(q)} T_x,$$

where $P_{w,w}(q) = 1$, $P_{x,w}(q) \in \mathbb{Z}[q]$, and $\deg P_{x,w}(q) \leq (\ell(x,w) - 1)/2$ whenever $x < w$.

The "change of basis" polynomials $P_{x,w}(q)$ are the **Kazhdan–Lusztig polynomials**, or **KL polynomials** for short. The cases when $x < w$ and $\deg P_{x,w}(q) = (\ell(x,w)-1)/2$ are especially interesting, as illustrated for type C_3 in 8.6 below; write $x \prec w$ when this happens. One can set $P_{x,w}(q) = 0$ in case $x \not\leq w$. It is conjectured in [**170**] (for arbitrary Coxeter groups) that all coefficients of these polynomials are nonnegative; this is still an open question in general. In our case the discussion in 8.5 will interpret these coefficients as dimensions of certain vector spaces.

We remark that Soergel [**239**] provides a streamlined development of these ideas, avoiding the R-polynomials entirely; his approach is motivated in part by the failure so far to find a straightforward interpretation of $R_{x,w}(q)$ in the framework of category $\mathcal{O}$ (Remark 8.11).

8.3. Examples

First we quote several elementary facts which follow directly from the algebraic theory in [**170**]:

(a) For all pairs $x \leq w$ in W, we have $P_{x,w}(0) = 1$. Combined with the degree bound for KL polynomials, this shows that $P_{x,w}(q) = 1$ whenever $\ell(x,w) \leq 2$. (See for example [**129**, Exer. 7.11].) These statements hold for arbitrary Coxeter groups.

(b) If W is *finite*, as in the case of a Weyl group, then $P_{x,w_\circ}(q) = 1$ for all $x \in W$. (See [**129**, Exer. 7.14].)

(c) If W is a *dihedral* group, such as the Weyl group in types A_2, B_2, G_2, then $P_{x,w}(q) = 1$ for all $x \leq w$. (See [**129**, 7.12(a)].)

While KL polynomials are recursively computable, a computer is soon needed: see [**129**, 7.12] for further discussion and references as of 1992. The combinatorial aspects of KL polynomials continue to be explored by many people in the wider context of Coxeter groups: see Brenti [**47**], Björner–Brenti [**29**, Chap. 5]. Here we sketch one small and one very large example.

(1) The first case in which KL polynomials of degree > 0 actually occur is $W = S_4$, the Weyl group of the Lie algebra $\mathfrak{sl}(4, \mathbb{C})$ of type A_3. Here $P_{x,w}(q) = 1 + q$ for two pairs satisfying $x \prec w$ [**170**, 6.1]; in turn, $P_{y,w}(q) = 1 + q$ holds whenever $y < x$. (This is confirmed by Goresky's machine

computations posted on his Web page [**114**].) Using the standard numbering $1, 2, 3$ of vertices in the Coxeter–Dynkin diagram, the two pairs are

$$s_2 < s_2 s_1 s_3 s_2 \text{ and } s_1 s_3 < s_1 s_3 s_2 s_3 s_1.$$

It is interesting to compare this configuration with the diagram of the Bruhat ordering of S_4 given by Björner–Brenti [**29**, Fig. 2.4], where permutation labelling $(abcd)$ is used. It is not difficult to locate the two special pairs. For example, the first pair is represented by the permutations (1324) and (3412).

(2) As $|W|$ increases, the difficulty of carrying out explicit computations of KL polynomials by a naive application of the original algorithm becomes evident. Among the exceptional types, E_8 illustrates well the challenge involved. Here $|W| = 2^{14}\, 3^5\, 5^2\, 7 = 696,729,600$.

A smaller, but still daunting, version of the problem for E_8 arises in the study of unitary representations of *real* semisimple Lie groups. This is explained informally by Vogan [**256**], following the successful computation for the split real group of type E_8 of related "Kazhdan–Lusztig–Vogan polynomials" attached to a geometrically defined $\mathcal{H}$-module. These polynomials encode essential character information and can be computed by an algorithm parallel to the original algorithm in [**170**]: see Lusztig–Vogan [**203**], Vogan [**255**]. The E_8 computation involved a team of Lie group specialists, including F. du Cloux and others with special programming skills. (Some algebraic background for Vogan's program is described below in the complex case: see 8.10–8.11.)

Whereas the KL polynomials are correlated with the geometry of the complex flag variety (8.5), notably the $|W|$ orbits of a Borel subgroup, the KLV polynomials involve a smaller number of orbits on this variety under the action of a (complexified) maximal compact subgroup of the real group. In the case at hand, this number is still impressively large: $320,206$. In fact, nontrivial local systems have to be paired with many orbits, for a total of $453,060$ pairs. Many reductions of the KLV algorithm are needed in order to make the computation feasible. Among the specific results obtained are the following:

- The actual number of *distinct* KLV polynomials for the $\mathbb{R}$-split group of type E_8 is $1,181,642,979$.

- The largest coefficient of any KLV polynomial is $11,808,808$, while the value of this polynomial at 1 is $60,779,787$.

- So far the largest value at 1 found for any KLV polynomial for E_8 is $62,098,473$.

8.4. Kazhdan–Lusztig Conjecture

In their 1979 paper, Kazhdan and Lusztig formulated a precise conjecture
[**170**, Conj. 1.5] for the formal characters of simple modules in the principal
block $\mathcal{O}_0$ of $\mathcal{O}$. Here there are $|W|$ simple modules. Since the antidominant
Verma module $M(-2\rho)$ is simple, it is natural to start with its highest weight
$\lambda := -2\rho$ and parametrize modules as $M_w := M(w \cdot \lambda), L_w := L(w \cdot \lambda)$.
With this notation, we know that $[M_w : L_x] \neq 0$ precisely when $x \leq w$ in
the Bruhat ordering (5.2).

Conjecture (Kazhdan–Lusztig). *Consider the principal block $\mathcal{O}_0$ of $\mathcal{O}$,
with simple modules L_w and Verma modules M_w indexed as above by $w \in W$.
Then the composition factor multiplicities of Verma modules are determined
in $K(\mathcal{O}_0)$ by the equations*

$$(1) \qquad [L_w] = \sum_{x \leq w} (-1)^{\ell(x,w)} P_{x,w}(1) \, [M_x].$$

In particular, the multiplicities depend only on W.

Thanks to a sort of "inversion" formula for the KL polynomials attached
to finite Coxeter groups ([**170**, §3]), the KL conjecture is easily seen to be
equivalent to:

$$(2) \qquad [M_w] = \sum_{x \leq w} P_{w_\circ w, w_\circ x}(1) \, [L_x], \quad \text{or} \quad [M_w : L_x] = P_{w_\circ w, w_\circ x}(1).$$

Notice that two extreme cases are consistent with the conjecture, in
view of the facts noted above in 8.3: On one hand, the antidominant Verma
module $M(\lambda)$ is already simple, in agreement with $P_{1,1}(q) = 1$. On the
other hand, the classical Weyl–Kostant formula developed in Chapter 2
corresponds to the case of $L(w_\circ \cdot \lambda)$. Here all $P_{x,w_\circ}(q) = 1$ and these occur
with alternating signs in the conjectured formula above as they do in the
classical formula.

One striking consequence of the conjecture is that the numerical data
involved must be the same for the Lie algebras of types B_ℓ and C_ℓ: these
have the same Weyl group, even though the Lie algebras are nonisomorphic.

Exercise. Let $\mathfrak{g} = \mathfrak{sl}(4, \mathbb{C})$. Deduce from the inverse formulation (2) of the
conjecture together with 8.3(1) that a Verma module such as $M(0)$ whose
highest weight is regular, dominant, and integral must have precisely 22
composition factors of multiplicity 1 and two of multiplicity 2.

Though the KL Conjecture looks at first somewhat opaque, the formulation does bring W to the forefront. Kazhdan and Lusztig were actually
motivated at first by the deep work of Springer on Weyl group representations in connection with the unipotent variety of a semisimple algebraic

group G over $\mathbb{C}$ having Lie algebra $\mathfrak{g}$. The individual unipotent classes are smooth (as group orbits) but their closures sometimes have singularities. A convenient desingularization involves the flag variety G/B, where B has Lie algebra $\mathfrak{b}$. This variety in turn has a nice cell decomposition, but here too the closures of cells may be singular. The geometry of the flag variety, and indirectly the Springer representations, turn out to be intimately related with KL polynomials, as we discuss next. It is somewhat miraculous that these geometric ideas lead to a solution of the multiplicity problem for Verma modules.

8.5. Schubert Varieties and KL Polynomials

As promised, we outline briefly one geometric interpretation of KL polynomials; see Kazhdan–Lusztig [**171**] (as well as the 1980 account by Gelfand–MacPherson [**109**], written just before the proof of the KL Conjecture).

Let G be a simply connected semisimple algebraic (or Lie) group over $\mathbb{C}$ whose Lie algebra is isomorphic to $\mathfrak{g}$. Fix a Borel subgroup B relative to Φ^+, with a maximal torus T. Then the Weyl group W of Φ is isomorphic to $N_G(T)/T$. Corresponding to this set-up is the *Bruhat decomposition* $G = \bigcup_{w\in W} BwB$ (disjoint union), where the double coset BwB is independent of the choice of coset representative for w in $N_G(T)$.

Write $X := G/B$, the *flag variety*. This is a smooth projective variety of dimension $m = |\Phi^+|$, which inherits from the Bruhat decomposition a decomposition into *Bruhat cells* $X_w := BwB/B$. Here X_w is isomorphic to an affine space of dimension $\ell(w)$. The closure $\overline{X_w}$ is called a *Schubert variety*. It is the union of X_w and some smaller Bruhat cells, giving a cell decomposition. Indeed, Chevalley showed that $X_x \subset \overline{X_w}$ if and only if $x \leq w$ in the Chevalley–Bruhat ordering; this is where the ordering originated. Schubert varieties include many special cases of interest, including flag varieties, but are not always smooth.

The cell decomposition of $\overline{X_w}$ makes its classical cohomology easy to compute: Since the real dimension of a cell X_x is $2\ell(x)$, all cohomology groups in odd degrees vanish while the even Betti number $\dim H^{2i}(\overline{X_w})$ is the number of $x \leq w$ for which $\ell(x) = i$. These data combine to give the *Poincaré polynomial*

$$\mathcal{P}_w(q) := \sum_{i=0}^{\ell(w)} \dim H^{2i}(\overline{X_w}) q^i.$$

In [**170**] (and its Appendix) it was already observed that there is a close connection between the failure of local Poincaré duality for $\overline{X_w}$ and the fact that certain associated $P_{x,w}(q)$ with $x < w$ fail to equal 1. This occurs in case $\overline{X_w}$ is not "rationally smooth" along some cells. Examples are seen, for instance, in two Schubert varieties when $\mathfrak{g} = \mathfrak{sl}(4, \mathbb{C})$ [**170**, 1.6]; as noted earlier, there are two $w \in W$ for which not all $P_{x,w}(q)$ are trivial.

In [**171**] Kazhdan and Lusztig went on to interpret their polynomials in terms of the newly perfected Deligne–Goresky–MacPherson theory of intersection cohomology and the Beilinson–Bernstein–Deligne theory of perverse sheaves on singular complex varieties (to which Gabber made important contributions). We use the notation *IH* for the associated intersection cohomology groups. (In [**171**] they actually work with varieties over an algebraic closure of $\mathbb{F}_p$ rather than over $\mathbb{C}$, but the cohomology calculations are done in characteristic 0.)

In this framework the KL polynomials turn out to be the Poincaré polynomials for *local* intersection cohomology of Schubert varieties. If $x \leq w$, the polynomial is 1 precisely when $\overline{X_w}$ is rationally smooth along X_y for all $x \leq y \leq w$. For Schubert varieties in general, intersection cohomology vanishes in odd degrees; this is consistent with the fact that KL polynomials involve the variable q rather than $q^{\frac{1}{2}}$.

Theorem (Kazhdan–Lusztig). *For each $x \leq w$ in W, we have*

$$P_{x,w}(q) = \sum_{i=0}^{\ell(w)} q^i \, \dim IH_x^{2i}(\overline{X_w}).$$

In particular, all coefficients of $P_{x,w}(q)$ are nonnegative.

The global Poincaré polynomial for intersection cohomology is given explicitly by:

$$\sum_{i=0}^{\ell(w)} \dim IH^{2i}(\overline{X_w}) q^i = \sum_{x \leq w} q^{\ell(x)} P_{x,w}(q).$$

Unlike the usual Poincaré polynomial, this one always exhibits duality (originating in Verdier duality): the coefficients of q^i and $q^{\ell(w)-i}$ agree.

8.6. Example: W of Type C_3

It is instructive to look at the case when W is of type C_3, using Goresky's tables [**114**] to track a couple of cases where singularities occur in Schubert varieties. (In this algebraic context, we use "singularity" as a shorthand for the failure of rational smoothness, ignoring other topological singularities.) Note that for C_3 and some other cases, his numbering of vertices in the Coxeter–Dynkin graph of W differs from that in Bourbaki [**45**]. Here his labels $1, 2, 3$ are the reverse of the standard ones, so $s_1 s_2$ has order 4 while $s_2 s_3$ has order 3 and $s_1 s_3 = s_3 s_1$.

For type C_3 (or B_3, where W is the same), we have $|W| = 48$. The elements are listed by indices 1–48, compatible with increasing length; the omitted index 1 refers to $1 \in W$. The length of each element is given, together with its "code": for example, 213 denotes the reduced decomposition $w = s_2 s_1 s_3$. (When many reduced decompositions of a single w occur, one gets into a typical computational problem in constructing such tables.) The indices of elements of length $\ell(w) - 1$ which lie below w in the Bruhat ordering (corresponding to cells of codimension 1 in $\overline{X_w}$) are listed; this permits a recursive determination of all $x < w$. From this in turn one can compute the Betti numbers, which are given explicitly in the table.

In many cases the Schubert varieties are (rationally) smooth. But when singularities occur along certain cells, this is indicated together with the associated KL polynomials. Take for example w with index 47; its length is 8 and its code is 21232123. The Betti numbers are $1, 3, 5, 7, 8, 8, 6, 3, 1$, corresponding to 42 cells in $\overline{X_w}$ belonging to 42 elements $x \le w$.

There is a singularity of minimal codimension along the 3-dimensional cell represented by $x \in W$ with index 15 and code 232 (the longest element in a subgroup of type A_2). Its closure is smooth and is in fact isomorphic to the flag variety of $\mathrm{SL}(3, \mathbb{C})$. The KL polynomial is $P_{x,w}(q) = 1 + q^2$, abbreviated in the table by 1 0 1. Moreover, $\overline{X_w}$ has singularities along the 6 cells X_y lying in $\overline{X_x}$, with corresponding KL polynomials $P_{y,w} = 1 + q^2$. (These are not listed individually but are easily recovered from the Bruhat graph.) Adding up all KL polynomials $P_{z,w}$ with $z \le w$, one gets the coefficients of the intersection cohomology Poincaré polynomial, called IH in the tables: $1, 3, 6, 9, 10, 9, 6, 3, 1$. Note that the total difference between these dimensions and the Betti numbers is 6, agreeing with the number of $z < w$ for which $P_{z,w} = 1 + q^2$.

Things get more complicated when unrelated singularities occur. Take for example w with index 43 and Betti numbers $1, 3, 5, 7, 8, 7, 4, 1$. There is a singularity along each of two 4-dimensional cells with indices 17 and 22; in both cases the KL polynomial is $1 + q$, which propagates to lower elements in the Bruhat ordering until one reaches a different singularity at the element of index 3 (the simple reflection s_2). Here and also at the identity element of W the KL polynomial changes to $1 + q + q^2$. The IH polynomial then has coefficients $1, 4, 9, 13, 13, 9, 4, 1$.

8.7. Jantzen's Multiplicity One Criterion

It is instructive to compare the KL Conjecture with the known multiplicity results summarized in 8.1. We already indicated in 8.3 that small rank cases computed earlier agree with results predicted by the conjecture. A more interesting example is the general criterion developed by Jantzen for

$[M(\lambda) : L(\mu)] = 1$. Its proof required a considerable amount of technical work. But the criterion itself (for the principal block) can be recovered from the KL Conjecture; this requires elementary combinatorial arguments together with the fact that KL polynomials have nonnegative coefficients when W is finite (8.5). On the other hand, the *proof* of the conjecture involves deep methods drawn from other parts of mathematics.

Here is an outline of steps involved, based on an unpublished letter from Lusztig to Jantzen early in 1980 as well as the paper of Gabber–Joseph [**106**] (which appeared in preprint form later in 1980). The reader is encouraged to fill in the omitted details.

(1) The key lemma involves further development of the combinatorics of the auxiliary polynomials $R_{x,w}(q)$ for $x \leq w$ (8.2). Here $R_{w,w}(q) = 1$, while the recursive recipe by which the polynomials are computed shows that $R_{x,w}(q) = q - 1$ in case $\ell(w) - \ell(x) = 1$ (which means $w = rx$ for some reflection r). Using prime to denote derivative, this case gives $R'_{x,w}(1) = 1$.

Lemma. *Let* $x < w$ *in* W. *Then*

$$R'_{x,w}(1) = \begin{cases} 1 & \text{if } \ell(w) - \ell(x) = 1, \\ 0 & \text{otherwise.} \end{cases}$$

The proof, given in Gabber–Joseph [**106**, 2.2], uses induction on $\ell(w)$ together with the recursive definition of R-polynomials in Kazhdan–Lusztig [**170**, 2.0.b–c]. Separate arguments are needed when s is a simple reflection such that $sw < w$, to deal with the cases $sx < x$ and $sx > x$.

(2) Now combine the lemma with a fundamental identity relating R-polynomials and KL polynomials for fixed $x \leq w$ (see [**170**, §2], [**171**, (2.1.5)], [**106**, 2.3], or [**129**, 7.10(20)]):

$$\sum_{x \leq y \leq w} R_{x,y}(q)P_{y,w}(q) = q^{\ell(x,w)} \, P_{x,w}(q^{-1}).$$

Just differentiate formally, set $q = 1$, and apply the lemma, to conclude:

Corollary. *If* $x < w$ *in the Weyl group* W, *then*

$$2P'_{x,w}(1) = \ell(x, w)P_{x,w}(1) - \sum_{x < rx \leq w} P_{rx,w}(1),$$

where r *runs over all reflections in* W.

(3) So far the arguments use only elementary combinatorics of the polynomials, but now we have to know that all coefficients of KL polynomials are nonnegative. For our Weyl group W, this follows from the geometric interpretation in 8.5. In particular, the statement $P_{x,w}(1) = 1$ is equivalent to $P_{x,w}(q) = 1$, since all KL polynomials have constant term 1.

If $x < w$, write $L := \ell(x, w)$ and let R be the number of reflections r for which $x < rx \le w$. Then we claim: Let $x < w$, and suppose that $P_{y,w}(q) = 1$ whenever $x < y \le w$. Then $P_{x,w}(q) = 1$ if and only if $R = L$.

By assumption, $P_{rx,w}(q) = 1$ for each of the R reflections r such that $x < rx \le w$, so step (2) yields:

$$2P'_{x,w}(1) = L P_{x,w}(1) - R.$$

Write $P_{x,w}(q) = \sum_i c_i q^i$, where $0 \le i \le (L-1)/2$ and all $c_i \ge 0$. Then differentiate and substitute to get

$$2\sum_{i>0} i c_i = L\sum_i c_i - R,$$

or

$$\sum_i (L - 2i) c_i = R - L.$$

Each term on the left side is nonnegative, so $R - L \ge 0$, while $R = L$ if and only if $c_i = 0$ for all $i \ge 1$, or if and only if $P_{x,w}(q) = 1$.

With this formalism in hand and assuming the truth of the KL Conjecture, we can recover Jantzen's multiplicity one theorem [147, 5.20] for the principal block $\mathcal{O}_0$. The general statement given at the end of 8.1 has to be adapted first, using the notation in 8.4: write $\lambda = w \cdot (-2\rho)$ and $\mu = x \cdot (-2\rho)$, where $x < w$. Thanks to Exercise 5.3, $|\Phi_\lambda^+| = \ell(w)$ and $|\Phi_\mu^+| = \ell(x)$, so the difference is $\ell(x, w) = L$. Recall that $\alpha \in \Phi_\lambda^+$ means $s_\alpha \cdot \lambda < \lambda$, which translates into $rw < w$ if $r = s_\alpha$. The number of such reflections for which $\mu \uparrow s_\alpha \cdot \lambda$ is then precisely the number of r for which $x \le rw < w$.

Theorem (Jantzen). *In $\mathcal{O}_0$, we have $[M_w : L_x] = 1$ for $x < w$ if and only if the following two conditions are satisfied:*

(a) *For all $y \in W$ such that $x \le y < w$, we have $[M_y : L_x] = 1$.*

(b) *$\ell(x, w)$ is the number of reflections r for which $x \le rw < w$.*

For the proof we assume the truth of the KL Conjecture, in the inverse format 8.4(2): $[M_w : L_x] = P_{w_\circ w, w_\circ x}(1)$. This equals 1 if and only if $P_{w_\circ w, w_\circ x}(q) = 1$, since all coefficients of the polynomial lie in $\mathbb{Z}^+$.

Given (a) and (b), we want to deduce $[M_w : L_x] = 1$ from the steps above. For this, set $w' := w_\circ w$ and $x' := w_\circ x$. Using the general fact that $\ell(w_\circ z) = \ell(w_\circ) - \ell(z)$ (0.3), we get $\ell(w', x') = \ell(x, w)$. Now $r' := w_\circ r w_\circ$ runs over all reflections as r does. Combined with the fact that $y < z$ if and only if $w_\circ z < w_\circ y$, this allows us to translate $x \le rw < w$ in (b) into $w' < r'w' \le x'$. From step (3) above, we obtain $P_{w',x'}(q) = 1$, which by the KL Conjecture proves that $[M_w : L_x] = 1$.

For the reverse implication, we have to invoke Verma's Theorem on embeddings of Verma modules. Given $[M_w : L_x] = 1$, the conditions $x \leq y < w$ in (a) ensure that $M_y \hookrightarrow M_w$ and thus $[M_y : L_x] = 1$ as required. Then part (3) above and the KL Conjecture imply (b).

8.8. Proof of the KL Conjecture

Once the KL Conjecture was announced, a number of independent attempts were made to understand it better and find a method of proof. Most of these attempts (especially those relying on algebraic methods alone) ran into insuperable barriers. But along the way some consequences of the conjecture as well as its relationship with the earlier Jantzen Conjecture were worked out. This work also suggested how one might interpret the *coefficients* of KL polynomials in terms of the module structure of Verma modules (including the Jantzen filtration). Some ideas due to Deodhar [**78**], Gabber–Joseph [**106**], and Gelfand–MacPherson [**109**] will be discussed below in the context of later refinements by Irving.

The first—and to date only—proofs of the KL Conjecture were published independently in 1981 by Beilinson–Bernstein [**16**] (in a detailed research announcement) and Brylinski–Kashiwara [**55**] (in a longer paper); see also Beilinson's 1983 ICM talk [**15**]. In each case the methods used, drawn from recent work on $\mathcal{D}$-modules and perverse sheaves, are essentially the same. In a Bourbaki seminar soon after these papers appeared, Springer [**242**] explained the ideas involved together with the conjectured applications to his own work on Weyl group representations. This seminar talk is still useful for its detailed overview of the deep mathematics here. (An extensive historical account is provided by Kleiman [**187**].) In the following section we outline the main steps in the proof of the KL Conjecture; filling in the details requires a considerable amount of background in modern sheaf theory and related matters.

For *nonintegral* (but compatible) weights, the analogue of the KL Conjecture does not follow straightforwardly from the machinery developed to prove the original conjecture. These methods only go as far as the case of weights with coordinates (relative to fundamental dominant weights) in $\mathbb{Q}$. Instead, one can transfer the problem for a block $\mathcal{O}_\lambda$ to an equivalent module category involving integral weights for a semisimple Lie algebra having Weyl group $W_{[\lambda]}$. This is worked out by Soergel [**237**, 2.5]: see Theorem 11 and the related Bemerkung, discussed below in 13.13. (A later approach by Kashiwara–Tanisaki [**169**] treats more generally the case of nonintegral weights for Kac–Moody algebras.) Now that the original KL Conjecture along with its analogue for arbitrary blocks of $\mathcal{O}$ have been established, it becomes important in further work on $\mathcal{O}$ to identify clearly where the statement—or the proof—of the conjecture is used.

Over the years the "localization" method articulated in the proof of the conjecture has continued to reverberate in neighboring parts of representation theory. (The method itself has also been revisited by Joseph–Perets–Polo [**163**].) This success in translating an intractable combinatorial problem in algebra into a problem in algebraic geometry has inspired many further developments in "geometric representation theory".

8.9. Outline of the Proof

Following the geometric interpretation of KL polynomials in the setting of Schubert varieties (8.5), it becomes clear that one wants to find a pathway from the representation theory of category $\mathcal{O}$ to the local geometry of the flag variety $X = G/B$. At the risk of oversimplifying matters, we indicate some of the main steps involved. The full proof requires far more background and details; see the recent book by Hotta–Takeuchi–Tanisaki [**124**].

(A) Rather than looking for an explicit category equivalence between the block $\mathcal{O}_0$ and a geometric category, it is best to start with the $U(\mathfrak{g})$-modules on which $Z(\mathfrak{g})$ acts trivially, i.e., by the character χ_0. These form a subcategory of $\operatorname{Mod} U(\mathfrak{g})$ whose intersection with $\mathcal{O}_0$ contains the M_w and L_w which figure in the original formulation of the KL Conjecture. If we denote by J_0 the two-sided ideal of $U(\mathfrak{g})$ generated by $\operatorname{Ker}\chi_0$ and set $U_0 := U(\mathfrak{g})/J_0$, we are therefore looking at the category of left U_0-modules.

If G is a semisimple group with Lie algebra $\mathfrak{g}$ and B a Borel subgroup with Lie algebra $\mathfrak{b}$, consider the flag variety $X = G/B$ (as in 8.5). The idea is to work with *differential operators*, not just functions, on X; locally, these arise from partial differentiation operators combined with multiplication by functions. Besides the sheaf $\mathcal{O}_X$ of regular functions on X (a notation unrelated to category $\mathcal{O}$), we have a larger sheaf $\mathcal{D}_X$ of differential operators on X. Here $\mathcal{D}_X$ is a quasi-coherent $\mathcal{O}_X$-module. The algebra $D_X := \Gamma(X, \mathcal{D}_X)$ of global sections is the algebra of (global) differential operators on X.

On the other hand, $\mathfrak{g}$ is viewed classically as the space of right-invariant tangent vector fields on G. This leads to an epimorphism $U(\mathfrak{g}) \to D_X$, whose kernel turns out to be J_0. So $U_0 \cong D_X$. This is derived indirectly using a well-known theorem of Kostant in the commutative setting: the ideal of polynomial functions on $\mathfrak{g}$ vanishing on the nilpotent variety $\mathcal{N} \subset \mathfrak{g}$ is generated by the G-invariant nonconstant polynomials. (The link with X is that the cotangent bundle T^*X projects naturally onto $\mathcal{N}$ and provides a resolution of singularities.)

(B) Working just in the category of D_X-modules for $U(\mathfrak{g})$ conceals too much of the local behavior relative to the stratification of X by Bruhat cells. To remedy this, one moves the problem into a category of sheaves on X, namely the category $\mathcal{C}$ of all $\mathcal{D}_X$-modules which are quasi-coherent over the structure sheaf $\mathcal{O}_X$. This has to be compared with the category $\mathcal{C}'$ of D_X-modules.

There are natural functors between $\mathcal{C}$ and $\mathcal{C}'$: In one direction, $\mathcal{F} \mapsto \Gamma(X, \mathcal{F})$, a module over D_X. In the other direction $M \mapsto \mathcal{D}_X \otimes_{D_X} M$. The fact that these define an equivalence of categories between $\mathcal{C}$ and $\mathcal{C}'$ is highly nontrivial; it rests on the proof that X is "$\mathcal{D}$-affine" (a subtle sheaf-theoretic analogue of the classical theorem of Borel–Weil), requiring that $\mathcal{D}_X$-modules be generated by their global sections and have zero sheaf cohomology in degrees > 0. This equivalence is a kind of noncommutative *localization*, inspired by the connection between the sheaf of regular functions on an affine variety and the local rings realized as global sections of the structure sheaf over affine open sets.

Our main concern is with the subcategory $\mathcal{C}_0'$ consisting of finitely generated, $U(\mathfrak{b})$-finite modules (which includes all M_w and L_w). Its counterpart $\mathcal{C}_0$ consists of coherent sheaves $\mathcal{M}$ killed by an ideal of finite codimension in $U(\mathfrak{b})$ and having a $U(\mathfrak{b})$-stable "good" filtration. Under the category equivalence, one sees that $\mathcal{C}_0$ is equivalent to $\mathcal{C}_0'$.

In this situation, it is shown that the $\mathcal{D}$-modules in $\mathcal{C}_0$ are *holonomic* and have *regular singularities.*

(C) Now it requires some close study to identify the specific sheaves $\mathcal{M}_w$ and $\mathcal{L}_w$ corresponding to M_w and L_w. Here one encounters local cohomology and complexes of sheaves.

To make contact with the local description of KL polynomials in 8.5, one needs to go beyond the techniques of classical sheaf theory such as derived functors. The (bounded) *derived category* $D_c^b(X)$ of constructible sheaves on X provides the correct setting. Here the objects are complexes of sheaves, but with a more subtle notion of morphism than usual in sheaf theory. As in the theory of $\mathcal{D}$-modules, one has to study delicate features of various functors associated to maps: surjections $G/B \to G/P$ with $\mathbb{P}^1$-fibers (where P is "minimal") or embeddings $X_w \hookrightarrow X$.

There is a sophisticated modern version of classical ideas about systems of differential equations, known as the *Riemann–Hilbert Correspondence*, which allows one to pass from $\mathcal{D}$-modules to objects in the derived category. Unfortunately $D_c^b(X)$ is not itself an abelian category. But the crucial objects corresponding to Verma modules and simple modules turn out to be *perverse sheaves*, which do form an abelian category. Here the objects have finite filtrations with simple quotients, much as in $\mathcal{O}_0$.

To a Verma module corresponds an easily constructed complex, which is nonzero in only one degree and is supported on a single Bruhat cell, whereas the simple objects are more elusive. But in the Grothendieck group one has (as in category $\mathcal{O}$) a unipotent change of basis matrix relating the two types of objects. Here one wants to express the simple objects in terms of the Verma objects. As indicated in 8.5, the local intersection cohomology groups recover the KL polynomials as Poincaré polynomials. Finally these yield the multiplicities in $K(\mathcal{O}_0)$ conjectured by Kazhdan–Lusztig.

8.10. Ext Functors and Vogan's Conjecture

Work of Vogan and others in the 1970s involving the broader study of unitary representations of real reductive Lie groups led him to a conjecture about wall-crossing functors which turned out to be equivalent to the KL Conjecture: see [**254**, §3] as well as [**253, 255**]. In particular, his viewpoint allows one to interpret the KL polynomials in terms of $\mathfrak{n}$-cohomology. This translates formally, as indicated in Theorem 6.15, into the language of the functors Ext^n.

Following fundamental work of Harish-Chandra and others, the description of unitary representations can to some extent be reduced to the algebraic study of "Harish-Chandra modules" for the complexified Lie algebra of the Lie group relative to a maximal compact subgroup. Assuming that G is semisimple (and linear), the Lie algebra $\mathfrak{g}$ is identified with the one we have been studying. If K is a maximal compact subgroup of G, the resulting Harish-Chandra modules are usually called $(\mathfrak{g}, K)$-modules. A prototype for the entire theory is the case of principal series representations for complex Lie groups, which translates into the language of category $\mathcal{O}$ (see Chapter 10 below). In the papers cited above, Vogan starts here. The idea is to compute characters of irreducibles in an algorithmic style; this depends heavily on getting information about $\mathfrak{n}$-cohomology at each step.

Using Zuckerman's version of translation and wall-crossing functors for Harish-Chandra modules, joint work of Speh and Vogan led to a general conjecture formulated by Vogan in [**253**, 3.15]. This encapsulates what is needed to carry out an inductive step in the algorithm. It become known as **Vogan's Conjecture** and can be described briefly in the general setting, using his notation. Given a simple $(\mathfrak{g}, K)$-module X and a simple root α for which translation to the α-wall fails to annihilate X, the associated wall-crossing functor $\psi_\alpha \varphi_\alpha$ yields a complex $0 \to X \to \psi_\alpha \varphi_\alpha(X) \to X \to 0$ with cohomology $U_\alpha(X)$.

Conjecture. *In the above notation, the $(\mathfrak{g}, K)$-module $U_\alpha(X)$ is semisimple.*

In the special case corresponding to the principal block $\mathcal{O}_0$, with modules M_w, L_w ($w \in W$) indexed by weights $w \cdot (-2\rho)$ as before, this translates as follows. Recall the discussion in 7.14–7.15. Setting $s = s_\alpha$ and assuming that $ws > w$, we know that $\Theta_s L_w$ has simple head and simple socle both isomorphic to L_w. This yields a complex:

$$0 \to L_w \to \Theta_s L_w \to L_w \to 0.$$

Then the conjecture is that $\operatorname{Rad}\Theta_s L_w / \operatorname{Soc}\Theta_s L_w$ is semisimple.

8.11. KLV Polynomials

How does Vogan's Conjecture relate to the KL Conjecture? Here is a brief heuristic explanation (which does not correlate precisely with the history). Recall Proposition 6.14, which was proved in a fairly elementary way (using BGG Reciprocity): in $K(\mathcal{O}_0)$ we have

$$[L_w] = \sum_{x \leq w} \sum_{i \geq 0} (-1)^i \dim \operatorname{Ext}^i_{\mathcal{O}}(M_x, L_w)\, [M_x].$$

Comparing this with the KL Conjecture and using $(-1)^i = (-1)^{-i}$, we see that the truth of the conjecture would imply

$$P_{x,w}(1) = \sum_{i \geq 0} (-1)^{\ell(x,w)-i} \dim \operatorname{Ext}^i_{\mathcal{O}}(M_x, L_w).$$

This makes it tempting to look at a sort of generating function in the variable $q^{\frac{1}{2}}$; the choice of variable ensures that we get a polynomial of degree at most $\ell(x, w)$ thanks to Theorem 6.11(b). Define for $x \leq w$ in W,

$$\widetilde{P}_{x,w} := \sum_{i \geq 0} (-1)^{\ell(x,w)-i} q^{(\ell(x,w)-i)/2}\, \dim \operatorname{Ext}^i_{\mathcal{O}}(M_x, L_w).$$

In the axiomatic setting of "BGG algebras", Irving [**137**, §4] calls this a **KLV polynomial**. If we assume the truth of the KL Conjecture, setting $q = 1$ recovers $P_{x,w}(1)$ in view of the remarks above.

Now Vogan's conclusions about the KL Conjecture can be summarized as follows.

Theorem. *With modules in the block $\mathcal{O}_0$ labelled as above, the following statements are equivalent:*

(a) *The KL Conjecture is true.*

(b) *For all $x \leq w$, the KLV polynomial $\widetilde{P}_{x,w}$ is a polynomial in q and coincides with $P_{x,w}(q)$. Thus*

$$P_{x,w}(q) = \sum_{i \geq 0} q^i \dim \operatorname{Ext}^{\ell(x,w)-2i}_{\mathcal{O}}(M_x, L_w).$$

(c) *For all $x \leq w$, $\mathrm{Ext}^i_{\mathcal{O}}(M_x, L_w) = 0$ unless $i \equiv \ell(x,w)$ (mod 2).*

Most of this is made explicit in [**254**, §3], notably the fact that (b) implies (a). Condition (c) mirrors in a way the vanishing of cohomology in odd degrees which appears in the geometric setting (8.5) and indirectly accounts for the fact that $P_{x,w}$ is a polynomial in q rather than just $q^{1/2}$.

Remark. In their study of the homological algebra associated with the KL Conjecture, a possible interpretation of the auxiliary polynomials $R_{x,w}(q)$ was proposed by Gabber–Joseph [**106**, §5] (just before the conjecture itself was settled). Somewhat in parallel with the above description of the coefficients of KL polynomials in terms of various $\dim \mathrm{Ext}^i_{\mathcal{O}}(M_x, L_w)$, they defined (using different notation) polynomials of the form

$$\widetilde{R}_{x,w}(q) := \sum_i (-1)^{\ell(x,w)-i} q^i \dim \mathrm{Ext}^i_{\mathcal{O}}(M_x, M_w).$$

The hypothesis that $\widetilde{R}_{x,w}(q) = R_{x,w}(q)$ became known as the *Gabber–Joseph Conjecture* and remained open for a decade. Its plausibility came from the fact that the $\widetilde{R}$-polynomials satisfy some of the same recursion relations used to compute R-polynomials. Some supporting evidence for the absolute values of coefficients predicted by the conjecture emerged from the methods developed by Carlin [**56**] to compute Ext^n between Verma modules with regular integral highest weights. This involves construction of a spectral sequence for derived functors of adjoint functors (here translation functors). In particular, Carlin proves that the highest nonvanishing Ext group has dimension 1:

$$\dim \mathrm{Ext}^{\ell(x,w)}_{\mathcal{O}}(M_x, M_w) = 1 \text{ if } x \leq w.$$

He also recovers known results for related $\mathrm{Ext}^i_{\mathcal{O}}(M_x, L_w)$, in the spirit of Vogan's work.

But eventually Boe [**31**] was able to use a computer to locate a specific counterexample to the Gabber–Joseph conjecture; his results were then replicated by hand. The idea behind this is fairly simple: since the coefficients of the $\widetilde{R}$-polynomials alternate in sign (by definition), it is enough to locate an R-polynomial having two successive nonzero coefficients with the same sign. The "smallest" possible example (relative to rank and $\ell(x,w)$) turns out to occur when W has type $\mathrm{B}_4 = \mathrm{C}_4$. Further examples were found in types D_4 and A_5, and presumably exist commonly in higher ranks.

In spite of this negative outcome, the work of Beilinson–Ginzburg–Soergel [**22**] on Koszul duality (discussed in 13.15 below) might lead to a more subtle version of the conjecture.

8.12. The Jantzen Conjecture and the KL Conjecture

As mentioned earlier, the KL Conjecture immediately raised questions about its possible relationship with the Jantzen filtration of a Verma module (5.3). The Jantzen filtration with its associated Sum Formula was introduced in Chapter 5 primarily as a tool in the proof of the BGG Theorem; but it was also expected to be helpful at least indirectly in the quest for composition factor multiplicities.

Recall the set-up, for an arbitrary Verma module $M(\lambda)$, $\lambda \in \mathfrak{h}^*$. The filtration $M(\lambda) = M(\lambda)^0 \supset M(\lambda)^1 \supset \cdots$ involves proper inclusions, with $M(\lambda)^1 = N(\lambda) = \operatorname{Rad} M(\lambda)$, and ends at 0. As observed in 5.3, the last nonzero term $M(\lambda)^n$ has index $n = |\Phi_\lambda^+|$, where $\Phi_\lambda^+ = \{\alpha > 0 \mid s_\alpha \cdot \lambda < \lambda\}$. Now the Sum Formula reads:

$$\sum_{i>0} \operatorname{ch} M(\lambda)^i = \sum_{\alpha \in \Phi_\lambda^+} \operatorname{ch} M(s_\alpha \cdot \lambda).$$

The question raised by Jantzen in [**147**] about whether this filtration is compatible in a natural way with embeddings $M(\mu) \hookrightarrow M(\lambda)$ became known as the **Jantzen Conjecture**:

(∗)   *Let $\mu \uparrow \lambda$ in $\mathfrak{h}^*$, with $\mu < \lambda$, and consider the associated embedding $M(\mu) \hookrightarrow M(\lambda)$. Set $r := |\Phi_\lambda^+| - |\Phi_\mu^+|$. Then $M(\mu) \cap M(\lambda)^i = M(\mu)^{i-r}$ for all $i \geq r$. In particular, $M(\mu) \subset M(\lambda)^r$.*

Thanks to the BGG Theorem 5.1, it would be enough to prove this when $\mu = s_\alpha \cdot \lambda < \lambda$ with $\alpha > 0$. The crux of the problem is how to characterize the filtration and Sum Formula intrinsically: the original construction required passage to $\mathbb{C}[T]$ and then specialization $T \mapsto 0$.

Various special cases have also been studied under the rubric "Jantzen Conjecture":

- In (∗), consider only the case $\mu = s_\alpha \cdot \lambda < \lambda$ with α *simple*; here $r = 1$.

- Let λ and μ be *regular*, with $\lambda = w \cdot \nu$ and $\mu = x \cdot \nu$ for some antidominant weight ν. By Exercise 5.3, $r = \ell_\lambda(x, w)$ (using the length function in $W_{[\lambda]} = W_{[\mu]}$), so condition (∗) becomes:

  $$M(x \cdot \nu) \cap M(w \cdot \nu)^i = M(x \cdot \nu)^{i - \ell(x,w)} \text{ for all } i \geq r.$$

- Let λ and μ be *integral*; for example, consider just the principal block $\mathcal{O}_0$ or other regular integral block.

In [**106**] Gabber and Joseph showed by algebraic methods that the truth of a weak version of the Jantzen conjecture would imply the truth of the KL Conjecture (not yet proved at the time). More precisely, they formulated (as did Deodhar and Gelfand–MacPherson, independently) a "generalized

KL Conjecture" based on the Jantzen filtration: the multiplicities of simple modules in the filtration layers should be given by the coefficients of a corresponding KL polynomial. By setting $q = 1$ one obtains the KL Conjecture. Here is a precise statement of their main theorem.

Theorem (Gabber–Joseph). *Let $\lambda \in \mathfrak{h}^*$ be regular and antidominant. If $\alpha \in \Delta_{[\lambda]}, w \in W_{[\lambda]}$, and $ws_\alpha > w$, assume that*

$$M(w \cdot \lambda)^i \subset M(ws_\alpha \cdot \lambda)^{i+1} \cap M(w \cdot \lambda) \text{ for all } i \geq 0.$$

Then

(a) *All layers $M(w \cdot \lambda)_i$ are semisimple.*

(b) *The (generalized) KL Conjecture holds for all weights linked to λ. Therefore the KL Conjecture is true for these weights.*

Notice here that the hypothesis is a weaker version of the Jantzen Conjecture, involving only embeddings associated with *simple* roots. The fact that this is enough to imply the KL Conjecture suggests that the Jantzen Conjecture may be significantly stronger and is therefore unlikely to provide a pathway to proving the KL Conjecture. This is reinforced by the work of Beilinson–Bernstein discussed in the following section. But in another direction, Barbasch [13] was able to draw further consequences from the work of Gabber–Joseph for the Loewy structure of Verma modules: see 8.16.

The proof of the theorem above relies on moderately complicated inductions involving indices in the Jantzen filtration, coupled with study of wall-crossing functors and exact sequences. This yields a proof that $U_\alpha\big(L(w \cdot \lambda)\big)$ is semisimple (using Vogan's notation as in 8.10) and thus that all layers in the filtration are semisimple. Induction comes heavily into play when computing the multiplicities of simple modules in layers and then comparing the KL polynomials. Along the way, it is shown that the multiplicity data are compatible with the Jantzen Sum Formula. It is also shown (without invoking the hypothesis of the theorem) that the methods of proof recover some known consequences of the KL Conjecture.

Remark. For each $\lambda \in \mathfrak{h}^*$, the Jantzen Conjecture locates a maximal vector of weight $s_\alpha \cdot \lambda$ in a definite submodule $M(\lambda)^i$. On the other hand, a Shapovalov element as in 4.12 produces such embeddings uniformly across a root hyperplane. Carlin [58] investigates (especially in type A_ℓ) the connection of Shapovalov elements with Jantzen filtrations as λ varies.

8.13. Weight Filtrations and Jantzen Filtrations

The proof of the Jantzen Conjecture by Beilinson–Bernstein required a significant extension of the geometric techniques used in the proof of the KL

Conjecture. While the main ideas were worked out early in 1981 and outlined in Beilinson's 1983 ICM talk [**15**], a detailed proof only appeared in 1993 [**18**, 5.3]. The basic idea is to endow the related category of perverse sheaves with enriched structure, allowing the Jantzen filtration to be interpreted there as a *weight filtration* in the sense of Beilinson–Bernstein–Deligne and Gabber [**17**, §5].

The notion of "weight" here is unrelated to the Lie-theoretic notion arising in representation theory. It comes instead from the study of sheaves over $\overline{\mathbb{Q}}_\ell$ (the algebraic closure of an ℓ-adic field) on a scheme defined over a finite field $\mathbb{F}_q$ of characteristic $p \neq \ell$. The ideas here arose in Deligne's proof of the Weil Conjectures. Without attempting to explain the details, we can suggest the flavor of the language employed. For each power q^n there is a *Frobenius operator* acting on points of the scheme over $\overline{\mathbb{F}}_q$. Given a fixed point x for this action, there is an induced action of the Frobenius on the stalk at x of the given sheaf $\mathcal{F}$. If $w \in \mathbb{Z}$ and the eigenvalues for this induced action are algebraic numbers whose complex conjugates all have absolute value $(q^n)^{w/2}$ for any fixed point x, then $\mathcal{F}$ is said to be *pointwise pure of weight w*. A sheaf is then said to be *mixed* if it has a finite filtration with pointwise pure quotients. These ideas carry over to complexes of sheaves as well.

It is shown in [**17**, 5.3] that every *simple* mixed perverse sheaf is pure and that any mixed perverse sheaf admits a unique finite increasing filtration (the "filtration par le poids") such that the associated ith graded piece is pure of weight i. Morphisms of sheaves then preserve such filtrations. Based on these results, the correspondence between Verma modules and perverse sheaves developed in the proof of the KL Conjecture can be strengthened:

Theorem (Beilinson–Bernstein). *If $\lambda \in \mathfrak{h}_{\mathbb{Q}}^*$, the perverse sheaf corresponding to the Verma module $M(\lambda)$ is mixed and the counterpart of the Jantzen filtration is a weight filtration. As a result, the Jantzen Conjecture is true.*

As in the case of the KL Conjecture, the passage to arbitrary weights can be made by using the later work of Soergel [**237**, Thm. 11].

8.14. Review of Loewy Filtrations

As observed in 8.5 above, the *coefficients* of KL polynomials have a natural interpretation in geometry as dimensions of certain cohomology groups. It is reasonable to ask whether there is also a representation-theoretic interpretation. The answer involves the submodule structure of Verma modules. While this structure can become extremely complicated, some aspects of it turn out to correlate well with KL theory (in Vogan's formulation).

First we have to review briefly some standard module theory, which applies quite generally to modules satisfying both chain conditions. Since notation and terminology vary somewhat in the literature, we have to make our own conventions precise. Consider all (ascending or descending) filtrations of a nonzero module M having successive quotients which are nonzero and semisimple. A composition series is the longest possibility, but we seek instead the shortest and call it a **Loewy filtration** (or **Loewy series**); the semisimple quotients are then called **Loewy layers** and the number of these the **Loewy length** of M, written $\ell\ell M$. (The length turns out to depend only on M.)

There are two extreme examples:

(1) The **radical filtration** is defined inductively starting with $\operatorname{Rad}^0 M := M$ and $\operatorname{Rad}^1 := \operatorname{Rad} M$ (the **radical** of M), the smallest proper submodule N such that M/N is semisimple (equal to the intersection of all maximal submodules). In general, $\operatorname{Rad}^{k+1} M$ is the smallest proper submodule of $\operatorname{Rad}^k M$ for which the quotient is semisimple. The chain conditions ensure that this filtration is well-defined and ends at 0. Write $\operatorname{Rad}_k M := \operatorname{Rad}^k M/\operatorname{Rad}^{k+1} M$. Sometimes $\operatorname{Rad}_0 M$ is called the **head** of M, denoted $\operatorname{Hd} M$.

(2) The **socle filtration** is defined inductively starting with $\operatorname{Soc}^0 M := 0$ and $\operatorname{Soc}^1 := \operatorname{Soc} M$ (the **socle** of M), the largest semisimple submodule of M (equal to the sum of all simple submodules). In general, $\operatorname{Soc}^k M$ is the (unique) submodule of M satisfying $\operatorname{Soc}(M/\operatorname{Soc}^{k-1} M) = \operatorname{Soc}^k M/\operatorname{Soc}^{k-1} M$. The chain conditions ensure that this filtration is well-defined and ends at M. Write $\operatorname{Soc}_k M := \operatorname{Soc}^k M/\operatorname{Soc}^{k-1} M$.

These Loewy filtrations have a natural hereditary property relative to a submodule $N \subset M$: for example, $\operatorname{Soc}^k N = N \cap \operatorname{Soc}^k M$. (In the case of the radical filtration, one has to shift indices to take account of the difference in Loewy lengths.)

Given any descending Loewy filtration $M = M^0 \supset M^1 \supset \cdots$, one shows inductively that $\operatorname{Rad}^{r-k} M \subset M^{r-k} \subset \operatorname{Soc}^k M$ if $r = \ell\ell M$. In case the radical and socle filtrations coincide, giving the unique Loewy filtration, we say that M is **rigid**. Thus the interesting questions about a module M include: What is its Loewy length? Is M rigid? What are the composition factors of each Loewy layer?

8.15. Loewy Filtrations and KL Polynomials

As a consequence of the KL Conjecture and Jantzen Conjecture (or their proofs), one gets considerable insight into the Loewy structure of Verma modules and associated extensions. This has implications for the structure of projective covers as well. For an arbitrary $\lambda \in \mathfrak{h}^*$, we know initially that $\operatorname{Hd} M(\lambda) \cong L(\lambda)$ and $\operatorname{Rad} M(\lambda) = N(\lambda)$, while $\operatorname{Soc} M(\lambda) \cong L(\mu)$, the simple module with antidominant highest weight in the block determined by λ.

To keep the formulations simple, we look first at the principal block $\mathcal{O}_0$, whose simple modules L_w, Verma modules M_w, and indecomposable projectives P_w correspond to the weights $w \cdot (-2\rho)$, $w \in W$. As we discuss in the following section, most of the results adapt to all integral weights via translation functors and also (sometimes in modified form) to arbitrary weights. Adaptations to parabolic categories are more complicated, however (see 9.17).

Recall from 8.2 that $x \prec w$ means $x < w$, with $\deg P_{x,w}(q)$ as large as possible: $(\ell(x,w) - 1)/2$. In this case, denote by $\mu(x,w)$ the leading coefficient of $P_{x,w}(q)$; in all other cases, set $\mu(x,w) = 0$. The inversion formula for KL polynomials ensures that $\mu(x,w) = \mu(w_\circ w, w_\circ x)$ [**170**, Cor. 3.2].

Now we can summarize in a portmanteau theorem the main structural results following from the KL Conjecture.

Theorem. *In the principal block $\mathcal{O}_0$, denote modules as above by L_w, M_w, P_w with $w \in W$. Assume the truth of the KL Conjecture for $\mathcal{O}_0$.*

(a) *Each Verma module M_w is rigid and has Loewy length $\ell(w) + 1$.*

(b) *Loewy layers of Verma modules are determined by coefficients of KL polynomials: If $x < w$, then*

$$P_{w_\circ w, w_\circ x}(q) = \sum_k [\operatorname{Soc}_{\ell(x)+1+2k} M_w : L_x] \, q^k$$

$$= \sum_k [\operatorname{Rad}_{\ell(x,w)-2k} M_w : L_x] \, q^k.$$

(c) *If $x < w$, then $\dim \operatorname{Ext}^1_{\mathcal{O}}(L_x, L_w) = \mu(x, w)$.*

(d) *The antidominant projective P_1 is rigid and has Loewy length equal to $2\ell(w_\circ) + 1$. Moreover,*

$$\operatorname{Soc}_k P_1 \cong \bigoplus_{w \in W} \operatorname{Soc}_{k - 2\ell(w_\circ w)} M_w.$$

(e) *In general, P_w is rigid if and only if $[M_{w_\circ} : L_w] = 1$. In that case its Loewy length is $2\ell(w_\circ) - \ell(w) + 1$.*

We defer to the following section a discussion of the history and proofs of these statements. First we comment on some parts of the theorem and illustrate them in special cases.

(1) Given part (a), the equivalence of the two sums in part (b) is a simple bookkeeping exercise: here $\operatorname{Soc}_k = \operatorname{Rad}_{\ell(w)-k+1}$. Notice that the

Loewy lengths given in parts (a), (d), and (e) are consistent with what we know about the extreme cases $w = 1$ and $w = w_\circ$.

(2) In (b) the placement of simple modules in Loewy layers is most readily visualized in terms of the radical filtration. Assume $x < w$. The default situation occurs when $P_{w_\circ w, w_\circ x}(q) = 1$, in which case $[M_w : L_x] = 1$ (by the KL Conjecture). Here the index $k = 0$ in the summation corresponds to the unique radical layer in which L_x lies; this is layer $\ell(x, w)$. In general, whenever $x < w$ one copy of L_x always occurs in this layer; but other copies may occur in other layers.

Consider for example the 8-dimensional Schubert variety $\overline{X_w}$ in type C_3 treated in 8.6 above. Here M_w has 42 distinct composition factors L_x with $x \leq w$ (corresponding to Bruhat cells), but 6 of them occur twice when $P_{x,w}(q) = 1 + q^2$. The elements x in question have lengths $3, 2, 2, 1, 1, 0$ in W. So their "default" radical layers are numbered $5, 6, 6, 7, 7, 8$. The respective extra composition factors lie in radical layers $1, 2, 2, 3, 3, 4$.

(3) We claim that (c) follows readily from (b). To see this, we just have to interpret Ext^1 in terms of radical layers. Recall from Proposition 3.1(c) and Theorem 3.2(e) that whenever $\mu < \lambda$ in $\mathfrak{h}^*$, we have

$$\mathrm{Hom}_{\mathcal{O}}\big(N(\lambda), L(\mu)\big) \cong \mathrm{Ext}_{\mathcal{O}}\big(L(\lambda), L(\mu)\big) \cong \mathrm{Ext}_{\mathcal{O}}\big(L(\mu), L(\lambda)\big).$$

Here $N(\lambda) = \mathrm{Rad}\, M(\lambda)$. On the other hand,

$$\mathrm{Hom}_{\mathcal{O}}\big(\mathrm{Rad}\, M(\lambda), L(\mu)\big) \cong \mathrm{Hom}_{\mathcal{O}}\big(\mathrm{Rad}_1 M(\lambda), L(\mu)\big),$$

whose dimension is $[\mathrm{Rad}_1 M(\lambda) : L(\mu)]$ since radical layers are semisimple.

In our situation, it follows from (b) that

$$\dim \mathrm{Ext}^1_{\mathcal{O}}(L_x, L_w) = [\mathrm{Rad}_1 M_w : L_x] = [\mathrm{Rad}_{\ell(x,w)-2k} M_w : L_x],$$

the coefficient of q^k in $P_{w_\circ w, w_\circ x}(q)$. This in turn forces $\ell(x, w) - 2k = 1$, or $k = (\ell(x, w) - 1)/2$, the largest possible degree of the polynomial. So the coefficient in question is $\mu(w_\circ w, w_\circ x) = \mu(x, w)$.

In the framework of part (c), the "default" simple modules L_x for which $\dim \mathrm{Ext}_{\mathcal{O}}(L_x, L_w) = 1$ correspond to elements $x < w$ with $\ell(x, w) = 1$. In the C_3 example just discussed, there are three such x. But there is also a nonzero Ext group involving a single L_x for which $\ell(x, w) = 5$; here $P_{x,w}(q) = 1 + q^2$, so $x \prec w$. All Ext groups here have dimension 1, but in general very large coefficients can occur in KL polynomials (as shown by computations) and might yield high dimensional Ext groups for large values of $\mu(x, w)$.

(4) In part (d), the reader can translate the result into the language of radical layers in the spirit of (b). The layer picture is readily visualized here and is consistent with the "inverted" arrangement of Verma modules

in a standard filtration of a projective module suggested formally by BGG Reciprocity. First place $M_{w_\circ}$, with its head in the middle of the picture and its radical layers numbered from top to bottom as 0 to $\ell(w_\circ) + 1$. Then place $M_{w_\circ w}$ exactly $\ell(w)$ layers higher. In this way the total number of layers becomes $2\ell(w_\circ) + 1$, with L_1 at the top and bottom of the picture and $L_{w_\circ}$ in the middle. Moreover, the self-dual feature of P_1 becomes visible when all the pieces are assembled.

Exercise. When $\mathfrak{g} = \mathfrak{sl}(3, \mathbb{C})$, use parts (d) and (e) of the theorem to work out the Loewy layers of all M_w and P_w.

8.16. Some Details

Obviously much more needs to be said about the history and proof of Theorem 8.15, as well as about the case of an arbitrary block $\mathcal{O}_\lambda$ with $\lambda \in \mathfrak{h}^*$. We conclude this chapter with an overview of these matters and references to the primary sources.

(1) The history is somewhat complicated, since some questions about the possible relationship between coefficients of KL polynomials and the structure of Verma modules were already raised in the immediate aftermath of the KL Conjecture. As already mentioned, independent work of Deodhar [78], Gelfand–MacPherson [109], and Gabber–Joseph [106] explored the idea that the coefficients might predict natural filtration layers.

Deodhar formulated this combinatorially in the Hecke algebra setting by introducing polynomials in q whose sum would be $P_{x,w}(q)$ and whose values at 1 should have this interpretation. The viewpoint of Gelfand–MacPherson was more topological: they set up a sort of dictionary between category $\mathcal{O}$ data and Schubert varieties, which turned out not to lead to a proof of the KL Conjecture. Even so, their dictionary also suggested connections between KL polynomials and layer structure in Verma modules.

As stated above in 8.12, Gabber–Joseph [106] were able to show that the truth of the Jantzen Conjecture would imply the KL Conjecture. Along the way they showed that the Jantzen Conjecture already implies the semisimplicity of layers in the Jantzen filtration. This was strengthened by Barbasch [13] to conclude that the Jantzen filtration of any Verma module must then coincide with the socle filtration. This implies as in Theorem 8.15(c) that $\dim \mathrm{Ext}_\mathcal{O}(L_x, L_w)$ is equal to the leading coefficient $\mu(x, w)$ of $P_{x,w}(q)$ when its degree is $(\ell(x, w) - 1)/2$ (and is 0 otherwise).

(2) Once it became clear from the work of Beilinson–Bernstein that the Jantzen Conjecture would be even harder to prove than the KL Conjecture, systematic study of Loewy filtrations by Irving led to a proof of other parts of Theorem 8.15, along with generalizations: see [134, 135, 136, 140]. While

many details are required, the essential methods are inductive in nature: a natural starting point is the case of an antidominant Verma module.

Working with Vogan's version of the KL Conjecture, Irving made a crucial observation about its relationship with Loewy lengths.

Proposition. *The KL Conjecture for $\mathcal{O}_0$ is equivalent to the statement: For all $M \in \mathcal{O}_0$ and any simple reflection s, the functor Θ_s satisfies $\ell\ell(\Theta_s M) \leq \ell\ell M + 2$.*

What can be said about the Loewy structure of Verma modules and projectives when $\lambda \in \mathfrak{h}$ is arbitrary? Thanks to the category equivalences induced by translation functors, which obviously respect socle and radical series, Theorem 8.15 carries over at once to blocks involving *regular integral* weights. In fact, the proofs work equally well for arbitrary *regular* weights.

The case of *singular* blocks is more delicate: see the papers by Irving [**136, 140**]. It turns out that the results for regular blocks can be adapted to the singular case by careful use of translation functors. In particular, Verma modules are again rigid and the multiplicities of composition factors in their Loewy layers are again given by coefficients of certain KL polynomials. The polynomials are of the type $P_{x,w}(q)$, where x and w are minimal length coset representatives for cosets W/W_μ° when μ is singular. Moreover, the radical series of projectives are described much as in the regular case. (In his first paper, Irving had to invoke some of the deeper geometric ideas involving weight filtrations. But the second paper shows how to proceed in a more self-contained manner.)

(3) One concluding remark may be appropriate. While the basic facts about Loewy structure of Verma modules and projective modules in $\mathcal{O}$ have by now been established, the *proofs* depend on knowing the truth of the KL Conjecture (in a formulation covering all weights). Serious attempts by Irving and others to find more elementary algebraic proofs have all foundered. Similarly, it seems unlikely at this point that the KL Conjecture or Jantzen Conjecture will be proved by purely algebraic methods. But in mathematics it is risky to make predictions.

Part II

Further Developments

Chapter 9

Parabolic Versions of Category $\mathcal{O}$

For applications to Lie group representations a relative version $\mathcal{O}^{\mathfrak{p}}$ of category $\mathcal{O}$ is often required. This is a subcategory determined by a parabolic subalgebra $\mathfrak{p} \supset \mathfrak{b}$ of $\mathfrak{g}$. Study of $\mathcal{O}^{\mathfrak{p}}$ is roughly parallel, in the Lie group setting, to the study of manifolds G/P generalizing the flag manifold G/B.

We have deferred this topic until now partly in order to avoid an overlay of extra notation, but also due to the greater complexity of some of the results. While a combined treatment might be more efficient, we now have the option of applying some of the earlier theory for $\mathcal{O}$ to $\mathcal{O}^{\mathfrak{p}}$ rather than starting again from first principles. In any case, the simple modules in $\mathcal{O}^{\mathfrak{p}}$ are already known from the study of $\mathcal{O}$; so the emphasis shifts to understanding the analogues of Verma modules.

We start with a quick review in 9.1 of parabolic subalgebras, along with related roots and weights. Then the main ideas in this chapter fall under a number of headings:

- Basic features of $\mathcal{O}^{\mathfrak{p}}$: definition and first properties (9.3), parabolic Verma modules (9.4).

- Formal characters and Grothendieck group (9.6): in principle the characters of parabolic Verma modules can be deduced from the known results about $\mathcal{O}$. A more intrinsic approach is based on a relative version of Kazhdan–Lusztig theory (9.7).

- Projective modules and BGG Reciprocity in $\mathcal{O}^{\mathfrak{p}}$ (9.8).

- Module structure: maps between parabolic Verma modules (9.10), simplicity criterion (9.12–9.13), socles, self-dual projectives (9.14), blocks (9.15), analogue of the BGG resolution (9.16), filtrations and rigidity (9.17).

Along the way, a number of special cases are examined involving most often "large" parabolics. To avoid too many technical detours, we often emphasize *integral* weights; side remarks will indicate what is true in general.

9.1. Standard Parabolic Subalgebras

First we recall from 0.1 the relevant subalgebra structure in $\mathfrak{g}$. Each subset $I \subset \Delta$ defines a root system $\Phi_I \subset \Phi$, with positive roots Φ_I^+, negative roots Φ_I^-, and Weyl group $W_I \subset W$ generated by all s_α with $\alpha \in I$. Associated with the root system Φ_I are a number of subalgebras of $\mathfrak{g}$:

$$
\begin{aligned}
&\mathfrak{p}_I && \text{standard parabolic subalgebra} \\
&\mathfrak{l}_I && \mathfrak{h} \oplus \textstyle\sum_{\alpha \in \Phi_I} \mathfrak{g}_\alpha \ \text{(Levi subalgebra of } \mathfrak{p}_I) \\
&\mathfrak{u}_I && \textstyle\bigoplus_{\alpha \in \Phi^+ \setminus \Phi_I^+} \mathfrak{g}_\alpha \ \text{(nilradical of } \mathfrak{p}_I) \\
&\mathfrak{u}_I^- && \textstyle\bigoplus_{\alpha \in \Phi^- \setminus \Phi_I^-} \mathfrak{g}_\alpha \\
&\mathfrak{g}_I && [\mathfrak{l}_I, \mathfrak{l}_I] \ \text{(semisimple subalgebra of } \mathfrak{g}) \\
&\mathfrak{h}_I && \textstyle\bigoplus_{\alpha \in I} \mathbb{C} h_\alpha \ \text{(Cartan subalgebra of } \mathfrak{g}_I) \\
&\mathfrak{n}_I && \textstyle\bigoplus_{\alpha \in \Phi_I^+} \mathfrak{g}_\alpha \\
&\mathfrak{n}_I^- && \textstyle\bigoplus_{\alpha \in \Phi_I^-} \mathfrak{g}_\alpha \\
&\mathfrak{z}_I && \textstyle\bigcap_{\alpha \in I} \operatorname{Ker} \alpha \ (= \text{center of } \mathfrak{l}_I)
\end{aligned}
$$

In the extreme case $I = \emptyset$, we have $\mathfrak{p}_I = \mathfrak{b}$; at the other extreme, $\mathfrak{p}_\Delta = \mathfrak{g}$. In all cases we have direct sum decompositions:

$$
\begin{aligned}
\mathfrak{p}_I &= \mathfrak{l}_I \oplus \mathfrak{u}_I \\
\mathfrak{l}_I &= \mathfrak{g}_I \oplus \mathfrak{z}_I \\
\mathfrak{g}_I &= \mathfrak{n}_I^- \oplus \mathfrak{h}_I \oplus \mathfrak{n}_I \\
\mathfrak{h} &= \mathfrak{h}_I \oplus \mathfrak{z}_I \\
\mathfrak{n} &= \mathfrak{n}_I \oplus \mathfrak{u}_I \\
\mathfrak{n}^- &= \mathfrak{n}_I^- \oplus \mathfrak{u}_I^- \\
\mathfrak{g} &= \mathfrak{u}_I^- \oplus \mathfrak{l}_I \oplus \mathfrak{u}_I = \mathfrak{u}_I^- \oplus \mathfrak{p}_I
\end{aligned}
$$

Here W_I is the Weyl group of $\mathfrak{g}_I$ (or of the reductive algebra $\mathfrak{l}_I$).

As before the reader has to be aware of divergent notation and terminology in the literature; for example, S is commonly used rather than I, while the letter $\mathfrak{m}$ often denotes a Levi subalgebra or its derived algebra in the Lie group context.

9.2. Modules for Levi Subalgebras

To deal with representations of $\mathfrak{g}$ arising from a parabolic subalgebra $\mathfrak{p} = \mathfrak{p}_I$, we also need to supplement the notation of the previous section by considering weights.

From $\mathfrak{h} = \mathfrak{h}_I \oplus \mathfrak{z}_I$ we get a corresponding decomposition $\mathfrak{h}^* \cong \mathfrak{h}_I^* \oplus \mathfrak{z}_I^*$, where for example $\lambda \in \mathfrak{h}_I^*$ is 0 on $\mathfrak{z}_I$. It is easy to check that the $\alpha \in I$ form a basis of $\mathfrak{h}_I^*$, while the fundamental weights ϖ_β with $\beta \in \Delta \setminus I$ form a basis of $\mathfrak{z}_I^*$. For each $\lambda \in \mathfrak{h}^*$, denote by λ_I its $\mathfrak{h}_I^*$ component, identifiable with its restriction to $\mathfrak{h}_I$. Then the set of λ for which λ_I is dominant integral on $\mathfrak{h}_I$ may be denoted by

$$\Lambda_I^+ := \{\lambda \in \mathfrak{h}^* \mid \langle \lambda, \alpha^\vee \rangle \in \mathbb{Z}^+ \text{ for all } \alpha \in I\}.$$

The condition on λ is equivalent to $\langle \lambda + \rho, \alpha^\vee \rangle \in \mathbb{Z}^{>0}$ for all $\alpha \in I$ (or all $\alpha \in \Phi_I^+$). Obviously $\Lambda^+ \subset \Lambda_I^+$.

Exercise. Let $\lambda \in \Lambda_I^+$. If $w \in W_I$ and $w \neq 1$, prove that $w \cdot \lambda \notin \Lambda_I^+$.

The λ_I with $\lambda \in \Lambda_I^+$ are precisely the highest weights of finite dimensional $\mathfrak{g}_I$-modules. Such a module has a natural structure of simple $\mathfrak{l}_I$-module, where $\mathfrak{z}_I$ acts by the restriction of λ to $\mathfrak{z}_I$. Denote this $\mathfrak{l}_I$-module (or its restriction to $\mathfrak{g}_I$) by $L_I(\lambda)$. Conversely, every finite dimensional simple $\mathfrak{l}_I$-module has this form: by Schur's Lemma, the elements of $\mathfrak{z}_I$ must act by scalars. But when $\mathfrak{z}_I \neq 0$, many simple $\mathfrak{l}_I$-modules have the same restriction to $\mathfrak{g}_I$. Observe also that Weyl's Complete Reducibility Theorem for finite dimensional $\mathfrak{g}_I$-modules (0.8) extends immediately to modules for the reductive Lie algebra $\mathfrak{l}_I$ on which $\mathfrak{h}$ acts semisimply.

Besides the simple modules $L_I(\lambda)$, we need the corresponding Verma modules for $\mathfrak{g}_I$; these too become $U(\mathfrak{l}_I)$-modules in a natural way. Since they will play only an auxiliary role below, we denote them in an *ad hoc* way by

$$V_I(\lambda) := U(\mathfrak{l}_I) \otimes_{U(\mathfrak{h} \oplus \mathfrak{n}_I)} \mathbb{C}_\lambda \text{ with } \lambda \in \mathfrak{h}^*.$$

As in 1.16, it is easy to write down the formal character as a function on $\mathfrak{h}$: letting p_I be the partition function for the root system Φ_I analogous to the Kostant function p, we get $\operatorname{ch} V_I(\lambda) = p_I * e(\lambda)$.

When $\lambda \in \Lambda_I^+$, the description of the maximal submodule of $V_I(\lambda)$ given in Theorem 2.6 extends readily from $\mathfrak{g}_I$ to $\mathfrak{l}_I$, yielding an exact sequence of $U(\mathfrak{l}_I)$-modules:

$$(1) \qquad \bigoplus_{\alpha \in I} V_I(s_\alpha \cdot \lambda) \to V_I(\lambda) \to L_I(\lambda) \to 0.$$

9.3. The Category $\mathcal{O}^{\mathfrak{p}}$

Fix a standard parabolic subalgebra $\mathfrak{p} = \mathfrak{p}_\mathrm{I} \supset \mathfrak{b}$, corresponding to $\mathrm{I} \subset \Delta$. One can introduce the **parabolic category** $\mathcal{O}^{\mathfrak{p}}$ (or $\mathcal{O}^\mathrm{I}$) from first principles (as for example in Rocha [**226**]) by axioms similar to those used for $\mathcal{O}$ in 1.1. In this spirit, $\mathcal{O}^{\mathfrak{p}}$ is defined to be the full subcategory of $\mathrm{Mod}\, U(\mathfrak{g})$ whose objects M satisfy:

($\mathcal{O}^{\mathfrak{p}}$1) M is a finitely generated $U(\mathfrak{g})$-module.

($\mathcal{O}^{\mathfrak{p}}$2) Viewed as a $U(\mathfrak{l}_\mathrm{I})$-module, M is a direct sum of finite dimensional simple modules.

($\mathcal{O}^{\mathfrak{p}}$3) M is locally $\mathfrak{u}_\mathrm{I}$-finite.

Obviously $\mathcal{O}^{\mathfrak{p}}$ is a full subcategory of $\mathcal{O}$ and contains all finite dimensional modules. To simplify notation, we write $\mathrm{Hom}_{\mathcal{O}}(M, N)$ rather than $\mathrm{Hom}_{\mathcal{O}^{\mathfrak{p}}}(M, N)$ when $M, N \in \mathcal{O}^{\mathfrak{p}}$. There are two extreme examples: the case $\mathrm{I} = \emptyset$ just recovers $\mathcal{O}$, while $\mathrm{I} = \Delta$ yields the (semisimple) category of finite dimensional $U(\mathfrak{g})$-modules.

Since we have already studied the basic properties of $\mathcal{O}$, it is more efficient at this point to rely instead on a single property to identify $\mathcal{O}^{\mathfrak{p}}$ as a subcategory of $\mathcal{O}$.

Lemma. *Let $M \in \mathcal{O}$ have set of weights $\Pi(M)$. The following conditions on M are equivalent:*

(a) *M is locally $\mathfrak{n}_\mathrm{I}^{-}$-finite.*

(b) *For all $\alpha \in \mathrm{I}$ and $\mu \in \Pi(M)$, we have $\dim M_\mu = \dim M_{s_\alpha \mu}$.*

(c) *For all $w \in W_\mathrm{I}$ and $\mu \in \Pi(M)$, we have $\dim M_\mu = \dim M_{w\mu}$.*

(d) *The set $\Pi(M)$ is stable under W_I.*

Proof. Assuming (a), the action on M_μ of the copy of $\mathfrak{sl}(2, \mathbb{C})$ generated by x_α and y_α for a fixed $\alpha \in \mathrm{I}$ produces a finite dimensional submodule of M; this is clearly stable under $\mathfrak{h}$. Then the standard theory in 0.9 yields (b). Since W_I is generated by the reflections s_α for $\alpha \in \mathrm{I}$, (c) follows at once and trivially implies (d).

Now assume (d). Given a vector $v \in M$ of weight μ, only finitely many standard PBW monomials in $U(\mathfrak{n}_\mathrm{I})$ can take v to a nonzero vector. Therefore only finitely many weights of the form $\mu + \nu$ with ν a $\mathbb{Z}^{+}$-linear combination of Φ_I^{+} can lie in $\Pi(M)$. Now the longest element w_I of W_I interchanges Φ_I^{+} and Φ_I^{-}. Thanks to (d), w_I stabilizes $\Pi(M)$, so only finitely many $\mathbb{Z}^{+}$-linear combinations of Φ_I^{-} can be added to $\mu' := w_\mathrm{I}\mu$ to yield a weight of M. Again by (d), μ' is a typical weight of M, forcing M to be locally $\mathfrak{n}_\mathrm{I}^{-}$-finite. $\quad\square$

With this in hand, it is easy to verify some elementary facts about $\mathcal{O}^{\mathfrak{p}}$:

Proposition. *Fix $\mathfrak{p} = \mathfrak{p}_I$ as above.*

(a) *$M \in \mathcal{O}$ lies in $\mathcal{O}^{\mathfrak{p}}$ if and only if M satisfies the equivalent conditions of the lemma.*

(b) *$\mathcal{O}^{\mathfrak{p}}$ is closed under duality in $\mathcal{O}$.*

(c) *$\mathcal{O}^{\mathfrak{p}}$ is closed under direct sums, submodules, quotients, and extensions in $\mathcal{O}$, as well as tensoring with finite dimensional $U(\mathfrak{g})$-modules.*

(d) *If $M \in \mathcal{O}^{\mathfrak{p}}$ decomposes as $M = \bigoplus M^{\chi}$ with M^{χ} in $\mathcal{O}_{\chi}$, then each M^{χ} lies in $\mathcal{O}^{\mathfrak{p}}$; this gives a decomposition $\mathcal{O}^{\mathfrak{p}} = \bigoplus_{\chi} \mathcal{O}^{\mathfrak{p}}_{\chi}$. As a result, translation functors preserve $\mathcal{O}^{\mathfrak{p}}$.*

(e) *If the simple module $L(\lambda)$ lies in $\mathcal{O}^{\mathfrak{p}}$, then $\lambda \in \Lambda_I^+$.*

Proof. (a) Any $M \in \mathcal{O}^{\mathfrak{p}}$ satisfies $(\mathcal{O}^{\mathfrak{p}}2)$, therefore also satisfies the conditions in the lemma.

In the other direction, suppose $M \in \mathcal{O}$ satisfies those conditions. Since any module in $\mathcal{O}$ satisfies $(\mathcal{O}^{\mathfrak{p}}1)$ and $(\mathcal{O}^{\mathfrak{p}}3)$, we just have to verify $(\mathcal{O}^{\mathfrak{p}}2)$. Thanks to part (a) of the lemma, any weight vector $v \in M$ generates a finite dimensional $U(\mathfrak{g}_I)$-module on which $\mathfrak{h}$ acts semisimply since $\mathfrak{h}$ normalizes $\mathfrak{n}_I$ and $\mathfrak{n}_I^-$. It follows that every element of M lies in such a finite dimensional $U(\mathfrak{l}_I)$-module. Using complete reducibility, a standard argument shows that M is a direct sum of simple $U(\mathfrak{l}_I)$-modules, as required.

(b) Since $M^{\vee}$ has the same formal character as M, part (d) of the lemma is satisfied by $M^{\vee}$ in place of M. Thus $M^{\vee}$ lies in $\mathcal{O}^{\mathfrak{p}}$, by (a).

(c) and (d) These assertions follow readily from the criteria in the lemma, coupled with (a).

(e) Suppose $L(\lambda) \in \mathcal{O}^{\mathfrak{p}}$. In particular, if v^+ is a maximal vector of weight λ, then for each $\alpha \in I$ some power of y_α kills v^+. As in 1.6, the study of $\mathfrak{sl}(2, \mathbb{C})$-modules shows that $\langle \lambda, \alpha^{\vee} \rangle \in \mathbb{Z}^+$; thus $\lambda \in \Lambda_I^+$. $\qquad\square$

It is natural to ask whether the converse of (e) also holds; but it is complicated to apply the criteria in the lemma directly to $L(\lambda)$, since $\Pi\big(L(\lambda)\big)$ is hard to describe. Instead the proof will rely on an analogue of Verma modules constructed in 9.4 below.

There are a couple of ways to define interesting functors from $\mathcal{O}$ to $\mathcal{O}^{\mathfrak{p}}$. Given $M \in \mathcal{O}$, define $\underline{M}$ to be the set of all $\mathfrak{n}_I^-$-finite vectors in M. This is clearly a submodule and is the unique maximal element among the submodules of M lying in $\mathcal{O}^{\mathfrak{p}}$. Taking advantage of the fact that duality preserves $\mathcal{O}^{\mathfrak{p}}$, we can then set $\overline{M} := (\underline{M^{\vee}})^{\vee}$. This is the largest quotient of M lying in $\mathcal{O}^{\mathfrak{p}}$ and may be dubbed the **truncation** of M relative to $\mathfrak{p}$.

Exercise. (1) Given an exact sequence $M \to N \to 0$ in $\mathcal{O}$, prove that the resulting homomorphism $\overline{M} \to \overline{N}$ is surjective.

(2) Given an exact sequence $0 \to N \to M$ in $\mathcal{O}$, must the resulting map $\overline{N} \to \overline{M}$ be injective?

9.4. Parabolic Verma Modules

What is the appropriate generalization of Verma modules in this setting? The basic idea is to induce finite dimensional modules from $\mathfrak{p}_I$ to $\mathfrak{g}$ as we did with 1-dimensional $\mathfrak{b}$-modules. Start with the finite dimensional simple $\mathfrak{l}_I$-module $L_I(\lambda)$ for $\lambda \in \Lambda_I^+$. Then inflate $L_I(\lambda)$ to a $\mathfrak{p}_I$-module via the projection $\mathfrak{p}_I \to \mathfrak{p}_I/\mathfrak{u}_I \cong \mathfrak{l}_I$, so that $\mathfrak{u}_I$ acts trivially. The resulting $\mathfrak{p}_I$-module is generated by a maximal vector v^+ of weight λ. Finally, define the **parabolic Verma module**

$$M_I(\lambda) := U(\mathfrak{g}) \otimes_{U(\mathfrak{p}_I)} L_I(\lambda).$$

Obviously $\mathfrak{u}_I^-$ acts freely on $L_I(\lambda)$, just as $\mathfrak{n}^-$ acted freely on a (simple!) 1-dimensional $\mathfrak{b}$-module of weight λ in the construction of Verma modules. Thus $M_I(\lambda)$ is generated as a $U(\mathfrak{g})$-module by the maximal vector $1 \otimes v^+$ of weight λ. It follows that $M_I(\lambda)$ is a quotient of $M(\lambda)$; in particular, $L(\lambda)$ is its unique simple quotient.

The construction of $M_I(\lambda)$ makes it easy to compute its formal character. By the PBW Theorem, the module $M_I(\lambda)$ when viewed as a $U(\mathfrak{u}_I^-)$-module is isomorphic to $U(\mathfrak{u}_I^-) \otimes L_I(\lambda)$. Accordingly,

(1) $$\operatorname{ch} M_I(\lambda) = p^I * \operatorname{ch} L_I(\lambda),$$

where p^I is defined like the function p in 1.16 but restricted to the sums of positive roots not in Φ_I^+. Note that $p = p^I * p_I$ with p_I as in 9.2.

While it is not straightforward here to write down explicitly a basis of $M_I(\lambda)$ in terms of PBW monomials applied to v^+, we can use the factorization $U(\mathfrak{n}^-) \cong U(\mathfrak{u}_I^-) \otimes U(\mathfrak{n}_I^-)$ to obtain a basis of the form

$$y_1^{i_1} \cdots y_r^{i_r}\, u \cdot v^+,$$

where the y_i correspond to an enumeration of $\Phi^+ \setminus \Phi_I$ and $u \cdot v^+$ runs over a basis of the finite dimensional space $U(\mathfrak{n}_I^-) \cdot v^+$.

The exact sequence in the following theorem originates in the work of Lepowsky [**201**, Prop. 2.1]:

Theorem. *Let* $\lambda \in \Lambda_I^+$.

(a) *The module* $M_I(\lambda)$ *lies in* $\mathcal{O}^\mathfrak{p}$; *so its quotient* $L(\lambda)$ *also does.*

(b) *There is an exact sequence*

$$(2) \qquad \bigoplus_{\alpha \in I} M(s_\alpha \cdot \lambda) \to M(\lambda) \to M_I(\lambda) \to 0.$$

(c) $M_I(\lambda)$ *coincides with the universal highest weight module* $\overline{M(\lambda)}$ *in* $\mathcal{O}^{\mathfrak{p}}$ *defined earlier in terms of truncation.*

Proof. (a) To show that $M_I(\lambda)$ lies in $\mathcal{O}^{\mathfrak{p}}$, it will be enough (by Proposition 9.3(a)) to check condition (d) in Lemma 9.3: the set of weights of $M_I(\lambda)$ is stable under W_I. From the character formula (1) we know that these weights are characterized as the differences $\mu - \nu$, where $\mu \in \Pi\big(L_I(\lambda)\big)$ and ν is a $\mathbb{Z}^+$-linear combination of $\Phi^+ \setminus \Phi_I$. Clearly W_I permutes $\Pi\big(L_I(\lambda)\big)$. On the other hand, each simple reflection s_α with $\alpha \in I$ permutes the positive roots other than α in both Φ^+ and Φ_I^+ (0.3(1)). So for all $w \in W_I$ the weight $w\mu - w\nu$ is again a weight of $M_I(\lambda)$ as required.

(b) As observed in 9.2(1), Theorem 2.6 extends to $\mathfrak{l}_I$ and gives an exact sequence of $U(\mathfrak{l}_I)$-modules:

$$(3) \qquad \bigoplus_{\alpha \in I} V_I(s_\alpha \cdot \lambda) \to V_I(\lambda) \to L_I(\lambda) \to 0.$$

As in the case of $L_I(\lambda)$, we can inflate $V_I(\lambda)$ to a $U(\mathfrak{p}_I)$-module by letting the nilradical $\mathfrak{u}_I$ act trivially. Then $M(\lambda) \cong U(\mathfrak{g}) \otimes_{U(\mathfrak{p}_I)} V_I(\lambda)$. This inflate-induce functor is exact and transforms the exact sequence into the desired exact sequence (2) in $\mathcal{O}$.

(c) It is clear that $s_\alpha \cdot \lambda \notin \Lambda_I^+$ for all $\alpha \in I$, so $L(s_\alpha \cdot \lambda) \notin \mathcal{O}^{\mathfrak{p}}$. It follows that $M(s_\alpha \cdot \lambda)$ lies in the kernel of the natural map $\varphi : M(\lambda) \to \overline{M(\lambda)}$. So the exact sequence in (b) implies that φ induces a surjection $M_I(\lambda) \to \overline{M(\lambda)}$. Thanks to (a), $M_I(\lambda) \in \mathcal{O}^{\mathfrak{p}}$. Since $\overline{M(\lambda)}$ is a universal highest weight module in $\mathcal{O}^{\mathfrak{p}}$, this surjection must be an isomorphism. $\square$

Exercise. On the level of formal characters, show that for $\lambda \in \Lambda_I^+$:

$$\mathrm{ch}\, M(\lambda) = p^I * \mathrm{ch}\, V_I(\lambda) = p^I * p_I * e(\lambda),$$

consistent with $\mathrm{ch}\, M(\lambda) = p * e(\lambda)$ (1.16).

In view of Proposition 9.3, part (a) of the theorem shows that $M \in \mathcal{O}$ *lies in* $\mathcal{O}^{\mathfrak{p}}$ *if and only if all its composition factors* $L(\lambda)$ *satisfy* $\lambda \in \Lambda_I^+$.

It is often convenient to parametrize weights in Λ_I^+ by Weyl group elements. When $\lambda \in \Lambda$ is *dominant* and *regular*, the weights $w \cdot \lambda$ are in natural bijection with the elements of W. If we then let W^I be the set of minimal length right coset representatives in $W_I \backslash W$, it is easy to check that

$$w \cdot \lambda \in \Lambda_I^+ \text{ if and only if } w \in W^I$$

(as noted by Boe [**30**, 2.4]). This uses an observation which follows readily from 0.3(4):

$$W^{\mathrm{I}} = \{w \in W \mid w^{-1}\Phi_{\mathrm{I}}^{+} \subset \Phi^{+}\}.$$

Question: What can be said if λ is no longer required to be regular?

Remarks. (1) Jantzen [**147**, 2.25] discusses the analogue for $\mathcal{O}^{\mathfrak{p}}$ of Theorem 7.6 on *translation functors* and their effect on Verma modules. As one might expect, things get more complicated than before: translating a parabolic Verma module to the closure of a facet may yield either the corresponding parabolic Verma module or else 0.

(2) The name **generalized Verma module** is more widely used; but its meaning has also been broadened (for example in work of Futorny, Mazorchuk, and others) to include modules induced from *infinite dimensional* simple $U(\mathfrak{p})$-modules. Another variant is the label **relative Verma module**. The reader needs to be aware of such differences in terminology, as well as notation, throughout the literature.

9.5. Example: $\mathfrak{sl}(3, \mathbb{C})$

It is easy to illustrate some features of category $\mathcal{O}^{\mathfrak{p}}$ when $\mathfrak{g} = \mathfrak{sl}(3, \mathbb{C})$, as in 4.11 and 5.4. (However, here as in other rank 2 cases the situation is artificially simplified by the fact that multiplicities in Verma modules are all 1.) We look only at integral weights.

Denoting the simple roots by α, β, set $\mathrm{I} := \{\alpha\}$. Then the weights $\lambda \in \Lambda_{\mathrm{I}}^{+}$ are those lying on or above the hyperplane orthogonal to α; the regular ones lie in the three Weyl chambers corresponding to $1, s_\beta, s_\beta s_\alpha$ (which are minimal length right coset representatives in $W_{\mathrm{I}} \backslash W$). Label the linked ones as $w \cdot \lambda$ for each fixed $\lambda \in \Lambda^{+}$.

What can be said about the parabolic Verma modules for regular weights in Λ_{I}^{+}? Since $|W_{\mathrm{I}}| = 2$, it is clear from the Weyl–Kostant character formula that the exact sequence 9.4(3) is actually a short exact sequence and similarly for the induced sequence 9.4(2). This makes the formal characters (hence composition factor multiplicities) easy to work out.

Consider $\mu := s_\beta s_\alpha \cdot \lambda$, which is minimal among linked weights lying in Λ_{I}^{+}; thus $M_{\mathrm{I}}(\mu) = L(\mu)$. This is a proper quotient of $M(\mu)$ by the submodule $M(s_\alpha \cdot \mu) = M(w_\circ \cdot \lambda)$.

Next consider $\mu := s_\beta \cdot \lambda$. In this case the kernel of $M(\mu) \to M_{\mathrm{I}}(\mu)$ is $M(s_\alpha \cdot \mu) = M(s_\alpha s_\beta \cdot \lambda)$, which has two composition factors. Thus $M_{\mathrm{I}}(\mu)$ has the remaining two composition factors $L(\mu) = L(s_\beta \cdot \lambda)$ and $L(s_\beta s_\alpha \cdot \lambda)$.

Finally, consider $M_{\mathrm{I}}(\lambda)$, which is the quotient of $M(\lambda)$ by $M(s_\alpha \cdot \lambda)$. The latter has four composition factors, leaving only two for $M_{\mathrm{I}}(\lambda)$ with highest weights λ and $s_\beta \cdot \lambda$. In particular, this gives an example in which $M_{\mathrm{I}}(\lambda) = \overline{M(\lambda)}$ fails to capture all composition factors of $M(\lambda)$ which lie in $\mathcal{O}^{\mathfrak{p}}$. (Compare part (b) of Exercise 9.3.)

Exercise. Work out the nonzero module homomorphisms among the various parabolic Verma modules in this example.

9.6. Formal Characters and Composition Factors

A central concern in our study of category $\mathcal{O}$ was the determination of formal characters of the modules $L(\lambda)$ for arbitrary $\lambda \in \mathfrak{h}^*$. This turned out to be a deep problem, leading to a solution (in KL theory) which is essentially algorithmic in nature. An equivalent formulation of the problem involves the two bases $\{[L(\lambda)]\}$ and $\{[M(\lambda)]\}$ of the Grothendieck group $K(\mathcal{O})$: expressing one basis in terms of the other involves a unipotent integral matrix, whose entries are to be calculated.

The problem for $\mathcal{O}^{\mathfrak{p}}$ is formally similar. Here it is clear that $K(\mathcal{O}^{\mathfrak{p}})$ may be viewed as a subgroup of $K(\mathcal{O})$ with two natural bases: one consisting of symbols $[L(\lambda)]$ and the other consisting of symbols $[M_{\mathrm{I}}(\lambda)]$ with λ running over Λ_{I}^+. Again the problem is to express one basis in terms of the other, which again involves a unipotent integral matrix.

Given what is already known about $\mathcal{O}$, the formal character problem for $\mathcal{O}^{\mathfrak{p}}$ is already solved in principle. We observed in 9.4(1) that $\operatorname{ch} M_{\mathrm{I}}(\lambda) = p^{\mathrm{I}} * \operatorname{ch} L_{\mathrm{I}}(\lambda)$ for all $\lambda \in \Lambda_{\mathrm{I}}^+$. This in turn is a $\mathbb{Z}^+$-linear combination of $\operatorname{ch} L(\lambda)$ (counted once) and various $\operatorname{ch} L(\mu)$ with $\mu \in \Lambda_{\mathrm{I}}^+$ linked to λ by W and $\mu < \lambda$. To compute the coefficients one can proceed indirectly, starting with the Weyl–Kostant formula for $\mathfrak{g}_{\mathrm{I}}$ (extended to $\mathfrak{l}_{\mathrm{I}}$):

$$(1) \qquad \operatorname{ch} L_{\mathrm{I}}(\lambda) = \sum_{w \in W_{\mathrm{I}}} (-1)^{\ell(w)} \operatorname{ch} V_{\mathrm{I}}(w \cdot \lambda).$$

Recall that the length function on W_{I} is just the restriction of the length function on W, so there is no conflict here in the use of $\ell(w)$: see 0.3. Also, Exercise 9.4 shows that $p^{\mathrm{I}} * \operatorname{ch} V_{\mathrm{I}}(\mu) = \operatorname{ch} M(\mu)$. To put the pieces together, multiply both sides of (1) (in the convolution sense) by p^{I} to get (as in Jantzen [**146**, Lemma 1]):

Proposition. *If $\lambda \in \Lambda_{\mathrm{I}}^+$, then*

$$\operatorname{ch} M_{\mathrm{I}}(\lambda) = \sum_{w \in W_{\mathrm{I}}} (-1)^{\ell(w)} \operatorname{ch} M(w \cdot \lambda).$$

Note that if we had developed the full BGG resolution of $L_{\mathrm{I}}(\lambda)$, this would correspond to an exact sequence completing the one in Theorem 9.4(b).

Now the idea is to compute (by the methods of KL theory) the multiplicities of composition factors of all $M(w \cdot \lambda)$ and add them up with alternating signs to get the list of composition factors (with multiplicity) of $M_{\mathrm{I}}(\lambda)$.

Example. In the case $\mathfrak{g} = \mathfrak{sl}(3, \mathbb{C})$ considered in 9.5 above, we were essentially following the method outlined above. Here it was reasonably straightforward, since the composition factor multiplicities of Verma modules are easy to work out and are all 1. In the inverse format, all coefficients are then ± 1. Even so, the end results in $\mathcal{O}^{\mathfrak{p}}$ are simpler yet. For example, for $\lambda \in \Lambda^+$ we arrive at

$$[L(\lambda)] = [M_{\mathrm{I}}(\lambda)] - [M_{\mathrm{I}}(s_\beta \cdot \lambda)] + [M_{\mathrm{I}}(s_\beta s_\alpha \cdot \lambda)].$$

In higher rank cases (especially when $\mathfrak{p}$ is taken to be a maximal parabolic subalgebra), the method has an obvious drawback: it requires one to work out a large number of composition factor multiplicities when the highest weights involved lie outside Λ_{I}^+, even though these ultimately cancel.

Exercise. In case $\lambda \in \Lambda^+$, combine Proposition 9.6 above with the Weyl–Kostant character formula (2.4) to express $\operatorname{ch} L(\lambda)$ as an alternating sum of the $\operatorname{ch} M_{\mathrm{I}}(w \cdot \lambda)$, with w ranging over the set W^{I} of minimal right coset representatives for $W_{\mathrm{I}} \backslash W$.

9.7. Relative Kazhdan–Lusztig Theory

A more intrinsic approach to computing formal characters within the framework of category $\mathcal{O}^{\mathfrak{p}}$ was developed by Deodhar [**79**] (with refinements in [**80**]) and (independently, but slightly later) by Casian–Collingwood [**61**]. Their goals and methods were different, but in each case the same kind of basic combinatorial framework is introduced: a *relative Hecke module*. This is a module with involution over the Hecke algebra $\mathcal{H}$ of W (8.2) whose basis is in bijection with the left (or right) coset space of W_{I} in W and can be labelled with minimal length elements of these cosets. Since the notational conventions in the two papers differ from ours and from each other, we just paraphrase briefly what is done. The common theme is a construction of *relative KL polynomials*.

Deodhar was looking for a treatment that would work for a general Kac–Moody variety of type G/P, not relying on a fibration $G/B \to G/P$. His goal was to compute the dimension of intersection cohomology groups of Schubert varieties in these analogues of parabolic flag varieties, using a version of KL theory involving mainly combinatorial techniques. Here the Hecke module and relative KL polynomials are essential. He does not treat representations in our category $\mathcal{O}^{\mathfrak{p}}$ explicitly but does point out that the relative KL polynomials can be expressed in terms of the usual KL polynomials for a finite Weyl group W.

On the other hand, Casian and Collingwood wanted a direct approach to representations in $\mathcal{O}^{\mathfrak{p}}$ which would permit the computation of intersection cohomology for P-orbits on the flag variety without embedding the problem in $\mathcal{O}$. This fed into the study of homomorphisms between parabolic Verma modules, as discussed below in 9.10. As a basic tool they formulate and prove a relative KL Conjecture for $\mathcal{O}^{\mathfrak{p}}$ involving regular integral weights (using the deep geometric methods employed in the proof of the original KL Conjecture). This is where the Hecke module and relative KL polynomials come in.

9.8. Projectives and BGG Reciprocity in $\mathcal{O}^{\mathfrak{p}}$

To round out the picture of category $\mathcal{O}^{\mathfrak{p}}$ for a fixed $\mathrm{I} \subset \Delta$ and $\mathfrak{p} = \mathfrak{p}_{\mathrm{I}}$, we have to look for projective (or injective) objects. By imitating closely the original arguments of BGG [**27**], Rocha [**226**] showed how to carry over all the essential ideas from $\mathcal{O}$ such as BGG Reciprocity; in this treatment the results for $\mathcal{O}$ are recovered as special cases when I is empty. To generalize the notion of standard filtration, one says that a module $M \in \mathcal{O}^{\mathfrak{p}}$ has a **standard filtration** (relative to I) if it has a filtration with subquotients isomorphic to various $M_{\mathrm{I}}(\lambda)$. As before, the multiplicity of each parabolic Verma module as a subquotient is well determined by M and written $\big(M : M_{\mathrm{I}}(\lambda)\big)$.

We can formulate the main conclusions as follows:

Theorem. *Fix* $\mathrm{I} \subset \Delta$.

(a) *The category $\mathcal{O}^{\mathfrak{p}}$ has enough projectives (and by duality enough injectives).*

(b) *Among the projectives are the parabolic Verma modules $M_{\mathrm{I}}(\lambda)$ with λ dominant.*

(c) *The tensor product of a projective module in $\mathcal{O}^{\mathfrak{p}}$ and a finite dimensional module is again projective.*

(d) *Every projective in $\mathcal{O}^{\mathfrak{p}}$ is a direct sum of indecomposables $P_{\mathrm{I}}(\lambda)$ indexed by Λ_{I}^{+}, where $P_{\mathrm{I}}(\lambda)$ is a projective cover of $L(\lambda)$.*

(e) *If $P \in \mathcal{O}^{\mathfrak{p}}$ is projective, it has a standard filtration (relative to I). In particular, $P_{\mathrm{I}}(\lambda)$ has such a filtration with $\big(P_{\mathrm{I}}(\lambda) : M_{\mathrm{I}}(\lambda)\big) = 1$ and $\mu > \lambda$ for all other subquotients $M_{\mathrm{I}}(\mu)$.*

(f) *The analogue of BGG Reciprocity holds in $\mathcal{O}^{\mathfrak{p}}$: if $\lambda, \mu \in \Lambda_{\mathrm{I}}^{+}$, then*

$$\big(P_{\mathrm{I}}(\lambda) : M_{\mathrm{I}}(\mu)\big) = [M_{\mathrm{I}}(\mu) : L(\lambda)].$$

Since the treatment of $\mathcal{O}$ in Chapter 3 differs significantly from that of BGG, we invite the reader at this point to revisit that chapter and verify that all the steps can indeed be generalized to $\mathcal{O}^{\mathfrak{p}}$. Here is a list of the main points to be checked:

(1) The observations about duality in Theorem 3.3 adapt to $\mathcal{O}^{\mathfrak{p}}$. In particular,

$$\dim \operatorname{Hom}_{\mathcal{O}}(M_{\mathrm{I}}(\mu), M_{\mathrm{I}}(\lambda)^{\vee}) = \delta_{\lambda\mu},$$

$$\operatorname{Ext}_{\mathcal{O}^{\mathfrak{p}}}(M_{\mathrm{I}}(\mu), M_{\mathrm{I}}(\lambda)^{\vee}) = 0 \text{ for all } \lambda, \mu \in \Lambda_{\mathrm{I}}^{+}.$$

(2) Either method of proof of Theorem 3.6 can be modified to show that the tensor product of $M_{\mathrm{I}}(\lambda)$ and a finite dimensional module M has a standard filtration with subquotients $M_{\mathrm{I}}(\lambda + \mu)$ as μ runs over the weights of M (counting multiplicities).

(3) The properties of standard filtrations in Proposition 3.7 carry over to $\mathcal{O}^{\mathfrak{p}}$. If M has a standard filtration here, it is $U(\mathfrak{u}_{\mathrm{I}}^{-})$-free. Similarly, Theorem 3.7 adapts to $\mathcal{O}^{\mathfrak{p}}$.

(4) The analogue of Proposition 3.8 is true in $\mathcal{O}^{\mathfrak{p}}$, leading to the existence of enough projectives as in Theorem 3.8.

(5) Theorem 3.9 adapts to $\mathcal{O}^{\mathfrak{p}}$. In particular, for all $M \in \mathcal{O}^{\mathfrak{p}}$ we have

$$\dim \operatorname{Hom}_{\mathcal{O}}(P_{\mathrm{I}}(\lambda), M) = [M : L(\lambda)].$$

(6) Any projective module in $\mathcal{O}^{\mathfrak{p}}$ has a standard filtration, as in Theorem 3.10.

(7) The analogue of BGG Reciprocity (3.11) in (f) above holds in $\mathcal{O}^{\mathfrak{p}}$.

Remark. Using the fact that $\mathcal{O}$ has enough projectives, one could apply the notion of "truncation" discussed at the end of 9.3 to prove the analogous result for $\mathcal{O}^{\mathfrak{p}}$: start by showing that $\overline{P}$ is projective in $\mathcal{O}^{\mathfrak{p}}$ whenever P is projective in $\mathcal{O}$. In particular, $\overline{P(\lambda)} \cong P_{\mathrm{I}}(\lambda)$ when $\lambda \in \Lambda_{\mathrm{I}}^{+}$.

Exercise. In the case $\mathfrak{g} = \mathfrak{sl}(3, \mathbb{C})$ discussed in 9.5, show that $P_{\mathrm{I}}(s_{\beta} \cdot \lambda)$ is self-dual, as is $P_{\mathrm{I}}(s_{\beta}s_{\alpha} \cdot \lambda)$.

9.9. Structure of Parabolic Verma Modules

In Chapters 4–5 we were able to answer definitively several related questions about the structure of Verma modules, using elementary methods:

- For all $\lambda, \mu \in \mathfrak{h}^{*}$, we have $\dim \operatorname{Hom}_{\mathcal{O}}\left(M(\mu), M(\lambda)\right) \leq 1$, while every nonzero homomorphism is injective (4.2).

- Each $M(\lambda)$ has a unique simple submodule (4.1), which is in fact isomorphic to a Verma module (4.2).

- The Verma module $M(\lambda)$ is simple if and only if λ is antidominant: $\langle \lambda + \rho, \alpha^{\vee} \rangle \notin \mathbb{Z}^{>0}$ for all $\alpha > 0$ (4.8). To prove this in the general case we had to show first the existence of embeddings $M(s_{\alpha} \cdot \lambda) \hookrightarrow M(\lambda)$ when $\alpha > 0$ and $s_{\alpha} \cdot \lambda < \lambda$ (4.6).

- As a consequence, we deduced that the blocks of $\mathcal{O}$ are parametrized naturally by antidominant weights (4.9).

- $[M(\lambda) : L(\mu)] \geq 1$ if and only if $\mu \uparrow \lambda$ (that is, μ is strongly linked to λ), which happens if and only if $\operatorname{Hom}_{\mathcal{O}}\big(M(\mu), M(\lambda)\big) \neq 0$ (5.1).

Similar questions arise in the category $\mathcal{O}^{\mathfrak{p}}$ but are much less straightforward to resolve; some remain open except in special cases. Where definitive answers are lacking, we will survey the current state of the literature. The questions often turn out to be interrelated in subtle ways, so the section divisions below are somewhat artificial.

- What can be said about homomorphisms $M_{\mathrm{I}}(\mu) \to M_{\mathrm{I}}(\lambda)$, including the dimension of the Hom space and the injectivity of nonzero maps?

- Which modules $M_{\mathrm{I}}(\lambda)$ are simple?

- Which simple modules $L(\mu)$ occur in $\operatorname{Soc} M_{\mathrm{I}}(\lambda)$, and with what multiplicity?

- When is $[M_{\mathrm{I}}(\lambda) : L(\mu)] \neq 0$ for $\lambda, \mu \in \Lambda_{\mathrm{I}}^{+}$?

- Is there an analogue of the Jantzen filtration for $M_{\mathrm{I}}(\lambda)$?

- Can the radical and socle series layers of $M_{\mathrm{I}}(\lambda)$ be determined explicitly? Is $M_{\mathrm{I}}(\lambda)$ rigid?

9.10. Maps between Parabolic Verma Modules

To study the possible module homomorphisms $M_{\mathrm{I}}(\mu) \to M_{\mathrm{I}}(\lambda)$ with λ, μ in Λ_{I}^{+}, it is natural to begin with the associated Verma modules. We know that $\operatorname{Hom}_{\mathcal{O}}\big(M(\mu), M(\lambda)\big) \neq 0$ just when $\mu \uparrow \lambda$, in which case this space has dimension 1. Moreover, any nonzero homomorphism of this type is injective. The case $\mathfrak{g} = \mathfrak{sl}(3, \mathbb{C})$ with $\mathrm{I} = \{\alpha\}$ studied in 9.5 shows already some complications in the parabolic setting: there is no nonzero map from $M_{\mathrm{I}}(s_{\beta}s_{\alpha} \cdot \lambda)$ to $M_{\mathrm{I}}(\lambda)$, while the nonzero map $M_{\mathrm{I}}(s_{\beta} \cdot \lambda) \to M(\lambda)$ has a nontrivial kernel.

In spite of these potential problems, something can always be done when $\lambda, \mu \in \Lambda_{\mathrm{I}}^{+}$ and $\mu \uparrow \lambda$. Start with an embedding $\varphi : M(\mu) \to M(\lambda)$, which is unique up to a nonzero scalar. Then compose φ with the natural projection $\pi : M(\lambda) \to M_{\mathrm{I}}(\lambda)$. If the image of $\pi \circ \varphi$ is nonzero, the universal property of $M_{\mathrm{I}}(\mu)$ in $\mathcal{O}^{\mathfrak{p}}$ permits us to factor this map through a nonzero homomorphism $\varphi_{\mathrm{I}} : M_{\mathrm{I}}(\mu) \to M_{\mathrm{I}}(\lambda)$. Otherwise set $\varphi_{\mathrm{I}} = 0$. The map φ_{I} is called the **standard map** associated with φ. But it may well be 0, as illustrated above when $\mathfrak{g} = \mathfrak{sl}(3, \mathbb{C})$: here the embedding $\varphi : M(s_{\beta}s_{\alpha} \cdot \lambda) \to M(\lambda)$ yields $\varphi_{\mathrm{I}} = 0$.

This notion was introduced by Lepowsky in [**202**, §3]. He observed that even when a standard map is 0, there may exist a nonzero homomorphism between the parabolic Verma modules in question. A little thought shows that this can happen only when $M(\lambda)$ has a repeated composition factor of type $L(\mu)$: indeed, when $\varphi_{\mathrm{I}} = 0$ the canonical submodule of type $M(\mu)$ in $M(\lambda)$ must be sent to 0 by the projection $M(\lambda) \to M_{\mathrm{I}}(\lambda)$. In his 1982 Yale thesis (see [**30**]) Boe refined the methods of Lepowsky. In particular, he worked out many further examples and gave precise conditions for a standard map to be 0. Combining their results, one gets:

Theorem. *Let $\lambda, \mu \in \Lambda_{\mathrm{I}}^{+}$ and consider a standard map $\varphi_{\mathrm{I}} : M_{\mathrm{I}}(\mu) \to M_{\mathrm{I}}(\lambda)$ in $\mathcal{O}^{\mathfrak{p}}$ associated with a nonzero map $\varphi : M(\mu) \to M(\lambda)$.*

- (a) *The map φ_{I} may be 0 or may fail to be injective even if it is nonzero.*

- (b) *If $\varphi_{\mathrm{I}} = 0$ but $[M(\lambda) : L(\mu)] > 1$, there might be a nonzero homomorphism between these parabolic Verma modules.*

- (c) *The map $\varphi_{\mathrm{I}} = 0$ if and only if μ is strongly linked to λ by a chain of higher weights with at least one of these not lying in Λ_{I}^{+}.*

Here are a few comments on the proof.

(a) This was already observed in connection with Example 9.5.

(b) Many examples emerge from the discussion in Lepowsky [**198**]: see pp. 225–226.

(c) Again this is illustrated in Example 9.5. Lepowsky's Proposition 3.9 in [**202**] goes in this direction, but was made more precise by Boe's Theorem 3.3 in [**30**].

As a corollary of (c), one recovers an earlier result of Lepowsky [**202**, 3.7]: if $\lambda \in \Lambda^{+}$ and $w' > w$ in the Bruhat ordering with $\ell(w') - \ell(w) = 1$, then a standard map $M_{\mathrm{I}}(w' \cdot \lambda) \to M_{\mathrm{I}}(w \cdot \lambda)$ must be nonzero. (This can be observed in Example 9.5.)

Having seen some of the obstacles that can arise in the study of maps between parabolic Verma modules, one has to approach with caution the further question:

$$\textit{What is the dimension of } \mathrm{Hom}_{\mathcal{O}}\left(M_{\mathrm{I}}(\mu), M_{\mathrm{I}}(\lambda)\right)?$$

For Verma modules this space always has dimension ≤ 1 (Theorem 4.2(b)), which is closely related to the fact that $\mathrm{Soc}\, M(\lambda)$ is always simple. As discussed further in 9.14 below, the simple socle property breaks down for parabolic Verma modules. This was first seen when λ is *singular*: an example in which $\mathrm{Soc}\, M_{\mathrm{I}}(\lambda)$ has a repeated summand $L(\mu)$ was found by

Irving [**133**, 9.6] in type D$_4$; here I $= \{\alpha\}$ with α corresponding to the vertex of the Dynkin diagram connected to all other vertices. It follows that $\dim \mathrm{Hom}_{\mathcal{O}}\big(M_{\mathrm{I}}(\mu), M_{\mathrm{I}}(\lambda)\big) > 1$.

All of the early examples worked out for *regular integral* weights produced Hom spaces of dimension ≤ 1, leading Casian–Collingwood to suggest tentatively at the end of [**61**, (4.18)] that this might be a general fact. The main result of their paper gives a computable lower bound for the dimension of $\mathrm{Hom}_{\mathcal{O}}\big(M_{\mathrm{I}}(\mu), M_{\mathrm{I}}(\lambda)\big)$ in case λ, μ are regular and integral. But a note added in proof points out an example found by Irving–Shelton [**142**, 5.3] which satisfies these conditions and has a Hom space of dimension 2; this involves the same D$_4$ set-up as above.

9.11. Parabolic Verma Modules of Scalar Type

The determination of Hom spaces between parabolic Verma modules is one of the open-ended problems arising in the study of $\mathcal{O}^{\mathfrak{p}}$. Typically the best results have been obtained under one or more restrictive hypotheses on the weights permitted (which may for example be assumed to be integral or regular or both) or on the type of parabolic subalgebra involved (for example, maximal). There has also been a lot of case-by-case study of the simple types.

Here we look at the best-studied special case. In case $\dim L_{\mathrm{I}}(\lambda) = 1$, Boe [**30**, 4.1] defines $M_{\mathrm{I}}(\lambda)$ to be of **scalar type**. In other words, $\langle \lambda, \alpha^{\vee} \rangle = 0$ for all $\alpha \in \mathrm{I}$. In particular, $L_{\mathrm{I}}(\lambda)$ is a trivial $U(\mathfrak{n}_{\mathrm{I}}^{-})$-module. Therefore $M_{\mathrm{I}}(\lambda)$ is isomorphic to $U(\mathfrak{u}_{\mathrm{I}}^{-})$ as a $U(\mathfrak{u}_{\mathrm{I}}^{-})$-module.

An obvious example is $M_{\mathrm{I}}(0)$, but there are many others. This special class of modules provides a good laboratory for the closer study of parabolic Verma modules and maps between them. Initial observations were already made by Lepowsky [**200**, Thm. 1.1], [**202**, Prop. 3.12]:

Proposition. *Let $\lambda, \mu \in \Lambda_{\mathrm{I}}^{+}$, and suppose $M_{\mathrm{I}}(\mu)$ is of scalar type. Then*

    (a) *$M_{\mathrm{I}}(\mu)$ has a unique simple submodule.*

    (b) *Any nonzero homomorphism from $M_{\mathrm{I}}(\mu)$ to $M_{\mathrm{I}}(\lambda)$ is injective. This applies in particular to any nonzero standard map.*

    (c) *If $M_{\mathrm{I}}(\lambda)$ is also of scalar type, then $\dim \mathrm{Hom}_{\mathcal{O}}\big(M_{\mathrm{I}}(\mu), M_{\mathrm{I}}(\lambda)\big) \leq 1$.*

Proof. As observed above, the assumption that $\dim L_{\mathrm{I}}(\mu) = 1$ shows that $M_{\mathrm{I}}(\mu)$ is free of rank one as a $U(\mathfrak{u}_{\mathrm{I}}^{-})$-module. The absence of zero-divisors in $U(\mathfrak{u}_{\mathrm{I}}^{-})$ then allows one to argue just as in 4.1 and 4.2. $\qquad\square$

It is quite possible in (b) to have $\varphi_{\mathrm{I}} = 0$, as seen in the example of 9.5. Part (c) is recovered by Boe [**30**, 4.5] in the special case of hermitian symmetric spaces: see 9.18 below.

Matumoto [**207, 208**] has explored in a more comprehensive way the question answered in special cases earlier: *When is there a nonzero homomorphism between two parabolic Verma modules of scalar type?* His 2006 paper includes a helpful survey of previous work on the problem, then proceeds in a mostly case-by-case fashion to classify homomorphisms. Translation functors, Jantzen's simplicity criterion, and Kazhdan–Lusztig theory all play important roles here. A unifying theme is the idea of building homomorphisms for arbitrary $\mathfrak{p}$ from "elementary" ones based on maximal parabolics.

Remark. Other work related to parabolic Verma modules of scalar type has been done by Gyoja [**118, 119**] and Marastoni [**205**]. These modules have been studied from other viewpoints as well. In particular, Oda–Oshima [**224**] have worked out good generating sets for their annihilators.

9.12. Simplicity of Parabolic Verma Modules

In category $\mathcal{O}$ there is a clearcut criterion for a Verma module $M(\lambda)$ to be simple: the highest weight λ must be *antidominant*. The proof is fairly direct for integral weights (4.4) but otherwise depends on a closer study of embeddings of Verma modules (4.8).

The question turns out to be far more complicated in the parabolic setting. But it is easy to get started by formulating a natural substitute for antidominance. Consider the following condition on a weight $\lambda \in \Lambda_{\mathrm{I}}^{+}$, which requires in effect that λ be "as antidominant as possible":

$$(*) \qquad\qquad \langle \lambda + \rho, \beta^{\vee} \rangle \notin \mathbb{Z}^{>0} \text{ for all } \beta \in \Phi^{+} \setminus \Phi_{\mathrm{I}}.$$

Theorem. *Assume $\lambda \in \Lambda_{\mathrm{I}}^{+}$.*

    (a) *If λ satisfies $(*)$, then $M_{\mathrm{I}}(\lambda)$ is simple.*

    (b) *If λ is regular and $M_{\mathrm{I}}(\lambda)$ is simple, then λ satisfies $(*)$.*

Although one might prefer to omit the regularity condition in (b), Theorem 9.13 below shows (as do Jantzen's examples) that a general simplicity criterion requires a more delicate formulation.

Proof. (a) Suppose on the contrary that $M_{\mathrm{I}}(\lambda)$ has a composition factor $L(\mu)$ with $\mu < \lambda$ and $\mu \in \Lambda_{\mathrm{I}}^{+}$. Thanks to the character formula (Proposition 9.6), $L(\mu)$ must occur as a composition factor of some $M(w \cdot \lambda)$, $w \in W_{\mathrm{I}}$. If $w \neq 1$, it is easy to see (Exercise 9.2) that $w \cdot \lambda \notin \Lambda_{\mathrm{I}}^{+}$. In particular, $\mu < w \cdot \lambda$; this is also true if $w = 1$.

The BGG Theorem 5.1, applied to $M(w \cdot \lambda)$, implies that μ is obtained from $w \cdot \lambda$ by a sequence of downward reflections. Let s_γ with $\gamma > 0$ be the first of these for which $\gamma \notin \Phi_I^+$ (it must exist since $\mu \in \Lambda_I^+$), while $x \in W_I$ is the product of the previous reflections with w. Now $\mu \leq s_\gamma x \cdot \lambda < x \cdot \lambda$, forcing $\langle x(\lambda + \rho), \gamma^\vee \rangle \in \mathbb{Z}^{>0}$. In turn, $\langle \lambda + \rho, (x^{-1}\gamma)^\vee \rangle \in \mathbb{Z}^{>0}$.

Now x is a product of simple reflections s_α with $\alpha \in I$. Since each s_α takes positive roots other than α to positive roots (0.3(1)), it follows that x^{-1} takes $\Phi^+ \setminus \Phi_I$ to itself. So the root $\beta := x^{-1}\gamma$ violates assumption $(*)$.

(b) To simplify the notation, we assume here that all weights are *integral*. The proof again involves the BGG Theorem, but now the regularity condition on λ allows us to shift the problem to the Bruhat ordering of W as in 5.2. Parametrize elements in the linkage class of λ in terms of a *dominant* weight λ_0, setting $\lambda = w \cdot \lambda_0$ for a (unique) $w \in W$. In turn, $\mu < \lambda$ will translate into $\mu = w_1 \cdot \lambda_0$ with $w_1 > w$ (hence $\ell(w_1) > \ell(w)$). By the BGG Theorem, this is equivalent to $M(\mu) \hookrightarrow M(\lambda)$.

Consider such a proper embedding. In view of Proposition 9.6, the simplicity of $M_I(\lambda)$ will then force $L(\mu)$ to be a composition factor of some $M(x \cdot \lambda)$ with $x \in W_I$ but $x \neq 1$. This translates into $w < xw \leq w_1$; unless $w_1 \in W_I w$, the last inequality is strict and $\ell(w_1) - \ell(w) \geq 2$.

Now the strategy for proving (b) is simple: Suppose there exists β in $\Phi^+ \setminus \Phi_I$ for which $(*)$ fails. Clearly $s_\beta \notin W_I$, where all reflections are with respect to roots in Φ_I. Set $w_1 := s_\beta w$. We now have $w_1 > w$ with $w_1 \notin W_I w$ as above. Take w_1 of minimal length with this property (and forget about β). If we can show that $\ell(w_1) = \ell(w) + 1$, we have contradicted the conclusion in the previous paragraph and thereby proved (b).

We just have to rule out $\ell(w_1) \geq \ell(w) + 2$. If this inequality holds, the fact that all adjacent elements in the Bruhat ordering differ in length by 1 (0.4(b)) yields some w_2 with $\ell(w_1) = \ell(w_2) + 2$ and $w_1 > w_2 \geq w$. In turn, 0.4(d) says that precisely two elements of W lie strictly between w and w_2. Since the length difference is 1, these have the form $s_\alpha w_2$ and $s_\beta w_2$ for distinct positive roots α, β (unrelated to the discarded β above).

By the minimality of w_1, we get $s_\alpha, s_\beta \in W_I$ and thus $\alpha, \beta \in \Phi_I^+$. On the other hand, the definition of the ordering shows the existence of $\gamma, \delta \in \Phi^+$ with $w_1 = s_\gamma s_\alpha w_2 = s_\delta s_\beta w_2$. In turn, $s_\gamma s_\alpha = s_\delta s_\beta$, or $s_\alpha s_\beta = s_\gamma s_\delta$. Applying both sides to a typical element in the root lattice $\mathbb{Q}\Phi$, we get

$$\mathbb{Q}\alpha + \mathbb{Q}\beta = \mathbb{Q}\gamma + \mathbb{Q}\delta.$$

This forces $\gamma, \delta \in \Phi \cap \mathbb{Q}\Phi_I = \Phi_I$. Finally we reach the contradiction $w_1 = s_\gamma s_\alpha w_2 \in W_I w$. $\qquad\square$

Part (a) was found independently by Wallach [258], Conze-Berline and Duflo [73], and Jantzen [146, Lemma 2]. (A less elegant but more elementary proof, not requiring the BGG Theorem, was later devised by Jantzen [147, 1.17].) The proof of part (b) follows suggestions by Jantzen.

9.13. Jantzen's Simplicity Criterion

Jantzen's criterion for simplicity of a parabolic Verma module relies heavily on a generalization of the determinant formula for a contravariant form worked out for Verma modules in 5.8. Recall from 3.15 that any highest weight module of weight λ, such as $M_\mathrm{I}(\lambda)$, has an essentially unique nonzero contravariant form. The radical of the form is the maximal submodule and is therefore 0 precisely when the module is simple. Using this criterion for simplicity in the case of Verma modules was unnecessary, but here there seems to be no obvious substitute (without using higher-powered methods based on Kazhdan–Lusztig theory).

The generalized determinant formula is a more elaborate version of the earlier one, with a correspondingly more elaborate proof. Once this is in hand, specializing to a concrete weight $\lambda \in \Lambda_\mathrm{I}^+$ leads quickly to the simplicity criterion. In order to state the theorem and its corollaries precisely, we have to introduce more notation.

Start with the character formula in Proposition 9.6 for parabolic Verma modules:

$$(1) \qquad \operatorname{ch} M_\mathrm{I}(\lambda) = \sum_{w \in W_\mathrm{I}} (-1)^{\ell(w)} \operatorname{ch} M(w \cdot \lambda).$$

It is convenient to denote by $\theta(\lambda)$ the right side of this equation when $\lambda \in \mathfrak{h}^*$ is arbitrary. It is easy to check that $\theta(w \cdot \lambda) = (-1)^{\ell(w)}\theta(\lambda)$ for all $w \in W_\mathrm{I}$; in turn, $\theta(\lambda) = 0$ if $w \cdot \lambda = \lambda$ for some $w \neq 1$ in W_I.

Next we single out certain sets of roots:

$$\begin{aligned}
\Psi &:= \Phi \setminus \Phi_\mathrm{I}, \\
\Psi_\lambda^+ &:= \{\beta \in \Psi^+ \mid \langle \lambda + \rho, \beta^\vee \rangle \in \mathbb{Z}^{>0}\}, \\
\Phi_\beta &:= (\mathbb{Q}\Phi_\mathrm{I} + \mathbb{Q}\beta) \cap \Phi \text{ if } \beta \in \Psi, \\
\Phi_\beta^+ &:= \Phi_\beta \cap \Phi^+.
\end{aligned}$$

We also need to formulate two conditions on λ:

(M) For all $\beta \in \Psi_\lambda^+$, there is a $\gamma \in \Phi_\beta$ with $\langle \lambda + \rho, \gamma^\vee \rangle = 0$.

(M+) For all $\beta \in \Psi_\lambda^+$, there is a $\gamma \in \Phi_\beta$ with $\langle \lambda + \rho, \gamma^\vee \rangle = 0$, $s_\beta \gamma \in \Phi_\mathrm{I}$.

With all this notation in place it is easy to state (but not prove!) Jantzen's theorem [146, Satz 4].

Theorem. *Let $\lambda \in \Lambda_{\mathrm{I}}^{+}$. Then $M_{\mathrm{I}}(\lambda)$ is simple if and only if for all $\beta \in \Psi_{\lambda}^{+}$,*

$$\sum_{\gamma} \theta(s_{\gamma} \cdot \lambda) = 0,$$

where the sum is over all $\gamma \in \Phi_{\beta}^{+} \setminus \Phi_{\mathrm{I}}$ for which $\langle \lambda + \rho, \gamma^{\vee} \rangle \in \mathbb{Z}^{>0}$.

Here are four corollaries which follow from the theorem; the last of these is contained in Theorem 9.12.

Corollary. *Let $\lambda \in \Lambda_{\mathrm{I}}^{+}$. Then:*

(a) *$M_{\mathrm{I}}(\lambda)$ is simple if and only if the sum of all $\theta(s_{\beta} \cdot \lambda)$ with $\beta \in \Psi_{\lambda}^{+}$ is 0.*

(b) *If $M_{\mathrm{I}}(\lambda)$ is simple, then condition (M) holds.*

(c) *If condition (M+) holds, then $M_{\mathrm{I}}(\lambda)$ is simple.*

(d) *Let λ be regular. Then $M_{\mathrm{I}}(\lambda)$ is simple if and only if $\Psi_{\lambda}^{+} = \emptyset$.*

In [**146**, Satz 4] Jantzen obtains a streamlined criterion when all components of Φ are of type A: *$M_{\mathrm{I}}(\lambda)$ is simple if and only if condition (M+) holds for λ.*

The reader might find it an interesting project to apply these criteria to a few concrete low rank cases in which λ is singular.

9.14. Socles and Self-Dual Projectives

In category $\mathcal{O}$ we found that $\operatorname{Soc} M(\lambda)$ is simple for all $\lambda \in \mathfrak{h}^{*}$. Moreover, the highest weight of this socle is the unique antidominant weight in the $W_{[\lambda]}$-orbit of λ (under the dot action). Nothing quite so straightforward can be said in the parabolic case. As remarked in 9.10 above, examples found in the 1980s showed that socles of parabolic Verma modules need not be simple even when the highest weights involved are well-behaved. In general the structure of $\operatorname{Soc} M_{\mathrm{I}}(\lambda)$ seems to be quite difficult to work out. But in one case it is easy to prove simplicity (following Irving [**133**, 4.1]):

Proposition. *Let $\lambda \in \mathfrak{h}^{*}$ be dominant and regular. If $\lambda \in \Lambda_{\mathrm{I}}^{+}$, then $\operatorname{Soc} M_{\mathrm{I}}(\lambda)$ is simple.*

Proof. To simplify notation we assume $\lambda \in \Lambda^{+}$ (but the result is general).

Look first at the special case when $\lambda = 0$ on $\mathfrak{h}_{\mathrm{I}}$: that is, $\langle \lambda, \alpha^{\vee} \rangle = 0$ for all $\alpha \in \mathrm{I}$. Then $M_{\mathrm{I}}(\lambda)$ is of scalar type, so its socle is simple by Proposition 9.11(a).

In general, define $\nu = \sum_{\alpha \in \Delta} c_{\alpha} \varpi_{\alpha}$ to be the unique weight for which $c_{\alpha} = \langle \lambda, \alpha^{\vee} \rangle$ when $\alpha \in \mathrm{I}$ and $c_{\alpha} = 0$ when $\alpha \notin \mathrm{I}$. Clearly $\mu := \lambda - \nu \in \Lambda^{+}$ and thus lies in Λ_{I}^{+}. It vanishes on $\mathfrak{h}_{\mathrm{I}}$, so $\operatorname{Soc} M_{\mathrm{I}}(\mu)$ is simple. On the other hand, λ and μ both lie inside the dominant W-chamber. This implies that T_λ^μ and T_μ^λ are inverse category equivalences (7.8); they interchange $M_{\mathrm{I}}(\lambda)$ and $M_{\mathrm{I}}(\mu)$. Such an equivalence preserves socles, so $\operatorname{Soc} M_{\mathrm{I}}(\lambda)$ is also simple. $\qquad\square$

It remains a problem here to specify the highest weight of the socle. The example for $\mathfrak{g} = \mathfrak{sl}(3,\mathbb{C})$ in 9.5 already illustrates the potential difficulty of identifying this weight in advance when $\lambda \in \Lambda_{\mathrm{I}}^+$ is specified.

Irving went on to raise a further question: *Which projective modules $P_{\mathrm{I}}(\lambda)$ in $\mathcal{O}^{\mathfrak{p}}$ are self-dual?* In category $\mathcal{O}$ the fact that only the antidominant simple module in a block can occur as the socle of a Verma module singled out this class of simple modules for special attention. It was shown in 4.10 for integral weights and in 7.16 for more general weights that the antidominant simples $L(\lambda)$ are precisely the heads (and socles) of self-dual projectives $P(\lambda)$.

In a similar spirit, Irving [133] formulated a reasonable conjecture and proved it in special cases. He defines $\lambda \in \mathfrak{h}^*$ to be a **socular weight** if $L(\lambda)$ occurs as a summand in the socle of some parabolic Verma module. The conjecture is that $P_{\mathrm{I}}(\lambda)$ is self-dual if and only if λ is socular. In an Addendum to that paper, following suggestions of D. Garfinkle, he filled in the outline of a general proof of this conjecture:

Theorem. *Let* $\lambda \in \Lambda_{\mathrm{I}}^+$. *Then* $P_{\mathrm{I}}(\lambda)$ *is self-dual if and only if* λ *is socular.*

The proof requires a lot of machinery, including wall-crossing functors (7.15) as well as one notion not yet introduced here which plays a crucial role in the study of primitive ideals in $U(\mathfrak{g})$: the *Gelfand–Kirillov dimension* of a module (13.3 below). In the proof of the theorem a key step is the characterization of socular simple modules as those of largest possible Gelfand–Kirillov dimension.

Exercise. In the case $\mathfrak{g} = \mathfrak{sl}(3,\mathbb{C})$ treated in 9.5, verify the criterion in the theorem.

9.15. Blocks of $\mathcal{O}^{\mathfrak{p}}$

Earlier we determined the blocks of $\mathcal{O}$: those involving just integral weights are the subcategories $\mathcal{O}_\chi$ (1.13), while those involving nonintegral weights require a more refinemed partition of the simple modules in $\mathcal{O}_\chi$ (4.9). Recall that we place two simple modules in the same block if they extend nontrivially; the smallest equivalence relation generated by this relation partitions the simple modules into blocks. In turn, an indecomposable module lies in a block provided all its composition factors do.

What can be said about $\mathcal{O}^{\mathfrak{p}}$? Since $\mathcal{O}^{\mathfrak{p}} = \bigoplus_{\chi} \mathcal{O}^{\mathfrak{p}}_{\chi}$, the subcategories $\mathcal{O}^{\mathfrak{p}}_{\chi}$ again provide at least a first approximation to blocks. For arbitrary $\mathfrak{p}$ the precise block decomposition of $\mathcal{O}^{\mathfrak{p}}_{\chi}$ has not yet been worked out completely, so we focus mainly on *integral* weights to keep the notation simple. (As in 1.13, the reader is cautioned to be aware of differing meanings assigned to the term "block" in the literature: in some papers "block" simply means a subcategory $\mathcal{O}^{\mathfrak{p}}_{\chi}$ of $\mathcal{O}^{\mathfrak{p}}$.)

It is easy to get started in the search for blocks lying in $\mathcal{O}^{\mathfrak{p}}_{\chi}$.

Proposition. *Given $\mathfrak{p}$ and a central character χ, each block $\mathcal{B}$ in $\mathcal{O}^{\mathfrak{p}}_{\chi}$ must contain at least one simple parabolic Verma module.*

Proof. Consider the finite partially ordered set of weights $\lambda \in \Lambda^{+}_{\mathrm{I}}$ for which $\chi = \chi_{\lambda}$ and $L(\lambda)$ lies in $\mathcal{B}$. If λ is *minimal* in this set, then no composition factor $L(\mu)$ of $M_{\mathrm{I}}(\lambda)$ with $\mu < \lambda$ can satisfy $\mu \in \Lambda^{+}_{\mathrm{I}}$. Thus $M_{\mathrm{I}}(\lambda)$ is simple and lies in $\mathcal{B}$. $\qquad\square$

As a result, the simplicity criteria discussed in 9.12–9.13 limit the possible blocks. Here is a summary of special cases where the blocks have been determined.

Examples. (1) There is one easy case: when $\chi = \chi_{\lambda}$ with λ *regular integral*, then $\mathcal{O}^{\mathfrak{p}}_{\chi}$ contains only one simple parabolic Verma module (Theorem 9.12) and is therefore already a block.

(2) Enright–Shelton [**92**, Cor. 12.14] explain in a particular (singular!) case how a category $\mathcal{O}^{\mathfrak{p}}_{\chi}$ splits into two blocks. (Their terminology is entirely different from ours, however.) This occurs for a Hermitian symmetric pair (see 9.18 below).

(3) In [**51**] Brundan studies the case $\mathfrak{g} = \mathfrak{gl}(n, \mathbb{C})$, whose representation theory is close to that of $\mathfrak{sl}(n, \mathbb{C})$ but is enriched by the presence of a center. His techniques are mostly special to this case, where partitions of n provide a natural parametrization of weights, and rely especially on the degenerate cyclotomic Hecke algebra attached to a related complex reflection group. In the introductory discussion following the statement of his Theorem 2, he explains how it implies for arbitrary integral weights that the blocks of $\mathcal{O}^{\mathfrak{p}}$ for $\mathfrak{gl}(n, \mathbb{C})$ coincide with the categories $\mathcal{O}^{\mathfrak{p}}_{\chi}$.

(4) Boe and his student K. Platt have worked out further special cases, in the context of the determination of the *representation type* of blocks: finite, tame, or wild. This builds on recent work discussed in 13.8 below.

9.16. Analogue of the BGG Resolution

In the special case when $\lambda \in \Lambda^{+} \subset \Lambda^{+}_{\mathrm{I}}$, it is easy to derive an analogue in $\mathcal{O}^{\mathfrak{p}}$ of the classical Weyl–Kostant character formula (2.4) for $L(\lambda)$, as indicated in Exercise 9.6. Start with the alternating sum formula

$$\operatorname{ch} L(\lambda) = \sum_{w \in W} (-1)^{\ell(w)} \operatorname{ch} M(w \cdot \lambda).$$

Then group the terms in the sum according to the right cosets $W_{\mathrm{I}} \backslash W$; an alternating sum over W_{I} occurs in each partial sum. Finally use Proposition 9.6 to obtain the rewritten formula

$$(1) \qquad \operatorname{ch} L(\lambda) = \sum_{w \in W^{\mathrm{I}}} (-1)^{\ell(w)} \operatorname{ch} M_{\mathrm{I}}(w \cdot \lambda).$$

Here the sum is over the set W^{I} of minimal length right coset representatives in $W_{\mathrm{I}} \backslash W$. For example, when $\mathfrak{g} = \mathfrak{sl}(3, \mathbb{C})$ and $\mathrm{I} = \{\alpha\}$ as in 9.5, we get

$$\operatorname{ch} L(\lambda) = \operatorname{ch} M_{\mathrm{I}}(\lambda) - \operatorname{ch} M_{\mathrm{I}}(s_\beta \cdot \lambda) + \operatorname{ch} M_{\mathrm{I}}(s_\beta s_\alpha \cdot \lambda).$$

As in Chapter 6 it is natural to raise the question of realizing (1) as the Euler character of a resolution of $L(\lambda)$ by parabolic Verma modules. The analogue of the BGG resolution is given as follows.

Theorem. *Let $\lambda \in \Lambda^+$ and let r be the greatest length of any element in W^{I}. Then there is an exact sequence*

$$0 \to C_r^{\mathrm{I}} \to \cdots \to C_1^{\mathrm{I}} \to C_0^{\mathrm{I}} = M_{\mathrm{I}}(\lambda) \to L(\lambda) \to 0,$$

where

$$C_k^{\mathrm{I}} := \bigoplus_{w \in W^{\mathrm{I}}, \, \ell(w) = k} M_{\mathrm{I}}(w \cdot \lambda).$$

Moreover, the map $C_k^{\mathrm{I}} \to C_{k-1}^{\mathrm{I}}$ is nonzero on each summand $M_{\mathrm{I}}(w \cdot \lambda)$, where it is a standard map.

In the example $\mathfrak{g} = \mathfrak{sl}(3, \mathbb{C})$ with $\mathrm{I} = \{\alpha\}$, the resolution looks like

$$0 \to L(s_\beta s_\alpha \cdot \lambda) = M_{\mathrm{I}}(s_\beta s_\alpha \cdot \lambda) \to M_{\mathrm{I}}(s_\beta \cdot \lambda) \to M_{\mathrm{I}}(\lambda) \to L(\lambda) \to 0.$$

The proof follows essentially the same lines as the proof for category $\mathcal{O}$. Inspired by an earlier joint paper with Garland, which led to a weak type of BGG resolution in a more general Kac–Moody setting, Lepowsky [**202**] was able to construct both weak and strong versions of the above "relative" resolution by adapting the original BGG arguments. Soon afterwards Rocha [**226**] streamlined the proof, showing the coincidence of weak and strong resolutions. This was done by first investigating extensions of parabolic Verma modules. As in the case of Verma modules, the key point is that $\operatorname{Ext}_{\mathcal{O}^{\mathfrak{p}}}\big(M_{\mathrm{I}}(w \cdot \lambda), M_{\mathrm{I}}(w' \cdot \lambda)\big) = 0$ if $\ell(w) = \ell(w')$.

9.17. Filtrations and Rigidity

As in category $\mathcal{O}$, it is natural to ask for more detailed structural information about parabolic Verma modules as well as projectives in $\mathcal{O}^{\mathfrak{p}}$. But the results on $\mathcal{O}$ summarized in 8.14–8.15 are strongly dependent on the KL Conjecture in Vogan's form (or even Jantzen's Conjecture). So one cannot expect easy answers for $\mathcal{O}^{\mathfrak{p}}$, where even the socles of Verma modules are elusive.

In spite of this, there are a number of positive results, especially on Loewy length, which generalize in a reasonable way the results for $\mathcal{O}$. The proofs tend to follow a similar path, relying especially on the way Loewy length changes under wall-crossing functors. While some steps are elementary, the most definitive statements rely on the deeper ideas outlined in Chapter 8. Most of the relevant work is due to Irving and his collaborators Collingwood and Shelton, in papers published from 1985 to 1990. Here is a brief overview of some of the main points:

- Irving's papers [**134**] and [**135**] deal mainly with category $\mathcal{O}$, as discussed in 8.15. The emphasis is on computing Loewy lengths for Verma modules and their projective covers, using wall-crossing functors and Vogan's form of the KL Conjecture. Special attention is paid to working out Loewy layers for the self-dual projective in a block. This suggests a similar description for self-dual projectives in $\mathcal{O}^{\mathfrak{p}}$, which are classified by "socular" weights (see 9.14).

- Irving–Shelton [**142**] focus on the use of *simple projective* modules in $\mathcal{O}^{\mathfrak{p}}$, in tandem with wall-crossing functors, to pass from singular weights to regular weights. This process in $\mathcal{O}$ is effective (say for integral weights) since $L(-\rho) = P(-\rho)$. In particular, simple projectives always exist in type A_ℓ, leading to explicit Loewy length formulas. But the D_4 example which leads to a non-simple socle for a parabolic Verma module also leads to non-existence of a simple projective module.

  In a related paper, Collingwood–Irving–Shelton [**70**] work out the ideas further for a special class of maximal parabolic subalgebras (discussed below in 9.18).

- In a longer paper (modified in the separate Errata), Irving [**137**] studies more broadly a filtered version of category $\mathcal{O}^{\mathfrak{p}}$. This requires further development of the earlier Hecke module formalism introduced by Deodhar and Casian–Collingwood (9.7). Here there are four natural bases, corresponding to four types of modules in $\mathcal{O}^{\mathfrak{p}}$: simple, parabolic Verma, projective covers, and self-dual modules filtered by parabolic Vermas (which reappear below in Chapter

11). This leads to "filtered" characters, which correspond to module filtrations described in §§7–9. The end results here are typically less definitive than those in $\mathcal{O}$. For example, it is shown in Proposition 7.4.1 that a parabolic Verma module (with regular integral highest weight) is rigid provided its socle is simple.

At first Irving had hoped to avoid invoking the deeper geometric ideas in KL theory, but in the Errata he notes some problems with the details leading up to his main results and suggests how to recover most of them in the framework of geometric methods introduced by Beilinson–Ginzburg.

As this quick summary indicates, there are still many uncertainties about Loewy structure for an arbitary category $\mathcal{O}^{\mathfrak{p}}$ as well as about the tools required.

Remark. Unlike the study of filtrations in $\mathcal{O}$, there is no parallel role here for a "Jantzen filtration" in parabolic Verma modules. But Jantzen does introduce such a filtration in [**146**] while developing his determinant formula for the contravariant form. This implicitly produces a "sum formula" as well.

On the other hand, it is possible to construct a filtration in $M_{\mathrm{I}}(\lambda)$ simply by applying the canonical map $M(\lambda) \to M_{\mathrm{I}}(\lambda)$ to the submodules $M(\lambda)^i$. But it is unclear how to understand this filtration intrinsically or relate it to Jantzen's construction.

9.18. Special Case: Maximal Parabolic Subalgebras

To round out the chapter we take a brief look at the special case when $\mathfrak{p}$ is a *maximal* (proper) subalgebra of $\mathfrak{g}$. The results here are often obtained in case-by-case fashion and are both simpler and richer in detail than one can expect for arbitrary categories $\mathcal{O}^{\mathfrak{p}}$. For example, in some categories $\mathcal{O}^{\mathfrak{p}}_{\chi}$ with $\mathfrak{p}$ maximal the parabolic Verma modules turn out to have no repeated composition factors.

The literature cited here is fairly representative, but not exhaustive. The work of Boe, Casian, Collingwood, Enright, Irving, Lepowsky, and Shelton has been especially influential.

(A) One case which has been well-studied (starting with papers of Lepowsky) involves a simple matrix Lie group G of *real rank one*: here $\mathfrak{g}$ is the complexified Lie algebra of G, while $\mathfrak{p}$ is the complexified Lie algebra of a suitable maximal parabolic subgroup. There are only a few cases of this type; in a standard classification, with G simply connected, these are labelled $\mathrm{Spin}(n,1)$, $\mathrm{SU}(n,1)$, $\mathrm{Sp}(n,1)$, $\mathrm{F}_{4(-20)}$. In his lecture notes Collingwood [**66**] works out carefully the connections between Lie group representations and categories of $U(\mathfrak{g})$-modules.

A starting point for the papers of Boe–Collingwood [**32, 33**] is the question: Can one generalize to the parabolic setting the BGG Theorem 5.1, relating maps between parabolic Verma modules to the existence of composition factors? As already discussed in 9.10, there are significant obstacles when $\mathfrak{p}$ is arbitrary. But in real rank one the authors are able to describe all Hom spaces between parabolic Verma modules; this generalizes Lepowsky's results for scalar type. (Gyoja [**120**] provides some follow-up to [**32**].)

In their second paper, which is largely independent of the first and recovers some of its results, they show that in fact all composition factors of parabolic Verma modules have multiplicity one; but these do not correlate precisely with homomorphisms. Using KL theory they list explicitly the composition factors for each real rank one type. An interesting consequence is that the number of factors in each case is bounded independently of the rank of $\mathfrak{g}$. (In all of this work the weights involved are assumed to be regular and integral.)

(B) Maximal parabolic subalgebras of another special type occur in the theory of *Hermitian symmetric pairs* $(\mathfrak{g}, \mathfrak{p})$. These arise from the study of Hermitian symmetric spaces G/P, which are interesting manifolds of relatively low dimension. There are five families of indecomposable classical Hermitian symmetric pairs, which are listed in [**92**, Table 8.1] according to the root systems of $\mathfrak{g}$ and $\mathfrak{p}$:

$$(A_\ell, A_{k-1} \times A_{\ell-k}), \ (B_\ell, B_{\ell-1}), \ (C_\ell, A_{\ell-1}), \ (D_\ell, D_{\ell-1}), \ (D_\ell, A_{\ell-1}).$$

There are also two exceptional pairs (E_7, E_6) and (E_6, D_5). The latter are studied in detail by Collingwood [**67**].

Enright–Shelton [**92**] go on to consider the indecomposable classical Hermitian symmetric pairs. After laying some general groundwork in Part I, they compute case-by-case for Hermitian symmetric pairs the precise composition factors of parabolic Verma modules in the "principal block" (where highest weights have the form $w \cdot 0$). All composition factors here occur with multiplicity one (see also Boe–Collingwood [**34**]). The computation is done by concrete methods, avoiding KL theory, which permits the recovery of the relevant KL polynomials here. From this data they also get explicit combinatorial formulas for dimensions of the $\mathrm{Ext}^i_{\mathcal{O}^\mathfrak{p}}\left(M_I(x \cdot 0), L(w \cdot 0)\right)$ (in our notation) for minimal coset representatives in W^I.

Shelton [**232**] follows up this work by studying the Ext^i spaces between pairs of parabolic Verma modules, again for regular integral weights. He obtains recursion relations for their dimensions, which are explored further by Biagioli [**28**] in connection with Deodhar's parabolic version of R-polynomials.

Collingwood–Irving–Shelton [**70**] go further with the principal block, studying filtrations on parabolic Verma modules. In this special situation they can prove a rigidity theorem, specify the composition factors which occur in the Loewy layers, and compute the Loewy lengths. They show moreover that the single Loewy filtration here coincides with Gabber's weight filtration (see 8.13).

(C) In a two-part paper, Boe–Collingwood [**35**] look more broadly at pairs $(\mathfrak{g}, \mathfrak{p})$ for which $\mathfrak{p}$ is a maximal parabolic. The Hecke module approach (9.7) figures here, along with a study of P-orbit closures in the flag variety G/B when G and P have respective Lie algebras $\mathfrak{g}, \mathfrak{p}$.

They first classify completely those pairs $(\mathfrak{g}, \mathfrak{p})$ whose regular integral blocks are *multiplicity free* in the sense that the composition factor multiplicities of all parabolic Verma modules are 1. The list comprises Hermitian symmetric pairs (and close relatives), real rank one pairs (and their close relatives), several further infinite families, and a handful of special cases in low ranks.

All multiplicity free cases turn out to share nice properties, observed earlier in some of the special cases: (1) All parabolic Verma modules are rigid and admit closed formulas for the simples occurring in Loewy layers. (2) All Hom spaces between parabolic Verma modules have dimension ≤ 1; the dimension one cases can then be specified in terms of the "Gabber lattice" introduced by Casian–Collingwood [**61**]; so their Conjecture 4.3 is verified here.

Notes

Category $\mathcal{O}^{\mathfrak{p}}$ was first studied systematically by Rocha [**226**].

Foundations for the study of parabolic Verma modules are discussed in detail by Jantzen [**147**, 1.14–1.20], who works in somewhat greater generality than we do. The widely used terminology "generalized Verma module" arose in a number of early papers such as the one by Conze-Berline and Duflo [**73**].

Khomenko, Mazorchuk, and others study "generalized Verma modules" in a wider sense: here one induces from simple (possibly infinite dimensional) $\mathfrak{l}_I$-modules after inflating to $\mathfrak{p}$. (Many of their papers can be found in the references.)

The truncation functor $\mathcal{O} \to \mathcal{O}^{\mathfrak{p}}$ defined in 9.3 was introduced by Irving [**137**, §5] under the name "residue". He states that it is an exact functor; in the Errata he corrects this but points out that exactness does hold for those short exact sequences involved in his arguments.

Chapter 10

Projective Functors and Principal Series

This chapter deals with the *projective functors* defined by Joseph Bernstein and Sergei Gelfand [**24**], together with the related category of principal series Harish-Chandra modules for a complex semisimple Lie group having $\mathfrak{g}$ as Lie algebra. We do not attempt to reproduce all of the somewhat intricate arguments in [**24**], where the setting is more general than category $\mathcal{O}$; instead we try to motivate and explain their main theorems, indicating connections to translation functors.

The following chapters will introduce further endofunctors of $\mathcal{O}$ and related new objects. To put all of this in perspective, we start off in 10.1 with an overview of questions arising in the study of such functors.

One main theme of [**24**] is the fine structure of functors related to tensoring with finite dimensional modules. Some essential ideas are gathered in 10.2–10.3. After defining relevant module categories in 10.4, we introduce the formal properties of projective functors in 10.5. Examples include the translation functors studied in Chapter 7. The classification of indecomposable projective functors is then explained in 10.7–10.8.

Underlying much of this work is the broader goal of understanding the category of Harish-Chandra modules for $\mathfrak{g}$, which carry essential information about Lie group representations. In 10.9 we provide a brief overview of the relevant ideas here. It turns out that category $\mathcal{O}$ is closely related to the subcategory of "principal series" modules (10.10). (This will play a further role in Chapter 12.) The work of Bernstein–Gelfand predates the Kazhdan–Lusztig Conjecture; indeed, their paper was motivated in part by the attempt to understand the multiplicity problem for Verma modules in the wider context of Harish-Chandra modules.

Notational conventions. In earlier chapters we usually emphasized *antidominant* weights: the corresponding Verma modules are simple and thus form a natural starting point for the study of other simple modules. But here the emphasis is placed instead on *dominant* weights (relative to the dot-action of W), which can equally well be used to parametrize blocks $\mathcal{O}_\lambda$ (1.13–4.9). This shift is motivated by the fact that dominant Verma modules are *projective*. In particular, Bernstein–Gelfand [**24**] take this as their starting point.

To be consistent with our own previous conventions, we depart in many ways from the notation of Bernstein–Gelfand. For example, they parametrize a module of highest weight λ by $\lambda + \rho$ and use the usual rather than the dot-action of W.

10.1. Functors on Category $\mathcal{O}$

Though we took a few detours along the way, the unifying goal of Chapters 1–8 has been the determination of formal characters of simple modules in $\mathcal{O}$. Since 1980 a number of other problems and applications involving $\mathcal{O}$ have attracted attention; but there is less unity to the resulting literature, which reflects a variety of agendas.

Most of the early work in $\mathcal{O}$ focused on special objects: highest weight modules or others (such as projectives) closely related to these. Recent work has called attention to other families of modules, sometimes motivated by ideas originating elsewhere in representation theory. Reinforcing the study of specific modules in $\mathcal{O}$ is the study of numerous endofunctors, which in some cases preserve blocks but in other cases relate one block to another. We have already seen a number of examples: duality, tensoring with finite dimensional modules, translation functors, and wall-crossing functors. Endofunctors are often related closely to the way families of modules are constructed and filtered (or graded). Some of the later work builds on the earlier study of Jantzen filtrations of Verma modules or standard filtrations in other modules.

Most of the ideas introduced here can be developed up to a point in the algebraic framework used in earlier chapters. But gradually the methods of "modern" homological algebra (as codified in Gelfand–Manin [**110**]) become indispensable, especially the notion of derived category. Connections with algebraic geometry and with the representation theory of Lie groups also become more prominent. Here are some typical questions about a functor $\mathcal{F}$ on $\mathcal{O}$.

- What basic properties does $\mathcal{F}$ have: left or right exactness? preservation of blocks $\mathcal{O}_\lambda$? interaction with other functors such as duality or tensoring with finite dimensional modules?

- Is there an axiomatic characterization of $\mathcal{F}$?

- If a functor $\mathcal{F}_s$ is initially defined for each simple reflection s, do the composites relative to different reduced expressions for $w \in W$ agree (braid relations)?

- If $\mathcal{F}$ is exact, what can be said about the homomorphism induced by $\mathcal{F}$ on $K(\mathcal{O})$ or the induced functor on the bounded derived category $D^b(\mathcal{O})$?

- How is $\mathcal{F}$ constructed (or its existence proved)?

- What does $\mathcal{F}$ do to "standard" modules: Verma modules and their duals, simple modules, projectives, etc.? This is an essential test question for any functor introduced.

- What applications does $\mathcal{F}$ have to problems in representation theory, geometry or topology, etc.?

When $\mathcal{F}$ is *exact*, some questions are simpler to deal with. In particular, part (c) of the following proposition will be helpful later on in the chapter.

Proposition. *Let $\mathcal{F}$ be an exact functor on $\mathcal{O}$ and consider its restriction to a single block $\mathcal{B}$.*

(a) *If $\mathcal{F}$ vanishes on all simple modules in $\mathcal{B}$, then $\mathcal{F}M = 0$ for all $M \in \mathcal{B}$.*

(b) *If $\mathcal{F}M = 0$ for a module $M \in \mathcal{B}$, then $\mathcal{F} = 0$ on every composition factor of M.*

(c) *If $\mathcal{F}M(\lambda) = 0$ when $M(\lambda) \in \mathcal{B}$ with λ dominant, then $\mathcal{F} = 0$ on $\mathcal{B}$.*

Proof. (a) This is proved by a straightforward induction on the length of M. The hypothesis takes care of length 1. For the induction step, let $L(\lambda)$ be any simple quotient of M and consider the short exact sequence

$$0 \to N \to M \to L(\lambda) \to 0.$$

By induction, $\mathcal{F}N = 0 = \mathcal{F}L(\lambda)$, so exactness forces $\mathcal{F}M = 0$.

(b) Use induction on the length of M. If the length exceeds 1, proceed as in step (a), using a short exact sequence

$$0 \to N \to M \to L(\lambda) \to 0.$$

By assumption $\mathcal{F}M = 0$, so exactness again forces $\mathcal{F}N = 0 = \mathcal{F}L(\lambda)$. All composition factors of M other than $L(\lambda)$ occur in N, so we can appeal to the induction hypothesis.

(c) Thanks to part (a), it is enough to show that $\mathcal{F}$ vanishes at all simple modules in $\mathcal{B}$. The hypothesis coupled with part (b) shows that $\mathcal{F}$ vanishes at all composition factors of $M(\lambda)$. Finally, we appeal to Proposition 4.3 (extended to nonintegral weights in Example 5.1): because λ is *dominant*, we have $[M(\lambda) : L(\mu)] > 0$ for all simple modules $L(\mu) \in \mathcal{B}$. $\square$

10.2. Tensoring With a Dominant Verma Module

In preparation for the study of "projective functors", we consider an important special case of tensoring with a finite dimensional module. The viewpoint is somewhat different from that in Chapter 7.

Here we shall make essential use of Remark 3.9: *Let* $\lambda, \mu \in \mathfrak{h}^*$, *with* λ *dominant, and suppose* $\dim L < \infty$. *Then*

$$(*) \qquad \dim \mathrm{Hom}_{\mathcal{O}}\big(M(\lambda) \otimes L, M(\mu)\big) = \dim L_{\mu-\lambda}.$$

The proof was based on the fact that $M(\lambda)$ is projective when λ is dominant, which in turn implies that $M(\lambda) \otimes L$ is projective (Proposition 3.8). In this situation, it is reasonable to ask (for fixed λ but variable L) which indecomposable projectives $P(\mu)$ can occur as direct summands of such tensor products. An obvious necessary condition is that λ and μ be compatible: $\mu - \lambda \in \Lambda$ (which is automatic when all weights are integral).

Theorem. *Let* $\lambda, \mu \in \mathfrak{h}^*$ *be compatible, with* λ *dominant. Then the following conditions are equivalent:*

(a) *There exists a finite dimensional module* L *for which* $P(\mu)$ *is isomorphic to a direct summand of* $M(\lambda) \otimes L$.

(b) *There exists a finite dimensional module* L *for which*

$$\mathrm{Hom}_{\mathcal{O}}\big(M(\lambda) \otimes L, L(\mu)\big) \neq 0.$$

(c) $\langle \mu + \rho, \alpha^\vee \rangle \leq 0$ *whenever* $\alpha > 0$ *and* $\langle \lambda + \rho, \alpha^\vee \rangle = 0$.

When these conditions are satisfied, L *can be chosen so that* $P(\mu)$ *occurs just once as a summand of the tensor product.*

Observe first that (a) and (b) are equivalent, thanks to Theorem 3.9(b).

Note that (c) makes sense, since $\alpha \in \Phi_{[\lambda]} = \Phi_{[\mu]}$. The condition on α means that $s_\alpha \in W_\lambda^\circ$. So (c) is automatically true in case λ is *regular*, that is, $W_\lambda^\circ = 1$. At the other extreme is the case $\lambda = -\rho$, which was treated earlier as an isolated example (4.10): here μ must be *antidominant*. The technique used there will resurface here in the proof that (c) implies (a): by choosing $L = L(\nu)$ with $\nu \in \Lambda^+$ in the W-orbit of $\mu - \lambda$, we will see that $P(\mu)$ occurs precisely once as a summand of the tensor product in (a).

Example. To understand what the theorem says, it is useful to consider the *integral* weights for $\mathfrak{sl}(3, \mathbb{C})$. Here there are three possibilities: the two extreme cases just indicated along with the intermediate case when λ lies in only one of the two hyperplanes bounding the dominant chamber (shifted as usual by $-\rho$). The condition on μ is that it lies on or below this hyperplane.

10.3. Proof of the Theorem

The proof of Theorem 10.2 involves a number of steps, starting with the easier implication.

(1) We want to show that (b) implies (c), or equivalently, that the failure of (c) implies the failure of (b). Accordingly, we are given $\alpha > 0$ for which $\langle \lambda + \rho, \alpha^\vee \rangle = 0$ while $\langle \mu + \rho, \alpha^\vee \rangle > 0$. This condition on μ and α ensures (by Verma's Theorem 4.6) that $M(s_\alpha \cdot \mu)$ embeds properly in $M(\mu)$. In turn, whenever $\dim L < \infty$ we get an embedding

$$\mathrm{Hom}_{\mathcal{O}}\left(M(\lambda) \otimes L, M(s_\alpha \cdot \mu) \right) \hookrightarrow \mathrm{Hom}_{\mathcal{O}}\left(M(\lambda) \otimes L, M(\mu) \right).$$

We claim this is a bijection, for which it suffices to show that the dimensions of the Hom spaces agree. This will follow from $(*)$ above and the assumption that $s_\alpha \cdot \lambda = \lambda$: the Hom space on the left side has dimension equal to

$$\dim L_{s_\alpha \cdot \mu - \lambda} = \dim L_{s_\alpha \cdot \mu - s_\alpha \cdot \lambda} = \dim L_{s_\alpha(\mu - \lambda)} = \dim L_{\mu - \lambda},$$

since W-conjugates of a weight of a finite dimensional module have the same multiplicity. But by $(*)$ again, this last dimension is the dimension of the Hom space on the right side.

(2) Because $P := M(\lambda) \otimes L$ is projective, $\mathrm{Hom}_{\mathcal{O}}(P, ?)$ is exact. If we set $Q := M(\mu)/M(s_\alpha \cdot \mu)$ and consider the short exact sequence

$$0 \to M(s_\alpha \cdot \mu) \to M(\mu) \to Q \to 0,$$

it follows that $\mathrm{Hom}_{\mathcal{O}}\left(M(\lambda) \otimes L, Q \right) = 0$. But $L(\mu)$ is a quotient of Q, forcing $\mathrm{Hom}_{\mathcal{O}}\left(M(\lambda) \otimes L, L(\mu) \right) = 0$ (again by the projective property of P). Thus (b) fails, as required.

(3) It remains to prove that (c) implies (b). The choice of L is easy, as indicated in remarks before the proof: take $\nu \in \Lambda^+$ in the W-orbit of $\mu - \lambda$ and set $L := L(\nu)$. Then set $P := M(\lambda) \otimes L(\nu)$. Since $\mu - \lambda$ is an extremal weight of $L(\nu)$, we get from $(*)$:

$$\dim \mathrm{Hom}_{\mathcal{O}}\left(P, M(\mu) \right) = \dim L(\nu)_{\mu - \lambda} = 1.$$

The goal is to show that $M(\mu)$ can be replaced here by $L(\mu)$. (Thanks to (a), $P(\mu)$ will then occur as a summand of P with multiplicity 1.)

(4) The existence of a nonzero homomorphism $P \to M(\mu)$ implies that $\mathrm{Hom}_{\mathcal{O}}\left(P, L(\mu') \right) \neq 0$ for some composition factor $L(\mu')$ of $M(\mu)$. If $\mu' = \mu$, we are done. Otherwise $\mu' < \mu$. Now Theorem 5.1 ensures that μ' is strongly

linked to μ. In particular, there exists $\beta > 0$ such that $\mu' \leq s_\beta \cdot \mu < \mu$. Moreover, there are corresponding embeddings $M(\mu') \hookrightarrow M(s_\beta \cdot \mu) \hookrightarrow M(\mu)$. Since the last embedding is proper, we have $\langle \mu + \rho, \beta^\vee \rangle \in \mathbb{Z}^{>0}$. In turn, the assumption in (c) implies that we cannot have $\langle \lambda + \rho, \beta^\vee \rangle = 0$. The remainder of the proof is devoted to contradicting this statement, from which it will follow that $\mu' = \mu$ as desired.

(5) Imitate the previous argument based on $(*)$ by taking $\nu' \in \Lambda^+$ in the W-orbit of $s_\beta \cdot \mu - \lambda$, so that

$$\mathrm{Hom}_{\mathcal{O}} \left(M(\lambda) \otimes L(\nu'), M(s_\beta \cdot \mu) \right) \neq 0.$$

Thanks to the embedding $M(s_\beta \cdot \mu) \hookrightarrow M(\mu)$, this also gives

$$\mathrm{Hom}_{\mathcal{O}} \left(M(\lambda) \otimes L(\nu'), M(\mu) \right) \neq 0.$$

But the nonzero dimension of this Hom space is equal to $\dim L(\nu')_{\mu - \lambda} = \dim L(\nu')_\nu$. Thus $\nu \leq \nu'$.

(6) From step (4) we have $\mathrm{Hom}_{\mathcal{O}} \left(P, L(\mu') \right) \neq 0$. The projective property of P forces $\mathrm{Hom}_{\mathcal{O}} \left(P, M(\mu') \right) \neq 0$. The embedding $M(\mu') \hookrightarrow M(s_\beta \cdot \mu)$ then implies that $\mathrm{Hom}_{\mathcal{O}} \left(P, M(s_\beta \cdot \mu) \right) \neq 0$. By $(*)$ this Hom space has dimension equal to $\dim L(\nu)_{s_\beta \cdot \mu - \lambda} = \dim L(\nu)_{\nu'}$. As a result, $\nu' \leq \nu$. Combined with step (5), we get $\nu' = \nu$.

(7) Now we can derive the contradiction promised in step (4) by some computations with roots and weights. From $\nu' = \nu$ we deduce $s_\beta \cdot \mu - \lambda \in W(\mu - \lambda)$; being an integral weight, its euclidean inner product with itself equals that of $\mu - \lambda$. Moreover, we know that $\langle \mu + \rho, \beta^\vee \rangle \in \mathbb{Z}^{>0}$; call it m. Since $s_\beta \cdot \mu = \mu - m\beta$, a quick calculation therefore yields $m = \langle \mu - \lambda, \beta^\vee \rangle$. Comparing the two formulas for m, we then conclude that $\langle \lambda + \rho, \beta^\vee \rangle = 0$, contradicting the assumption in (c). $\qquad\square$

10.4. Module Categories

Now we turn to the the paper by Bernstein–Gelfand [**24**], in which a deeper study is made of tensoring $U(\mathfrak{g})$-modules with finite dimensional modules. (To be consistent with our earlier notational conventions, we have to depart considerably from their notation in what follows.) Although our main emphasis is still on category $\mathcal{O}$, their broader viewpoint places the problems studied in a more natural setting. In particular, they focus on the role played by the center $Z := Z(\mathfrak{g})$ of $U(\mathfrak{g})$ and the associated central characters $\chi : Z \to \mathbb{C}$.

Tensoring with a finite dimensional $U(\mathfrak{g})$-module L makes sense in a number of full subcategories of $\mathcal{M} := \mathrm{Mod}\, U(\mathfrak{g})$, for example the category $\mathcal{M}_f$ whose objects are the finitely generated $U(\mathfrak{g})$-modules and of course its subcategory $\mathcal{O}$. Less obvious is the case of the category $\mathcal{M}_{Zf}$ whose objects are the Z-finite $U(\mathfrak{g})$-modules. Here a standard result of Kostant has to be invoked [**24**, 2.5] to show that $\mathcal{M}_{Zf}$ is stable under this kind of tensoring. Thanks to Theorem 1.1(e), $\mathcal{O}$ lies in both $\mathcal{M}_f$ and $\mathcal{M}_{Zf}$.

In Chapter 7 we used the projection pr_λ from $\mathcal{O}$ to its direct summand $\mathcal{O}_\chi$ when $\chi = \chi_\lambda$. Such projections also play a useful role here, but with emphasis on the central character χ rather than an individual weight; so we write pr_χ. It makes sense to work in the larger category $\mathcal{M}_{Zf}$, which also decomposes naturally into a direct sum of subcategories indexed by central characters where Z acts as the sum of χ and a nilpotent operator.

In more detail, let J_χ be the two-sided ideal of $U = U(\mathfrak{g})$ generated by $\mathrm{Ker}\,\chi$ and set $U_\chi := U/J_\chi$. Define a full subcategory of $\mathcal{M}_{Zf}$ by $\mathcal{M}(\chi) := \mathrm{Mod}\,U_\chi$. The objects here are modules on which Z acts by the character χ: for example, Verma modules or simple modules in $\mathcal{O}_\chi$. More generally, when J_χ is replaced by its nth power, the quotient algebra is written U_χ^n and its module category is written $\mathcal{M}^n(\chi)$. Then

$$\mathcal{M}(\chi) \subset \mathcal{M}^2(\chi) \subset \cdots \subset \mathcal{M}^\infty(\chi),$$

where

$$\mathcal{M}^\infty(\chi) := \{M \in \mathcal{M} \mid \text{for each } v \in M, J_\chi^n \cdot v = 0 \text{ for } n \gg 0\}.$$

With this set-up, it follows that each $M \in \mathcal{M}_{Zf}$ decomposes uniquely into a direct sum of submodules $M^\chi \in \mathcal{M}^\infty(\chi)$ (with χ running over all central characters). In turn, we get a projection operator $\mathrm{pr}_\chi : \mathcal{M}_{Zf} \to \mathcal{M}^\infty(\chi)$ extending the operator on $\mathcal{O}$.

10.5. Projective Functors

For a fixed finite dimensional $U(\mathfrak{g})$-module L, denote by $\mathcal{F}_L$ the functor on $\mathcal{M}$ which sends M to $M \otimes L$ and takes a $U(\mathfrak{g})$-homomorphism $\varphi : M \to N$ to the map $\varphi \otimes \mathrm{id}$. These functors preserve various subcategories of $\mathcal{M}$, notably $\mathcal{O}$ and $\mathcal{M}_f$. As remarked above, Bernstein–Gelfand verify that $\mathcal{M}_{Zf}$ is also stable under $\mathcal{F}_L$.

A number of properties of the functors follow readily from the definition:

- $\mathcal{F}_L$ is exact and commutes with arbitrary direct sums and products.
- $\mathcal{F}_{L'} \circ \mathcal{F}_L = \mathcal{F}_{L \otimes L'}$.
- If L^* is the usual dual module, then $\mathcal{F}_{L^*}$ is a left and right adjoint to $\mathcal{F}_L$.
- $\mathcal{F}_L$ takes projectives to projectives in the categories mentioned above.

In a category whose objects are functors (say on $\mathcal{M}$ or one of its subcategories) and whose morphisms are natural transformations of functors, it makes sense to talk about *direct sums* of functors. (If the sum is infinite, the term "coproduct" may be used in place of "direct sum" to contrast with "product".) In turn, there is a notion of *indecomposable* functor. Define a **projective functor** on $\mathcal{M}_{Zf}$ or $\mathcal{O}$ to be a direct summand of some $\mathcal{F}_L$. The rationale for this label will soon emerge. It is clear from the properties of the $\mathcal{F}_L$ that direct sums, or direct summands, of projective functors, are again projective functors. Similarly, composites of projective functors are projective. Projective functors take projective modules to other projective modules in $\mathcal{M}_{Zf}$ (or $\mathcal{O}$).

A central issue in the paper [**24**] is the classification of all indecomposable projective functors, where as usual the zero functor does not qualify as indecomposable. It will be seen below that Theorem 10.2 embodies much of the necessary work for category $\mathcal{O}$. Some care has to be taken to verify that new functors arising as direct summands are in fact nonzero; for this Proposition 10.1 is a useful tool in $\mathcal{O}$.

The search for indecomposables can be started easily enough by composing a given projective functor $\mathcal{F}$ on $\mathcal{M}_{Zf}$ (resp. $\mathcal{O}$) in either order with the projection to a subcategory $\mathcal{M}^{\infty}(\chi)$ (resp. a block $\mathcal{O}_\lambda$). This decomposes $\mathcal{F}$ into a direct sum of projective functors, each vanishing outside the chosen subcategory and/or taking nonzero values only in that subcategory. For example, when L is the trivial 1-dimensional $U(\mathfrak{g})$-module, the identity functor $\mathcal{F}_L$ decomposes into a direct sum of projections in $\mathcal{M}_{Zf}$.

Example. In the case of category $\mathcal{O}$, the translation functors studied in Chapter 7 provide natural examples of projective functors, though they do not exhaust all possibilities. The initial definition of T_λ^μ in 7.1 required only that λ and μ be compatible; then a finite dimensional module $L = L(\overline{\nu})$ was specially chosen to have highest weight in the W-orbit of $\nu = \mu - \lambda$. Letting $\chi := \chi_\lambda$ and $\chi' := \chi_\mu$, we defined T_λ^μ to be $\mathrm{pr}_{\chi'} \circ \mathcal{F}_L \circ \mathrm{pr}_\chi$ followed by inclusion into $\mathcal{O}$. In particular, T_λ^μ is a direct summand of $\mathcal{F}_L$.

In this construction the choice of L depended on the given pair λ, μ. If instead L is given at the outset, the resulting functor $\mathcal{F}_L$ might or might not be a direct sum of translation functors. In any case, it is usually difficult to analyze the effect of T_λ^μ on a Verma module or simple module without imposing further constraints on λ, μ (as we did in Chapter 7). The reader might find it useful at this point to consider which functors T_λ^μ are indecomposable, keeping in mind Proposition 10.1 and Theorem 10.2.

To get started toward the classification theorem, we have to ask how far a projective functor can be decomposed into a direct sum (possibly infinite); in particular, is there always a decomposition into *indecomposable* projective functors? The answer turns out to be affirmative for the category $\mathcal{M}_{Zf}$

[**24**, 3.3(i)], though the proof is somewhat indirect. Here we give a straight-forward argument in the case of $\mathcal{O}$, where such a decomposition must be finite:

Proposition. *Given a projective functor $\mathcal{F}$ on category $\mathcal{O}$, its nonzero restriction to a block $\mathcal{O}_\lambda$ decomposes into a finite direct sum of indecomposable functors.*

Proof. In the block $\mathcal{O}_\lambda$, where λ can be assumed to be *dominant*, there is a unique projective Verma module $M(\lambda)$. Thanks to Proposition 10.1, $\mathcal{F}M(\lambda)$ must be nonzero. Since $\mathcal{F}M(\lambda)$ is again projective, it decomposes into a finite direct sum of indecomposable projectives in an essentially unique way; say there are n summands. If $\mathcal{F}$ fails to be indecomposable on $\mathcal{O}_\lambda$, it can be written as $\mathcal{F}_1 \oplus \mathcal{F}_2$. Neither functor can kill $M(\lambda)$ (again by Proposition 10.1); so each $\mathcal{F}_i M(\lambda)$ is a direct sum of fewer than n indecomposable projectives. This makes it clear that $\mathcal{F}$ can be decomposed into a direct sum of at most n indecomposable projective functors. $\qquad\square$

It is not obvious that a decomposition of the sort obtained here is unique; for this the detailed classification in Theorem 10.8 seems to be needed.

10.6. Annihilator of a Verma Module

The proof of Theorem 10.7 below requires one fundamental fact about $U(\mathfrak{g})$ which we have not so far introduced. For any $U(\mathfrak{g})$-module M, it is reasonable to ask for a description of the *annihilator* $\operatorname{Ann} M$ in $U(\mathfrak{g})$; this is the two-sided ideal consisting of all elements which act as 0 on M (in other words, the kernel of the representation $U(\mathfrak{g}) \to \operatorname{End} M$).

Theorem (Duflo). *Let $\lambda \in \mathfrak{h}^*$, with associated central character $\chi = \chi_\lambda$. Then $\operatorname{Ann} M(\lambda)$ is the two-sided ideal J_χ of $U(\mathfrak{g})$ generated by $\operatorname{Ker}\chi$. In particular, this ideal depends only on χ. Therefore $\operatorname{Ann} M(\mu) = J_\chi$ for all weights μ linked to λ.*

One inclusion is obvious: $J_\chi \subset \operatorname{Ann} M(\lambda)$. This was already used in 10.4. But the other inclusion is far more subtle. Duflo's theorem originates in his 1971 lecture in a Budapest summer school, published in [**86**] (along with the BGG paper [**26**] and others). It relies in an essential way on earlier work of Kostant ("separation of variables" theorem), which involves not just $U(\mathfrak{g})$ but also the nilpotent variety in $\mathfrak{g}$. For a thorough account, see Dixmier [**84**, 8.4].

Taking λ here to be *antidominant*, the fact (4.8) that $M(\lambda) = L(\lambda)$ then implies that J_χ is a *primitive* ideal: the kernel of an irreducible representation of $U(\mathfrak{g})$. Moreover, it can be shown in this setting (since we are working over an algebraically closed field) that the ideals J_χ are precisely the *minimal* primitive ideals of $U(\mathfrak{g})$. This opens the way to a deeper study of the primitive ideal spectrum and its relationship with category $\mathcal{O}$: see 13.1–13.3 below.

Remark. In his 1997 Montreal lectures, Joseph [162] provides an algebraic (rather than geometric) pathway to the proof, motivated by analogous issues for Kac–Moody algebras and quantum groups.

10.7. Comparison of Hom Spaces

A fundamental obstacle to classifying indecomposable projective functors is the difficulty in recognizing when a given functor is actually indecomposable. In the case of a module, the algebra of endomorphisms is a key tool: for example, $M \in \mathcal{O}$ is indecomposable if and only if $\mathrm{End}_\mathcal{O} M$ has precisely two involutions 0 and 1 (in other words, no nontrivial projections onto proper direct summands). The same test applies to a functor $\mathcal{F}$ and its algebra $\mathrm{End}\,\mathcal{F}$ of endomorphisms.

Our goal is to reduce the study of morphisms of functors to the study of morphisms of certain modules. In general, a morphism of functors $\varphi : \mathcal{F} \to \mathcal{G}$ on a module category $\mathcal{C}$ induces for each module $M \in \mathcal{C}$ a homomorphism $\varphi_M : \mathcal{F}M \to \mathcal{G}M$. In turn, $\varphi \mapsto \varphi_M$ defines a natural linear map $\mathrm{Hom}\,(\mathcal{F}, \mathcal{G}) \to \mathrm{Hom}_\mathcal{C}(\mathcal{F}M, \mathcal{G}M)$. When $\mathcal{F} = \mathcal{G}$, the functorial properties ensure that this becomes an algebra homomorphism $\mathrm{End}\,\mathcal{F} \to \mathrm{End}_\mathcal{C}\,\mathcal{F}M$.

To get more detailed information in this direction, we focus at first on the subcategory $\mathcal{M}(\chi)$ attached to a central character χ; its intersection with $\mathcal{O}_\chi$ contains for example all Verma modules $M(\lambda)$ with $\chi = \chi_\lambda$ as well as their simple quotients. Define a **projective χ-functor** to be a direct summand of some $\mathcal{F}_L(\chi)$, the restriction of $\mathcal{F}_L$ to the category $\mathcal{M}(\chi)$. The main technical tool for the classification of indecomposable projective functors is the following theorem, drawn from [24, 3.5, 3.7]. This reduces matters largely to the study of projectives in category $\mathcal{O}$.

Theorem. *Fix a central character χ.*

(a) *Let $\mathcal{F}, \mathcal{G}$ be projective χ-functors. If $\chi = \chi_\lambda$, then the natural map*
$$i_\lambda : \mathrm{Hom}\,(\mathcal{F}, \mathcal{G}) \to \mathrm{Hom}_\mathcal{O}\big(\mathcal{F}M(\lambda), \mathcal{G}M(\lambda)\big)$$
is a monomorphism.

(b) *The map i_λ is an isomorphism in case λ is either dominant or antidominant.*

(c) *The restriction $\mathcal{F}^\infty(\chi)$ of a projective functor $\mathcal{F}$ to $\mathcal{M}^\infty(\chi)$ is completely determined by its restriction $\mathcal{F}(\chi)$ to $\mathcal{M}(\chi)$. Any morphism $\varphi : \mathcal{F}(\chi) \to \mathcal{G}(\chi)$ extends to a morphism $\widehat{\varphi} : \mathcal{F}^\infty(\chi) \to \mathcal{G}^\infty(\chi)$; this* *is an isomorphism if φ is and can be chosen to be idempotent if φ is idempotent.*

Here we outline the methods used in the proof, referring to [**24**] for full details. Duflo's Theorem 10.6 plays an essential role in the proof, together with some classical results of Kostant. For the proofs of (a) and (b), one first makes an easy reduction to the case when $\mathcal{F} = \mathcal{F}_L$ and $\mathcal{G} = \mathcal{F}_{L'}$ for some finite dimensional modules L, L'.

(a) To show that i_λ is injective, we exploit the $U(\mathfrak{g})$-module $U_\chi = U/J_\chi$, where as in 10.4 J_χ is the ideal in U generated by $\operatorname{Ker}\chi$. The category of U_χ-modules is just $\mathcal{M}(\chi)$, with U_χ itself playing a sort of universal role in this category (though it need not lie in $\mathcal{O}$). Now a morphism $\varphi : \mathcal{F} \to \mathcal{G}$ induces a module homomorphism

$$\varphi_{U_\chi} : \mathcal{F}U_\chi \to \mathcal{G}U_\chi.$$

By choosing ordered bases (v_j) in L and (v_i') in L', we can express this map concretely as $1 \otimes v_j \mapsto \sum_i u_{ij} \otimes v_i'$ for some $u_{ij} \in U_\chi$. In turn, for an arbitrary $M \in \mathcal{M}(\chi)$, the U-map $U_\chi \to M$ defined by $u \mapsto u \cdot v$ allows us to write the map $\varphi_M : M \otimes L \to M \otimes L'$ explicitly as $v \otimes v_j \mapsto \sum_i u_{ij} \cdot v \otimes v_i'$. In the special case $M = M(\lambda)$, the assumption $i_\lambda(\varphi) = 0 = \varphi_{M(\lambda)}$ translates into $u_{ij} \cdot v = 0$ for all $v \in M(\lambda)$. In other words, u_{ij} annihilates $M(\lambda)$, so $u_{ij} = 0$ in U_χ by Duflo's Theorem. Now $\varphi_M = 0$ for all M, forcing $\varphi = 0$.

(b) Thanks to (a), we only have to show that

$$\dim \operatorname{Hom}_{\mathcal{O}}\big(\mathcal{F}M(\lambda), \mathcal{G}M(\lambda)\big) \leq \dim \operatorname{Hom}\big(\mathcal{F}, \mathcal{G}\big).$$

The first step is to prove, using a standard result of Kostant, that $\dim \operatorname{Hom}\big(\mathcal{F}, \mathcal{G}\big) = \dim(L \otimes (L')^*)_0$, the dimension of the 0-weight space. This involves some study of the (U_χ, U)-bimodules $U_\chi \otimes L$ and $U_\chi \otimes L'$, using the natural adjoint representation of $\mathfrak{g}$ on U. (Again the argument departs from category $\mathcal{O}$.)

What can be said about the other dimension? Consider the case when λ is *dominant*. Using the fact that $\dim L < \infty$, a standard vector space identification yields

$$\operatorname{Hom}_{\mathcal{O}}\big(M(\lambda) \otimes L, M(\lambda) \otimes L'\big) \cong \operatorname{Hom}_{\mathcal{O}}\big(M(\lambda), M(\lambda) \otimes (L^* \otimes L')\big).$$

By Theorem 3.6, the tensor product with $M(\lambda)$ has a standard filtration with quotients $M(\lambda + \nu)$; here ν ranges over the weights of $L^* \otimes L'$, taken with multiplicity. Since λ is dominant and therefore maximal in its linkage class, $\operatorname{Hom}_{\mathcal{O}}\big(M(\lambda), M(\lambda + \nu)\big) = 0$ unless $\nu = 0$ (and then the dimension is 1). So the Hom space has dimension at most $\dim(L^* \otimes L')_0$, which is readily seen to agree with $\dim(L \otimes (L')^*)_0$.

The argument when λ is *antidominant* is essentially the same, but based instead on the identification

$$\mathrm{Hom}_{\mathcal{O}}\left(M(\lambda) \otimes L, M(\lambda) \otimes L'\right) \cong \mathrm{Hom}_{\mathcal{O}}\left(M(\lambda) \otimes \left((L')^* \otimes L\right), M(\lambda)\right).$$

(c) The argument here is in the spirit of other lifting arguments in algebra, using the inverse limit of spaces $\mathrm{Hom}\left(\mathcal{F}^n(\chi), \mathcal{G}^n(\chi)\right)$ for the restrictions of functors to $\mathcal{M}^n(\chi)$, together with the "J_χ-adic topology".

By combining the parts of the theorem, one sees when λ is dominant that the indecomposability of a projective functor on $\mathcal{M}^\infty(\chi)$ is equivalent to indecomposability of the module $\mathcal{F}M(\lambda)$. In general, a decomposition of $\mathcal{F}M(\lambda)$ into indecomposable summands will yield a corresponding decomposition of $\mathcal{F}$.

10.8. Classification Theorem

A central result of [**24**] is the classification of all indecomposable projective functors on $\mathcal{M}_{Zf}$. To formulate the theorem we need a little more notation.

The group W acts on the set of all ordered pairs of compatible weights, with w sending (λ, μ) to $(w \cdot \lambda, w \cdot \mu)$. Denote the set of orbits for this action by Ξ. For each $\xi \in \Xi$, there clearly exists at least one pair (λ, μ) in this orbit for which λ is dominant and μ is minimal in its dot-orbit under the isotropy group W_λ°. Call such a pair *proper* for ξ. There may be many such pairs, one for each block if $\lambda \notin \Lambda$. These blocks $\mathcal{O}_\lambda$ are indexed by dominant weights λ lying in different cosets modulo Λ_r. The extreme case involves singleton blocks; it is in a sense the most "generic" (but least interesting) situation.

Theorem 3.3 in [**24**] is formulated in the wider context of $\mathcal{M}_{Zf}$ but then adapts readily to $\mathcal{O}$ if we take blocks into account when dealing with nonintegral weights.

Theorem. *There is a natural bijection $\xi \mapsto \mathcal{F}_\xi$ between the set Ξ and the set of (isomorphism classes of) indecomposable projective functors on $\mathcal{M}_{Zf}$, satisfying the condition: If (λ, μ) is a proper pair for $\xi \in \Xi$, then $\mathcal{F}_\xi$ sends $M(\lambda)$ to $P(\mu)$.*

Proof. (1) Given an indecomposable projective functor $\mathcal{F}$ on $\mathcal{O}$, it follows from the discussion in 10.5 that it must be zero outside a single block $\mathcal{O}_\lambda$ (with λ dominant). By Proposition 10.1, the projective module $\mathcal{F}M(\lambda)$ must be nonzero. Applying Theorem 10.7, we see that $\mathcal{F}M(\lambda)$ is *indecomposable*, thus isomorphic to some $P(\mu)$. By Theorem 10.2, the pair (λ, μ) properly represents its W-orbit $\xi \in \Xi$. So we can set $\mathcal{F}_\xi := \mathcal{F}$.

(2) In the other direction, let (λ, μ) be a proper pair for $\xi \in \Xi$. By Theorem 10.2, $P(\mu)$ is a direct summand of $\mathcal{F}_L M(\lambda)$ for some finite dimensional module L; indeed, L can be chosen so that $P(\mu)$ occurs just once

as a summand. Use Proposition 10.5 (generalized to $\mathcal{M}_{Zf}$) to decompose $\mathcal{F}_L$ into a direct sum of indecomposable functors. One (but only one!) of these, call it $\mathcal{F}_\xi$, must take $M(\lambda)$ to $P(\mu) \oplus P$ with P projective. Since $\mathcal{F}_\xi$ is indecomposable, Theorem 10.7 forces $\mathcal{F}_\xi M(\lambda)$ to be indecomposable as well. Thus $P = 0$ and $\mathcal{F}_\xi M(\lambda) = P(\mu)$.

Combining these steps yields the desired bijection. $\qquad\square$

Example. Recall from 7.15 the discussion of *wall-crossing functors* Θ_s attached to simple reflections in W (or $W_{[\lambda]}$). Given antidominant weights $\lambda, \mu \in \mathfrak{h}^*$ with λ regular and $\mu^\natural$ lying just in the s-wall of the Weyl chamber containing $\lambda^\natural$, one defines Θ_s to be the composite $T_\mu^\lambda T_\lambda^\mu$. This is a projective functor, 0 outside the block of λ. Combining Theorem 7.6 with Proposition 7.13, we get $\Theta_s M(w_\lambda \cdot \lambda) \cong P(w_\lambda s \cdot \lambda)$; here $w_\lambda \cdot \lambda$ is the *dominant* weight in the block. In particular, this result is independent of μ. It follows from the above theorem that the projective functor Θ_s is also independent of μ.

10.9. Harish-Chandra Modules

As indicated at the outset, an underlying motive in [**24**] for the study of projective functors is the theory of Harish-Chandra modules. Indeed, the main external motivation for the study of category $\mathcal{O}$ comes from the representation theory of semisimple (or more generally, reductive) Lie groups.

Start with a semisimple Lie group G having finite center and a maximal compact subgroup K, then complexify their respective Lie algebras to get Lie algebras $\mathfrak{g}$ and $\mathfrak{k}$ over $\mathbb{C}$. For example, if $G = \mathrm{SL}(2, \mathbb{R})$ and we take $K = \mathrm{SO}(2, \mathbb{R})$, then $\mathfrak{g} \cong \mathfrak{sl}(2, \mathbb{C})$ while $\mathfrak{k}$ is a Cartan subalgebra $\mathfrak{h}$ (in general, we would get a reductive Lie algebra).

Now if (π, V) is a well-behaved irreducible representation of G on a Banach space V, such as a unitary representation on a Hilbert space, then the subspace V° of K-finite vectors is in a natural way a simple $\mathfrak{g}$-module. Moreover, V° is the direct sum of finite dimensional simple $\mathfrak{k}$-modules, only finitely many from each isomorphism class. These are the "$\mathfrak{k}$-types", whose multiplicities go a long way toward determining π (or its "character", in the distribution sense). Here V° is a *Harish-Chandra module*. To develop the theory axiomatically in the Lie algebra setting, some treatments are more general in scope.

This theory is rich and well-organized but still incomplete in the original context of representations of real Lie groups. Here we outline the basic constructions in the best understood special case: *complex* semisimple groups. Such a group is actually an algebraic group, whose Lie algebra $\mathfrak{g}$ over $\mathbb{C}$ is of the type we have been studying. The construction leads in 10.10 below to a comparison of category $\mathcal{O}$ with the "principal series" Harish-Chandra modules. The foundations are explained in a number of sources, notably Duflo [**87**], Dixmier [**84**, Chap. 9], Jantzen [**148**, Kap. 6]. But the reader has to be aware of considerable variation in notation in the literature; we follow mainly the version of Bernstein–Gelfand [**24**] treated by Jantzen.

When a complex group G with Lie algebra $\mathfrak{g}$ over $\mathbb{C}$ is viewed as *real*, the complexified Lie algebra becomes $\mathfrak{g} \times \mathfrak{g}$. Its enveloping algebra is isomorphic to $U \otimes U$, where $U := U(\mathfrak{g})$. In turn, the center of $U \otimes U$ is $Z \otimes Z$, with $Z := Z(\mathfrak{g})$. Central characters here correspond naturally to pairs of characters of Z. The anti-involution τ of $\mathfrak{g}$ (see 0.5) extends to an anti-automorphism of U and leaves Z pointwise fixed. Now $\mathfrak{k}$ can be realized as a copy of $\mathfrak{g}$ embedded into $\mathfrak{g} \times \mathfrak{g}$ by $x \mapsto (x, -\tau(x))$.

Using τ, we get an isomorphism between U and its opposite algebra, which allows us to convert left U-modules into right U-modules. Thus a $\mathfrak{g} \times \mathfrak{g}$-module can be identified with a (U, U)-*bimodule*. Given this set-up, the linear space $\mathrm{Hom}(M, N)$ with $M, N \in \mathrm{Mod}\,U(\mathfrak{g})$ acquires a natural (U, U)-bimodule structure, as does $M \otimes N$. Setting $U_\chi := U/J_\chi$ as before, we can also specialize to (U, U_χ)-bimodules where part of the central action is fixed in advance.

For our limited purposes, a **Harish-Chandra module** for $\mathfrak{g} \times \mathfrak{g}$ is defined to be a finitely generated module whose restriction to $\mathfrak{k}$ decomposes into a direct sum of finite dimensional simple modules, each repeated only finitely many times. Write $\mathcal{HC}$ for the category of all such modules. Basic examples can be constructed by assigning to arbitrary modules $M, N \in \mathcal{O}$ the subspace of the linear space $\mathrm{Hom}(M, N)$ spanned by all $\mathfrak{k}$-finite vectors. The resulting (U, U)-bimodule $\mathcal{L}(M, N)$ then lies in $\mathcal{HC}$. A sort of dual construction is based on the tensor product $M \otimes N$: let $\mathcal{D}(M, N)$ be the span of all $\mathfrak{k}$-finite vectors in $(M \otimes N)^*$. The two constructions are related by

$$\mathcal{D}(M, N) \cong \mathcal{L}(N, M^\vee) \text{ for } M, N \in \mathcal{O}.$$

Remark. Observe that U or its quotient U/I by any two-sided ideal is naturally a (U, U)-bimodule. When $\mathfrak{g} = \mathfrak{k}$ is embedded as above in $\mathfrak{g} \times \mathfrak{g}$, it is easy to check that the resulting action of $x \in \mathfrak{g}$ is the adjoint action $u \mapsto xu - ux$. Thanks to the PBW filtration of U, all vectors are $\mathfrak{k}$-finite. In this way we get a large family of Harish-Chandra modules in the above sense. In case M is a U-module and I is the kernel of the associated representation, we conclude moreover that U/I embeds in $\mathcal{L}(M, M)$.

One major objective is the classification and description of simple Harish-Chandra modules. In the special case of complex groups, this rests on the study of category $\mathcal{O}$. Following work of Duflo and others, the connections between $\mathcal{O}$ and Harish-Chandra modules in this case were worked out independently by Bernstein–Gelfand [**24**], Enright [**90**], Joseph [**155**]. We continue to follow the version of [**24**], which is also developed in detail by Jantzen [**148**, Kap. 6].

For a fixed central character χ, the main interest lies in the category of (U, U_χ)-bimodules. Here the (isomorphism classes of) indecomposable projective objects turn out to be in natural 1–1 correspondence with the elements of the set Ξ used in 10.8 to classify indecomposable projective functors on $\mathcal{O}$. If $\xi \mapsto P_\xi$ and (λ, μ) is a proper pair for ξ, then

$$P_\xi \otimes_U M(\lambda) \cong P(\mu) \text{ as left } U(\mathfrak{g})\text{-modules,}$$

illustrating the intimate connection with category $\mathcal{O}$. In turn, one can show that P_ξ has a unique simple quotient, call it L_ξ. This leads to a precise classification:

Theorem. *With notation as above, assign to each $\xi \in \Xi$ the unique simple quotient L_ξ of the projective module P_ξ. This induces a bijection between Ξ and the isomorphism classes of simple Harish-Chandra modules for $\mathfrak{g} \times \mathfrak{g}$. Moreover, every Harish-Chandra module has finite length.*

To describe L_ξ in more detail would require a determination of multiplicities of the $\mathfrak{k}$-types occurring in its restriction to $\mathfrak{k}$. This is by no means an easy problem.

In general, a Harish-Chandra module X has a filtration of Jordan–Hölder type with quotients of the type L_ξ which are uniquely determined up to isomorphism and order of occurrence. Write $[X : L_\xi]$ for the resulting multiplicity. It is another basic problem to determine these multiplicities. We turn now to a special case where a reduction can be made to the analogous multiplicity problem for category $\mathcal{O}$.

10.10. Principal Series Modules and Category $\mathcal{O}$

In the Lie group theory which underlies the general formalism of Harish-Chandra modules, some group representations are easy to construct by induction. Though these representations are typically not irreducible, they provide a natural starting point. In the category $\mathcal{HC}$ a related class of modules can be studied under the rubric **principal series modules**. In the framework of [**24**], these are easily specified: associate to any compatible pair $\lambda, \mu \in \mathfrak{h}^*$ the Harish-Chandra module $X(\lambda, \mu) := \mathcal{D}\big(M(\lambda), M(\mu)\big)$.

Here are the main features, the first result being due to Duflo:

- If $\lambda, \mu \in \mathfrak{h}^*$ are compatible, then all modules $X(w \cdot \lambda, w \cdot \mu)$ with $w \in W_{[\lambda]} = W_{[\mu]}$ have the same composition factors, counting multiplicities.

- Let $\lambda, \nu \in \mathfrak{h}^*$ be compatible, with λ dominant. If $\xi \in \Xi$, then

$$[X(\lambda, \nu) : L_\xi] = \begin{cases} [M(\nu) : L(\mu)] \text{ if } (\lambda, \mu) \text{ is a proper pair for } \xi, \\ 0 \text{ otherwise.} \end{cases}$$

Taken together, these properties reduce the problem of computing composition factor multiplicities for principal series modules to the analogous problem for Verma modules; there the KL Conjecture in its extended form provides an answer.

Underlying the multiplicity comparison is a category equivalence [**24**, 5.9]. Suppose λ is *dominant*, with $\chi = \chi_\lambda$. In case λ is also *regular*, the functor sending X to $X \otimes_U M(\lambda)$ takes a principal series module X to a module in category $\mathcal{O}$. This gives an equivalence between the finitely generated (U, U_χ)-bimodules and the subcategory of $\mathcal{O}$ consisting of modules whose weights are compatible with λ. In case λ is irregular, the subcategory of $\mathcal{O}$ is more complicated to characterize. It consists of the modules M which are "projectively presentable" by direct sums of projectives $P(\mu)$ with $\xi = (\lambda, \mu)$ a proper pair for some $\xi \in \Xi$. If $\mathcal{P}$ is the collection of such projectives, the condition on M is that there exists a short exact sequence $P' \to P \to M \to 0$ with $P, P' \in \mathcal{P}$. (This generalizes the case when λ is regular, since $\mathcal{P}$ is then the collection of all projectives having weights compatible with λ.)

We remark that principal series modules and the category equivalence just described also play a significant role in Chapter 12 below.

Notes

The material in this chapter is drawn primarily from Bernstein–Gelfand [**24**], but with many differences in notation and emphasis. Theorem 10.2 reformulates some of the ideas in [**24**] along the lines of Jantzen's treatment in [**148**, 6.18–6.26]. The proof given here follows his suggestions.

Some later developments involving projective functors can be found in papers by Khoroshkin [**186**], Backelin [**12**], Khomenko [**174**].

Chapter 11

Tilting Modules

Next we introduce some objects in $\mathcal{O}$ called *tilting modules*. Such a module is characterized by the condition that both the module and its dual have standard filtrations. Motivation arises indirectly from the study of fusion rules in parallel module categories for semisimple algebraic groups in characteristic p, quantum enveloping algebras at a root of unity, or affine Kac–Moody algebras. In our setting, tilting modules play a role which is in a subtle sense dual to that of projectives.

After recalling known examples and working out some elementary properties of these modules in 11.1, we show how to parametrize the indecomposable ones by their highest weights (11.2). The effect of translation functors on certain tilting modules is explained in 11.3, leading to the development of "fusion rules" in 11.6. Then we discuss the deeper question of determining the formal characters of tilting modules (11.7): this cannot be understood properly without appeal to Kazhdan–Lusztig theory. Adaptations to parabolic categories $\mathcal{O}^{\mathfrak{p}}$ are outlined in 11.8.

Early work in this direction is due to Enright–Shelton [**91**], Collingwood–Irving [**69**], and Irving [**138**, §9]. Our treatment follows mainly that of Andersen–Paradowski [**6**, §1]. (However, they use dominant rather than antidominant weights to index linkage classes as well as translation functors.) As in [**6**], we simplify notation by limiting the discussion to *integral* weights (from 11.2 on). This amounts to working in the full subcategory $\mathcal{O}_{\mathrm{int}}$ of $\mathcal{O}$ whose objects are modules having weights in Λ. But the ideas can be generalized to arbitrary weights.

11.1. Tilting Modules

We have already encountered most of the "interesting" indecomposable objects in $\mathcal{O}$, but here we introduce a further class of modules which are characterized in terms of standard filtrations. The indecomposable ones turn out to be parametrized by highest weights but are rarely highest weight modules. While they do not appear to play as central a role in $\mathcal{O}$ as Verma modules and projective modules, their counterparts in related module categories (notably for affine Lie algebras, algebraic groups in prime characteristic, and quantum groups at a root of unity) have been influential.

We say that M is a **tilting module** if both M and $M^\vee$ have standard filtrations in the sense of 3.7. Therefore a *self-dual* module with a standard filtration is automatically a tilting module; whether all tilting modules are self-dual is less obvious at this point but will be proved as a consequence of the construction in 11.2. Following Donkin, the label "tilting module" is applied here by analogy with the partial tilting modules which are important in the study of representations of finite dimensional algebras by Brenner–Butler, Auslander–Reiten, Ringel, and others.

Among the modules in $\mathcal{O}$ already studied, obvious examples of tilting modules are the self-dual Verma modules $M(\lambda) = L(\lambda)$ with λ antidominant. Thanks to Theorem 4.8, there are no other self-dual Verma modules. At the other extreme, the projective cover of an antidominant Verma module is self-dual and has a standard filtration (7.16); it is therefore a tilting module.

Exercise. In the case $\mathfrak{g} = \mathfrak{sl}(2, \mathbb{C})$, show that every indecomposable tilting module is of one of these two types. (It is enough to consider integral weights.)

Given a tilting module, here are some ways to obtain others:

Proposition. *Let M be a tilting module. Then:*

    (a) *$M^\vee$ is also a tilting module.*

    (b) *If N is a tilting module, so is $M \oplus N$.*

    (c) *Any direct summand of M is a tilting module.*

    (d) *If $\dim L < \infty$, then $L \otimes M$ is a tilting module.*

    (e) *Applying any translation functor to M produces another tilting module.*

    (f) *If N is also a tilting module, then $\operatorname{Ext}^n_{\mathcal{O}}(M, N) = 0$ for all $n > 0$.*

Proof. Parts (a) and (b) are obvious, while (c) follows from Proposition 3.7(b).

Part (d) follows from Theorem 3.6, coupled with Exercise 3.2. Part (e) is a consequence of Theorem and Corollary 7.6. For (f), apply Exercise 6.12, using the fact that $N^\vee$ has a standard filtration. $\qquad\square$

It turns out that there is a hidden duality between tilting and projective modules in $\mathcal{O}$, which will be made explicit in a number of the results proved below. Roughly speaking, the role of the dominant and antidominant chambers is being reversed. This is already seen in the fact that dominant Verma modules are projective, while antidominant ones are tilting. (For a geometric perspective, see Beilinson–Bezrukavnikov–Mirković [**19**].)

Remark. As we remarked earlier, the origins of "tilting" theory lie in the representation theory of finite dimensional algebras. While that theory can sometimes be applied to blocks of $\mathcal{O}$ by taking advantage of Proposition 3.13, it is easy in the current setting to work out the initial steps directly.

Background on "tilting theory" in general, with special attention to modular representations of algebraic groups, is provided by Jantzen [**152**, II.E]; this builds on the work of Ringel and Donkin. In the case of category $\mathcal{O}$, the search for self-dual modules with standard filtrations began in the work of Collingwood–Irving [**69**]. The theory was later reformulated and extended by Andersen–Paradowski [**6**]. Analogues were then treated by Soergel [**239, 240**], inspired in part by Arkhipov's ideas.

11.2. Indecomposable Tilting Modules

From now on we work in $\mathcal{O}_{\mathrm{int}}$, whose Grothendieck group $\mathcal{K} := K(\mathcal{O}_{\mathrm{int}})$ is a subgroup of $K(\mathcal{O})$. In order to classify all tilting modules, it is enough (by Proposition 11.1(c)) to determine the indecomposable ones. It turns out that these are neatly parametrized (up to isomorphism) by their highest weights. Existence is proved using translation functors (Proposition 11.1(e)). For the isomorphism statement we have to invoke Proposition 3.7(a):

(∗)   *If M has a standard filtration and λ is maximal among the weights of M, then M has a submodule isomorphic to $M(\lambda)$, while the quotient $M/M(\lambda)$ has a standard filtration.*

Theorem. *Let $\lambda \in \Lambda$.*

> (a) *There exists an indecomposable tilting module $D(\lambda)$ in $\mathcal{O}_{\mathrm{int}}$ such that $\dim D(\lambda)_\lambda = 1$ and all weights μ of $D(\lambda)$ satisfy $\mu \le \lambda$. (In particular, $M(\lambda)$ occurs just once in any standard filtration of $D(\lambda)$ and then occurs as a submodule.)*

> (b) *Every indecomposable tilting module is isomorphic to some $D(\lambda)$.*

(c) *Up to isomorphism, $D(\lambda)$ is the only tilting module in $\mathcal{O}_{\text{int}}$ having the properties in* (a). *It is uniquely determined by its formal character.*

(d) *The symbols $[D(\lambda)]$ with $\lambda \in \Lambda$ form a basis of the group $\mathcal{K}$.*

Proof. (a) Existence is shown by an induction on length in W, using translation functors to and from walls of Weyl chambers. So we fix a regular antidominant weight λ and consider the linked weights $w \cdot \lambda$ as well as translated weights in walls of Weyl chambers. To get started, simply define $D(\lambda) := M(\lambda)$. Translating to any weight μ in the closure of the antidominant Weyl chamber then produces another simple Verma module $M(\mu)$ which is again a tilting module and can be labelled $D(\mu)$.

Now use induction on $\ell(w)$, assuming that $D(w \cdot \mu)$ exists for all weights $w \cdot \mu$ in the closure of the chamber containing $w \cdot \lambda$. Thanks to Proposition 11.1(e), translating any tilting module produces another.

Having defined $D(w \cdot \lambda)$, let s be any simple reflection for which $ws > w$ and choose an antidominant weight μ for which $W^\circ_\mu = \{1, s\}$. Then $D(w \cdot \mu) = D(ws \cdot \mu)$ is also defined. By Theorem 7.14, the tilting module $T^\lambda_\mu D(w \cdot \mu)$ involves in a standard filtration a nonsplit extension of two Verma modules, a submodule $M(ws \cdot \lambda)$ and a quotient $M(w \cdot \lambda)$; this module is indecomposable, with head $L(w \cdot \lambda)$. Moreover, $ws \cdot \lambda$ is the unique highest weight of $T^\lambda_\mu D(w \cdot \lambda)$, occurring with multiplicity 1. So the indecomposable summand involving this weight has the required properties of $D(ws \cdot \lambda)$.

Once $D(ws \cdot \lambda)$ is defined, it can be translated to the closure of the ws-chamber. For weights ν in walls of this chamber which are not walls of the w-chamber already considered, an application of Theorem 7.6 shows that the translated module has a unique occurrence of ν. So we can extract the desired indecomposable tilting module $D(\nu)$.

(b) Let D be an arbitrary indecomposable tilting module having λ as one of its maximal weights. By $(*)$ above, there is an embedding $M(\lambda) \hookrightarrow D$ as part of a standard filtration, while $D/M(\lambda)$ also has a standard filtration. Then Proposition 11.1(f) implies that $\mathrm{Ext}_\mathcal{O}\big(D/M(\lambda), D(\lambda)\big) = 0$. Thus the natural map

$$\mathrm{Hom}_\mathcal{O}\big(D, D(\lambda)\big) \to \mathrm{Hom}_\mathcal{O}\big(M(\lambda), D(\lambda)\big)$$

is surjective. This implies that the embedding $M(\lambda) \hookrightarrow D(\lambda)$ lifts to a homomorphism $\varphi : D \to D(\lambda)$. By the same kind of reasoning, there is a homomorphism $\psi : D(\lambda) \to D$ which is an isomorphism on the embedded copies of $M(\lambda)$. It follows that the endomorphism $\psi \circ \varphi$ is an isomorphism on D_λ, while $\varphi \circ \psi$ is an isomorphism on $D(\lambda)_\lambda$. Because the modules are indecomposable (and of finite length), it is a standard result that any endomorphism must be either nilpotent (which is impossible here) or invertible. In turn, it follows that both φ and ψ must be isomorphisms.

(c) This is an immediate consequence of part (b), when D is taken to be a module with the same properties as $D(\lambda)$ in (a).

(d) This follows from part (a), coupled with the easy Exercise 1.12. $\square$

Corollary. *Every tilting module is self-dual.*

Proof. It is enough to consider indecomposable tilting modules. Thanks to the theorem, such a module is isomorphic to $D(\lambda)$ for some λ. But $D(\lambda)^\vee$ is also an indecomposable tilting module. Since it has the same formal character as $D(\lambda)$, its unique maximal weight λ has multiplicity 1. Then the uniqueness assertion in (b) ensures that $D(\lambda)^\vee \cong D(\lambda)$. $\square$

Remarks. (1) Starting with our definition of tilting module, there seems to be no shorter way to obtain the corollary. The earliest work along this line, by Collingwood–Irving [**69**], proceeds in a contrary direction but leads to essentially the same results: their goal was to determine self-dual modules having standard filtrations.

(2) The direct sum of all $D(\lambda)$ in a given block corresponds to a "tilting module" in the original sense for a finite dimensional algebra whose module category is equivalent to the block (see 3.13).

At this point the initial examples in 11.1 fit easily into the classification. At one extreme, $D(\lambda) = M(\lambda) = L(\lambda)$ whenever λ is antidominant. Moreover, $L(-\rho)$ is also projective and injective. As a result, $T_{-\rho}^{w_\circ \cdot \lambda} M(-\rho)$ is both projective and tilting; its highest weight is $w_\circ \cdot \lambda$, so $D(w_\circ \cdot \lambda)$ must be a direct summand. It is also projective, with simple quotient $L(\lambda)$, hence coincides with $P(\lambda)$ (and is self-dual). More precisely, our direct construction in 4.10 shows that $D(w_\circ \cdot \lambda) \cong T_{-\rho}^\lambda L(-\rho)$. In these cases the formal character of the tilting module is therefore known. But in general it becomes more complicated to describe: see 11.7 below.

11.3. Translation Functors and Tilting Modules

As one might expect from the case of projective modules, it is nontrivial to work out the effect of translation functors on arbitrary tilting modules. Under the standard assumptions on λ, μ, Proposition 11.1 ensures that $T_\lambda^\mu D(w \cdot \lambda)$ is isomorphic to a direct sum of various $D(w' \cdot \mu)$; but the result could be complicated, for example if we are translating out from a wall. Theorem 7.11 does have a sort of mirror image here, which we work out only in the easiest case (following Andersen–Paradowski [**6**, Prop. 1.11]). This will be essential for the discussion of "fusion rules" in 11.6 below.

Fix antidominant weights $\lambda, \mu \in \Lambda$, with λ regular and $W_\mu^\circ = \{1, s\}$ for some simple reflection s. We want to analyze $T_\mu^\lambda D(w \cdot \mu)$ when $ws > w$, so $ws \cdot \lambda > w \cdot \lambda$. In this situation (or more generally, when μ is just assumed to lie in the closure of the facet of λ), Theorem 7.12 ensures that $\operatorname{ch} T_\mu^\lambda M(w \cdot \mu) = \operatorname{ch} M(w \cdot \lambda) + \operatorname{ch} M(ws \cdot \lambda)$. Since $D(w \cdot \mu)$ has a standard filtration, this guarantees at least that $T_\mu^\lambda D(w \cdot \mu) \neq 0$. More precisely, Theorem 7.14 gives a nonsplit short exact sequence:

$$(1) \qquad 0 \to M(ws \cdot \lambda) \to T_\mu^\lambda M(w \cdot \mu) \to M(w \cdot \lambda) \to 0.$$

Theorem. *Let $\lambda, \mu \in \Lambda$ be antidominant, with λ regular and $W_\mu^\circ = \{1, s\}$ for a simple reflection s. Assume that $w \in W$ satisfies $ws > w$, so $ws \cdot \lambda > w \cdot \lambda$. Then $T_\mu^\lambda D(w \cdot \mu) \cong D(ws \cdot \lambda)$.*

Proof. Since $w \cdot \mu$ is the unique maximal weight of $D(w \cdot \mu)$ and occurs with multiplicity 1, applying (1) above to the various Verma modules including $M(w \cdot \mu)$ in a standard filtration of $D(w \cdot \mu)$ shows that $T_\mu^\lambda D(w \cdot \mu)$ has $ws \cdot \lambda$ as its unique maximal weight (with multiplicity 1). Thus $T_\mu^\lambda D(w \cdot \mu) \cong D(ws \cdot \lambda) \oplus D$ for some tilting module D having all weights $< ws \cdot \lambda$. We just need to show that $D = 0$. This is equivalent to $T_\lambda^\mu D = 0$, in view of Theorem 7.6.

Now the strategy is to apply T_λ^μ to $T_\mu^\lambda D(w \cdot \mu)$. According to Corollary 7.12, the formal character of the resulting module is $2 \operatorname{ch} D(w \cdot \mu)$. But since it is a tilting module, we deduce

$$T_\lambda^\mu T_\mu^\lambda D(w \cdot \mu) \cong D(w \cdot \mu) \oplus D(w \cdot \mu).$$

In particular, this module has $w \cdot \mu$ as its unique maximal weight, with multiplicity 2. If we can show that $T_\lambda^\mu D(ws \cdot \lambda)$ already has $w \cdot \mu$ as a weight with multiplicity 2, it will follow that $T_\lambda^\mu D = 0$ as desired.

Start with an embedding $\varphi : M(ws \cdot \lambda) \to D(ws \cdot \lambda)$ (Proposition 3.7(a)). Dualize Theorem 6.13(c) to get $\operatorname{Ext}_{\mathcal{O}}\big(M(w \cdot \lambda), D(ws \cdot \lambda)\big) = 0$. Applying the functor $\operatorname{Hom}_{\mathcal{O}}\big(?, D(ws \cdot \lambda)\big)$ to (1) above, this allows us to extend φ to a homomorphism $\varphi : T_\mu^\lambda M(ws \cdot \mu) \to D(ws \cdot \lambda)$. Here $\operatorname{Ker} \varphi$ cannot involve the weight $w \cdot \lambda$, lest the sequence (1) split. So φ induces a nonzero homomorphism $M(w \cdot \lambda) \to D(ws \cdot \lambda)/M(ws \cdot \lambda)$. In particular, $w \cdot \lambda$ is a weight (obviously maximal) of the quotient module. This module has a standard filtration, so $M(w \cdot \lambda)$ must occur in such a filtration of $D(ws \cdot \lambda)$ along with $M(ws \cdot \lambda)$. It follows from Theorem 7.6 that $w \cdot \mu$ is a weight of $T_\lambda^\mu D(ws \cdot \lambda)$ with multiplicity (at least) 2, as claimed. $\qquad\square$

Corollary. *Under the hypotheses of the theorem, $\big(D(ws \cdot \lambda) : M(xs \cdot \lambda)\big) = \big(D(ws \cdot \lambda) : M(x \cdot \lambda)\big)$ for all $x \in W$.*

Proof. Thanks to the theorem, $D(ws \cdot \lambda) = T_\mu^\lambda D(w \cdot \mu)$. Since tilting modules (hence their filtration multiplicities) are determined by their formal characters, the corollary follows from the analogue of (1) above when w is replaced by x if $xs > x$ (or by xs if $x > xs$). $\qquad\square$

Notice how the theorem and corollary fail when $ws < w$, by taking the extreme case $w = s$: here $D(ws \cdot \lambda) = M(\lambda)$.

Remark. The classification by Collingwood–Irving of self-dual modules in $\mathcal{O}$ having a standard filtration [**69**, Thm. 3.1] leads to precisely the tilting modules as we have defined them. Their inductive construction relies heavily on translation functors, refined by the Bernstein–Gelfand theory of projective functors: the idea is to start with an antidominant Verma module, cross a wall, then extract the desired indecomposable summand, etc.

11.4. Grothendieck Groups

There is a natural action of W on $K(\mathcal{O})$ or its subgroup $\mathcal{K}$: Using the $\mathbb{Z}$-basis consisting of all symbols $[M(\lambda)]$ with $\lambda \in \mathfrak{h}^*$, an element $w \in W$ sends $[M(\lambda)]$ to $[M(w \cdot \lambda)]$. This action obviously respects the blocks. On the other hand, the Grothendieck group $\mathcal{R}$ of the subcategory $\mathcal{O}_{\mathrm{fd}}$ of $\mathcal{O}_{\mathrm{int}}$ whose objects are finite dimensional modules has a natural ring structure. Since tensoring with a finite dimensional module preserves $\mathcal{O}_{\mathrm{int}}$, the additive group $\mathcal{K}$ is then a module over $\mathcal{R}$.

Proposition. *The natural action of W on $\mathcal{K}$ commutes with the $\mathcal{R}$-module structure.*

Proof. Theorem 3.6 allows us to write for all $\mu \in \Lambda^+$ and $\lambda \in \mathfrak{h}^*$:

$$(1) \qquad [L(\mu)] \cdot [M(\lambda)] = \sum_{\pi \in \Lambda} \dim L(\mu)_\pi [M(\lambda + \pi)].$$

To prove the proposition we have to manipulate this formula, using the fact that $\dim L(\mu)_\pi = \dim L(\mu)_{w\pi}$ for all $w \in W$. Here each sum is taken over $\pi \in \Lambda$:

$$
\begin{aligned}
w \cdot \big([L(\mu)] \cdot [M(\lambda)]\big) &= \sum_\pi \dim L(\mu)_\pi [M(w \cdot (\lambda + \pi))] \\
&= \sum_\pi \dim L(\mu)_\pi [M(w \cdot \lambda + w\pi)] \\
&= \sum_\pi \dim L(\mu)_\pi [M(w \cdot \lambda + \pi)] \\
&= [L(\mu)] \cdot (w \cdot [M(\lambda)]).
\end{aligned}
$$

Here the formula (1) was applied in the first and last equality. $\qquad\square$

11.5. Subgroups of $\mathcal{K}$

Given a finite dimensional module $L(\mu)$ with $\mu \in \Lambda^+$, the tensor product $D(\lambda) \otimes L(\mu)$ decomposes into the direct sum of tilting modules $D(\nu)$ with various multiplicities. To compute these multiplicities explicitly is a difficult problem, tied to the computation of all formal characters $\operatorname{ch} D(\nu)$. (See 11.7 below.) But the task becomes reasonable when λ is regular and antidominant and we look only for the contributions of other regular antidominant weights ν to the direct sum. Indeed, the multiplicities here turn out to be given in terms of classical formulas for decomposing finite dimensional tensor products.

Exercise. Let $\mathfrak{g} = \mathfrak{sl}(2, \mathbb{C})$, with $\Lambda \cong \mathbb{Z}$. Work out the tensor products of the form $M(\lambda) \otimes L(\mu)$ with $\lambda \leq -2$ and $\mu \geq 0$, as direct sums of tilting modules. Compare the occurrence of regular antidominant summands with the classical Clebsch–Gordan formula for $L(w_\circ \lambda) \otimes L(\mu)$.

We can formulate the problem conveniently in the Grothendieck group language. It was already observed in Theorem 11.2(d) that the symbols $[D(\lambda)]$ with $\lambda \in \Lambda$ form a $\mathbb{Z}$-basis of $\mathcal{K}$. Denote by $\mathcal{K}^\circ$ the subgroup of $\mathcal{K}$ spanned by all $[D(\lambda)] = [M(\lambda)]$ with λ *regular* and *antidominant*; such a weight can be expressed uniquely as $w_\circ \cdot \lambda$ for some $\lambda \in \Lambda^+$. All other $[D(\lambda)]$ then span a complementary subgroup $\mathcal{K}'$ of $\mathcal{K}$; thus $\mathcal{K} = \mathcal{K}^\circ \oplus \mathcal{K}'$. How do these Grothendieck groups interact?

Proposition. *Consider the decomposition* $\mathcal{K} = \mathcal{K}^\circ \oplus \mathcal{K}'$.

(a) *The subgroup $\mathcal{K}'$ is equal to the $\mathbb{Z}$-span $\mathcal{K}''$ of all elements of $\mathcal{K}$ which are fixed by at least one reflection s_α with $\alpha > 0$.*

(b) *The subgroup $\mathcal{K}'$ of $\mathcal{K}$ is an $\mathcal{R}$-submodule; so $\mathcal{K}^\circ \cong \mathcal{K}/\mathcal{K}'$ inherits a natural $\mathcal{R}$-structure.*

Proof. (a) First consider $\lambda \in \Lambda$ which is *irregular*: $W_\lambda^\circ \neq 1$. In the Verma basis of $\mathcal{K}$, we can write $[D(\lambda)]$ as a $\mathbb{Z}^+$-linear combination of symbols $[M(\mu)]$ with μ linked to λ (thus also irregular). Since $W_\mu^\circ \neq 1$ is generated by reflections, there is some s_α fixing $[M(\mu)]$ for each μ which occurs. It follows that these symbols and therefore also $[D(\lambda)]$ belong to $\mathcal{K}''$.

Suppose on the other hand that λ is regular but not antidominant. In this case the reflection s in at least one wall of the chamber containing λ satisfies $s \cdot \lambda > \lambda$. It then follows from Corollary 11.3 that we can write

$$[D(\lambda)] = \sum_{w\cdot\lambda < ws\cdot\lambda} \left(D(\lambda) : M(w \cdot \lambda)\right) \left([M(w \cdot \lambda) + M(ws \cdot \lambda)]\right),$$

where the sum is taken over $w \in W$ and s a simple reflection. Since the reflection wsw^{-1} fixes the expression in parentheses, we are again in $\mathcal{K}''$. This shows that $\mathcal{K}' \subset \mathcal{K}''$, as required.

In the other direction, let λ be regular and antidominant; so $s \cdot \lambda > \lambda$ for all reflections s. In particular, s cannot fix $[D(\lambda)]$. The same is then true of any element of $\mathcal{K}^\circ$, so this subgroup of $\mathcal{K}$ intersects $\mathcal{K}''$ trivially. Conclusion: $\mathcal{K}'' = \mathcal{K}'$.

(b) Combine Proposition 11.4 with part (a). $\qquad\square$

Part (a) shows in particular that for any $\lambda \in \Lambda$ and $\alpha > 0$, we have

$$[M(\lambda)] + [M(s_\alpha \cdot \lambda)] \equiv 0 \mod \mathcal{K}'.$$

By iteration, we get:

(1) $\qquad [M(\lambda)] \equiv (-1)^{\ell(w)}[M(w \cdot \lambda)] \mod \mathcal{K}'$ for all $\lambda \in \Lambda$, $w \in W$.

Note in particular that $[M(\lambda)] \equiv 0 \mod \mathcal{K}'$ whenever λ fails to be regular.

11.6. Fusion Rules

For the proof of the main result, we need to recall a classical rule for decomposing the tensor product of two *finite dimensional* simple modules when the weights of one are explicitly known (see [**6**, 1.15]):

(1) $\qquad \mathrm{ch}\left(L(\mu) \otimes L(\lambda)\right) = \sum_{\nu \in \Lambda^+} \sum_{w \in W} (-1)^{\ell(w)} \dim L(\mu)_{w\cdot\nu - \lambda}\, \mathrm{ch}\, L(\nu).$

It will be convenient here to denote by λ a typical element of Λ^+, so that $w_\circ \cdot \lambda$ denotes a typical regular antidominant weight in Λ.

Theorem. *The additive isomorphism $\mathcal{K}^\circ \to \mathcal{R}$ given by $[M(w_\circ\cdot\lambda)] \mapsto [L(\lambda)]$ for $\lambda \in \Lambda^+$ is an isomorphism of $\mathcal{R}$-modules.*

Proof. If $\lambda, \mu \in \Lambda^+$, so $w_\circ \cdot \lambda$ is regular and antidominant, we have to rewrite the corresponding product in $\mathcal{K}$ (modulo $\mathcal{K}'$) in terms of the symbols $[M(w_\circ \cdot \nu)]$ with $\nu \in \Lambda^+$. Start with 11.4(1) above:

$$[L(\mu)] \cdot [M(w_\circ \cdot \lambda)] = \sum_{\pi \in \Lambda} \dim L(\mu)_\pi [M(w_\circ \cdot \lambda + \pi)].$$

Keeping in mind that irregular weights disappear modulo $\mathcal{K}'$, we claim that this sum is equivalent to

$$\sum_{\nu \in \Lambda^+} \sum_{w \in W} (-1)^{\ell(w)} \dim L(\mu)_{w\cdot\nu - \lambda}\, [M(w_\circ \cdot \nu)],$$

which will prove the theorem in view of (1) above.

This just requires a little bookkeeping. We may assume that $w_\circ \cdot \lambda + \pi$ in the first double sum is *regular*, so there is a unique $x \in W$ with $x \cdot (w_\circ \cdot \lambda + \pi) = w_\circ \cdot \nu$, where $\nu \in \Lambda^+$. Solving, we get $\pi = x^{-1} w_\circ \cdot \nu - w_\circ \cdot \lambda$. Thus $w_\circ \pi = (w_\circ x^{-1} w_\circ) \cdot \nu - \lambda$. Since $w_\circ \pi$ has the same multiplicity in $L(\mu)$ as π, while $w := w_\circ x^{-1} w_\circ$ has the same parity as x^{-1} or x, we get a term (modulo $\mathcal{K}'$) of the desired form by applying 11.5(1). $\square$

With the theorem in hand, we can see explicitly how to answer the question raised earlier about extracting from the tensor product of a (regular) antidominant Verma module and a finite dimensional module the tilting summands which belong to other regular antidominant weights. If λ and μ lie in Λ^+, define the **reduced tensor product** $M(w_\circ \cdot \lambda) \underline{\otimes} L(\mu)$ to be the direct sum of those tilting summands (counting multiplicity) $D(w_\circ \cdot \nu) = M(w_\circ \cdot \nu)$ with $\nu \in \Lambda^+$ which occur in the ordinary tensor product. According to the theorem, the multiplicities can be recovered from the classical ones for $L(\lambda) \otimes L(\mu)$ given by (1). (Note that in the proof of the theorem, we have added some details omitted in [6], where there is also a sign error in the subscript attached to $L(\mu)$.)

Remark. There is a formally parallel development for projective modules, which we sketch briefly. The group $\mathcal{K}$ becomes the $\mathbb{Z}$-span in $K(\mathcal{O})$ of the symbols $[P(\lambda)]$ with $\lambda \in \Lambda$. Here $\mathcal{K}'$ and $\mathcal{K}^\circ$ are the complementary subgroups spanned respectively by the $[P(\lambda)]$ with $\lambda \notin \Lambda^+$ and $\lambda \in \Lambda^+$. In the latter case, $P(\lambda) = M(\lambda)$. As above, $\mathcal{K}'$ is stable under the natural action of $\mathcal{R}$, inducing an $\mathcal{R}$-module structure on the quotient $\mathcal{K}/\mathcal{K}' \cong \mathcal{K}^\circ$. There is again a reduced tensor product: If $\mu \in \Lambda^+$, then

$$P(\lambda) \otimes L(\mu) \cong \bigoplus_{\nu \in \Lambda} n^\nu_{\lambda\mu} \, P(\nu).$$

If $\lambda \in \Lambda^+$, define

$$P(\lambda) \underline{\otimes} L(\mu) := \bigoplus_{\nu \in \Lambda^+} n^\nu_{\lambda\mu} \, P(\nu) \ \text{ and } \ [P(\lambda)] \cdot [L(\mu)] := [P(\lambda) \underline{\otimes} L(\mu)].$$

Again the resulting structure constants $n^\nu_{\lambda\mu}$ mirror the classical decomposition of $L(\lambda) \otimes L(\mu)$ into simple summands.

11.7. Formal Characters

So far we have sidestepped one of the most natural questions about the modules $D(\lambda)$: What are their formal characters? Equivalently, what are the standard filtration multiplicities $\big(D(\lambda) : M(\mu)\big)$?

This question was not addressed directly by Collingwood–Irving [69] (or by Andersen–Paradowski [6]). But the comments in [69, §3] are quite suggestive. They point out a sort of duality between tilting modules and projectives in $\mathcal{O}$, manifested in the similar methods of construction based on projective functors (in concrete terms, extracting appropriate direct summands after applying wall-crossing functors). In the case of tilting modules, however, one starts with an antidominant Verma module; in the case of projective modules, one starts with a dominant Verma module. This kind of duality between tilting and projective objects is also observed in parallel module categories, notably in the case of Kac–Moody algebras.

To make these ideas precise, recall the proof of BGG Reciprocity in 3.11. The strategy was to relate a filtration multiplicity and a composition factor multiplicity by equating each to the dimension of a Hom space. Here the goal can be restated as follows for integral weights (the general case is similar). Fix an antidominant weight $\lambda \in \Lambda$. Then for all $x \le w$ in W, the question becomes:

$$(1) \qquad \big(D(w \cdot \lambda) : M(x \cdot \lambda)\big) \overset{?}{=} [M(w_\circ x \cdot \lambda) : L(w_\circ w \cdot \lambda)].$$

By Theorem 3.9, the left side of (1) is equal to

$$\dim \mathrm{Hom}_{\mathcal{O}}\big(D(w \cdot \lambda), M(x \cdot \lambda)^{\vee}\big) = \dim \mathrm{Hom}_{\mathcal{O}}\big(M(x \cdot \lambda), D(w \cdot \lambda)\big),$$

using the fact that tilting modules are self-dual.

Does this agree with the right side of (1)? In the proof of BGG Reciprocity, the analogous argument was elementary. Here that no longer seems to be the case (but see the remarks below). The work of Soergel [**241**] implies the desired result but uses deeper geometric methods related to KL theory; this is inspired by earlier work of Andersen in the setting of quantum groups and characteristic p representations. Andersen's idea was to introduce a filtration in the indicated Hom space, somewhat analogous to the Jantzen filtration. In our situation, Soergel is able to analyze the resulting filtration layers and describe their dimensions in terms of coefficients of KL polynomials [**241**, Thm. 4.4]. Specializing the parameter to 1 then yields:

Theorem (Soergel). *With the above notation, for all $x \le w$ in W we have:*

$$\dim \mathrm{Hom}_{\mathcal{O}}\big(M(x \cdot \lambda), D(w \cdot \lambda)\big) = P_{x,w}(1).$$

Comparing this with the inverse form of the KL Conjecture 8.4(2), we see that the Hom space has dimension equal to the right side of (1). Thus the KL Conjecture coupled with BGG Reciprocity shows:

Corollary. *Equality holds in (1) above. Thus*

$$\big(D(w \cdot \lambda) : M(x \cdot \lambda)\big) = P_{x,w}(1).$$

Moreover, $\big(D(w \cdot \lambda) : M(x \cdot \lambda)\big) = \big(P(w_\circ w \cdot \lambda) : M(w_\circ x \cdot \lambda)\big).$

Soergel has pointed out that (1) can be proved without use of the KL Conjecture. The idea is to set up (using more homological algebra than we have developed) a self-equivalence of the subcategory of $\mathcal{O}$ whose objects are modules with a standard filtration. This interchanges projective modules and tilting modules while preserving filtration multiplicities as in the corollary. (In an unpublished later German version of [**240**], the Kac–Moody case in §2 is supplemented in an added §3 by an explicit discussion of the finite dimensional case.) See also Beilinson–Ginzburg [**21**, Thm. 6.10].

11.8. The Parabolic Case

The treatment of tilting modules in this chapter extends for the most part to a parabolic subcategory $\mathcal{O}^{\mathfrak{p}}$, defined in Chapter 9 relative to a subset $I \subset \Delta$; but some proofs become more complicated. We continue to focus just on *integral* weights and outline briefly what can be done, referring to Andersen–Paradowski [**6**] for details.

By a *tilting module* in $\mathcal{O}^{\mathfrak{p}}$ we mean a module M for which both M and $M^{\vee}$ have filtrations whose quotients are parabolic Verma modules $M_I(\lambda)$. For each integral weight $\lambda \in \Lambda_I^+$, the earlier arguments can be adapted to show the existence of a unique (up to isomorphism) tilting module $D_I(\lambda)$ having λ as highest weight.

Which parabolic Verma modules are also tilting modules in $\mathcal{O}^{\mathfrak{p}}$? In $\mathcal{O}$ the antidominant Verma modules play this role (being simple as well). Here the answer is similar: $M_I(\lambda)$ *is a tilting module if and only if* $M_I(\lambda) = L(\lambda)$. The weights $\lambda \in \Lambda_I^+$ satisfying this condition are somewhat complicated to characterize, but are specified by Jantzen's Theorem 9.13. In case λ is *regular*, the condition for simplicity of $M_I(\lambda)$ is just that λ be as antidominant as possible:

$$\langle \lambda + \rho, \beta^{\vee} \rangle < 0 \text{ for all } \beta \in \Phi^+ \setminus \Phi_I.$$

At the other extreme, we can ask which tilting modules in $\mathcal{O}^{\mathfrak{p}}$ are also projective: that is, when is $D_I(\lambda) \cong P_I(\lambda)$? For this it is necessary that $P_I(\lambda)$ be self-dual. On the other hand, all projectives in $\mathcal{O}^{\mathfrak{p}}$ have "standard filtrations" by parabolic Verma modules (9.8). So the self-dual ones are tilting modules. Whereas the self-dual projectives $P(\lambda)$ in $\mathcal{O}$ are just those with λ antidominant, the criterion for self-duality in $\mathcal{O}^{\mathfrak{p}}$ is more complicated (Theorem 9.14): *The projective tilting modules $P_I(\lambda)$ in $\mathcal{O}^{\mathfrak{p}}$ are those for which $L(\lambda)$ is a summand of the socle of some parabolic Verma module.*

The fusion rules in $\mathcal{O}$ also have a close analogue in $\mathcal{O}^{\mathfrak{p}}$ and lead back in a similar way to classical multiplicity formulas; however, the proof is more intricate than before: see [**6**, 1.20–1.23].

To work out the formal characters of the $D_I(\lambda)$ in terms of parabolic KL data, one can again apply Soergel's methods [**239**, **240**, **241**] to get a duality between projective and tilting modules.

Chapter 12

Twisting and Completion Functors

Here we introduce briefly some further objects of $\mathcal{O}$ and associated functors. There is a continuing flow of research papers involving these ideas, which we cite as we go along. As in Chapters 10 and 11, much of the original motivation comes from other areas of representation theory or neighboring subjects.

The sources we draw on vary considerably in notation as well as in the generality of weights considered. To streamline our account we assume that *all weights are integral* and often *regular* as well; the principal block $\mathcal{O}_0$ is convenient for illustrative purposes. For consistency we label blocks by *antidominant* weights, though some papers cited opt for dominant weights. When using translation functors T_λ^μ, we usually take λ to be regular and antidominant, while μ lies just in a single wall of the antidominant Weyl chamber. (But some results in the literature are valid more generally.)

Here we survey three main topics, which have originated independently over several decades but ultimately turn out to be intimately related:

- We start with *shuffled Verma modules* (later called *twisted Verma modules*): see 12.2–12.5. Such a module has the same formal character as a Verma module, but with the composition factors rearranged. There are several approaches to the construction of these modules, including *shuffling functors* derived from wall-crossing functors.

- Next we define (following Arkhipov) *twisting functors* (12.6). These endofunctors on $\mathcal{O}$ originate quite differently from the shuffling functors, but when applied systematically to Verma modules produce the same families.

- Earlier work of Enright and others involves *completion functors* (12.9). These were first developed in a limited framework as a tool for constructing "discrete series" modules for Lie groups algebraically. But a broader approach eventually leads to interesting interactions with twisting functors (12.12).

As in the case of wall-crossing functors, some constructions begin with a functor attached to a simple reflection (or simple root), followed by iteration relative to a given reduced expression for each $w \in W$. Then it is a problem to decide whether the resulting functor is independent of the choice of reduced expression: if so, we say the *braid relations* are satisfied. This in turn may permit applications to the study of knot and link invariants.

12.1. Shuffling Functors

Irving [**141**] studied certain modules in $\mathcal{O}$ which have the same formal characters as Verma modules but whose composition factors occur in other arrangements. He coined the term **shuffled Verma modules**. (The related name **twisted Verma modules** emerged in some of the later literature.) The prototype is a dual Verma module $M(\lambda)^\vee$ (3.3), but in general many new modules occur. On both philosophical and practical grounds, one should not expect to focus attention on just one Verma module but rather on all those lying in a block; this shows up immediately in our use of wall-crossing functors below.

A parallel family of modules occurs naturally in the setting of *principal series* Harish-Chandra modules for a complex semisimple Lie group whose Lie algebra is isomorphic to $\mathfrak{g}$: see 10.10. Without first establishing a category equivalence relating principal series modules to $\mathcal{O}$, one can attempt a direct functorial construction in each regular block $\mathcal{O}_\lambda$. This starts with the *wall-crossing functor* $\Theta_s = \Theta_s^\lambda$ (for a simple reflection s) defined in 7.14 for a fixed block $\mathcal{O}_\lambda$ with λ antidominant and regular. Here $\Theta_s = T_\mu^\lambda T_\lambda^\mu$ for a suitable weight μ in the s-wall of the antidominant Weyl chamber. Recall that the adjoint property of translation functors yields for each $M \in \mathcal{O}_\lambda$ a natural isomorphism

$$(1) \qquad \mathrm{Hom}_{\mathcal{O}}(T_\lambda^\mu M, T_\lambda^\mu M) \cong \mathrm{Hom}_{\mathcal{O}}(M, T_\mu^\lambda T_\lambda^\mu M).$$

The identity map on $T_\lambda^\mu M$ then corresponds to an *adjunction morphism* $\kappa_s : M \to \Theta_s M$, which may or may not be injective. (For the general notion, see Gelfand–Manin [**110**, II.3.24].)

The **shuffling functor** $\mathrm{Sh}_s = \mathrm{Sh}_s^\lambda$ takes M to the cokernel of κ_s:

$$M \to \Theta_s M \to \mathrm{Sh}_s M \to 0.$$

(Irving writes C_s rather than Sh_s, perhaps to suggest "cokernel"; however, we are reserving the letter C for completion functors.) It is routine to check that maps $M \to N$ induce maps $\mathrm{Sh}_s M \to \mathrm{Sh}_s N$ having the usual functorial properties.

Before considering the effect of shuffling functors on Verma modules, we observe that these functors are somewhat problematic in general. For one thing, they do not commute with arbitrary translation functors. Moreover, it is usually hard to determine whether or not κ_s is injective for a given M; this certainly fails if $M \neq 0$ but $T_\lambda^\mu M = 0$. One special case can be handled more easily. From (1) above we get immediately:

Lemma. *Suppose* $\dim \mathrm{End}_\mathcal{O} T_\lambda^\mu M = 1$. *Then* κ_s *is nonzero and is up to scalars the unique nonzero homomorphism* $M \to \Theta_s M$. $\qquad\square$

12.2. Shuffled Verma Modules

Fix an *antidominant* weight $\lambda \in \Lambda$, which for convenience we assume to be *regular* in order to simplify the use of translation functors. Thus λ lies inside the antidominant Weyl chamber for W. As above, we use an arbitrary μ lying in just one s-wall of that chamber to define a wall-crossing functor Θ_s and shuffling functor Sh_s. Here we consider the effect of Sh_s on a typical Verma module $M(w \cdot \lambda)$ in $\mathcal{O}_\lambda$.

Theorem. *Fix a regular antidominant weight* $\lambda \in \Lambda$. *Let* $w \in W$ *and let* s *be a simple reflection.*

    (a) *The adjunction morphism* $\kappa_s : M(w \cdot \lambda) \to \Theta_s M(w \cdot \lambda)$ *is nonzero.*

    (b) *If* $ws > w$, *then* $\mathrm{Sh}_s M(ws \cdot \lambda) \cong M(w \cdot \lambda)$. *On the other hand,* $\mathrm{Sh}_s M(w \cdot \lambda)$ *is an indecomposable module with simple head* $L(w \cdot \lambda)$ *and formal character equal to* $\mathrm{ch}\, M(ws \cdot \lambda)$.

    (c) *For all* w, *the formal character of* $\mathrm{Sh}_s M(w \cdot \lambda)$ *equals* $\mathrm{ch}\, M(ws \cdot \lambda)$.

Proof. (a) Recall from 7.6 that $T_\lambda^\mu M(w \cdot \lambda) \cong M(w \cdot \mu)$. On the other hand, $\dim \mathrm{End}_\mathcal{O} M(w \cdot \mu) = 1$. It follows from Lemma 12.1 that κ_s is nonzero and is up to scalars the unique nonzero homomorphism $M(w \cdot \lambda) \to \Theta_s M(w \cdot \lambda)$.

(b) Since $T_\lambda^\mu M(w \cdot \lambda) \cong M(w \cdot \mu)$, Theorem 7.14(a) provides a nonsplit short exact sequence:

$$(1) \qquad 0 \to M(ws \cdot \lambda) \to \Theta_s M(w \cdot \lambda) \to M(w \cdot \lambda) \to 0.$$

In particular, $\mathrm{ch}\, \Theta_s M(w \cdot \lambda) = \mathrm{ch}\, M(ws \cdot \lambda) + \mathrm{ch}\, M(w \cdot \lambda)$.

The obvious fact that $\Theta_s M(w \cdot \lambda) \cong \Theta_s M(ws \cdot \lambda)$ allows us to deduce from (1) coupled with part (a) that $\kappa_s : M(ws \cdot \lambda) \to \Theta_s M(ws \cdot \lambda)$ is proportional to the injection in this short exact sequence. In turn, we get $\mathrm{Sh}_s M(ws \cdot \lambda) \cong M(w \cdot \lambda)$.

On the other hand, the fact that κ_s is nonzero on $M(w \cdot \lambda)$ implies that $\kappa_s M(w \cdot \lambda)$ is a highest weight submodule in $\Theta_s M(w \cdot \lambda)$ of weight $w \cdot \lambda$. If its projection to $M(w \cdot \lambda)$ in (1) were nonzero, the projection would be surjective, contrary to the fact that the short exact sequence is nonsplit. Thus κ_s maps $M(w \cdot \lambda)$ into the submodule $M(ws \cdot \lambda)$ of $\Theta_s M(w \cdot \lambda)$. As a nonzero map between Verma modules, κ_s is therefore injective.

By definition there is another short exact sequence:

$$(2) \qquad 0 \to M(w \cdot \lambda) \to \Theta_s M(w \cdot \lambda) \to \mathrm{Sh}_s M(w \cdot \lambda) \to 0.$$

Character comparison with (1) shows that $\mathrm{ch}\,\mathrm{Sh}_s M(w \cdot \lambda) = \mathrm{ch}\, M(ws \cdot \lambda)$.

By Theorem 7.14(b), $\Theta_s M(w \cdot \lambda)$ has a simple head isomorphic to $L(w \cdot \lambda)$. So its nonzero quotient $\mathrm{Sh}_s M(w \cdot \lambda)$ must have the same simple head and therefore be indecomposable.

(c) In case $ws > w$, apply the second statement in (b). But if $ws < w$, apply the first statement in (b) to $w' := ws$. $\qquad\square$

The ultimate objective here is to start with the family $\{M(w \cdot \lambda)\}$ of Verma modules in $\mathcal{O}_\lambda$ indexed by W and obtain by some iteration process a larger family of shuffled modules indexed by pairs $(w, w') \in W \times W$ which can be described intrinsically. Although we can get started with a single functor Sh_s, there is immediately a problem in composing such functors relative to a reduced expression of a Weyl group element. How does one maintain control of $\mathrm{Sh}_t \mathrm{Sh}_s M(w \cdot \lambda)$ and further iterations in the spirit of the above theorem? And will the reduced expression chosen affect the outcome? On the level of formal characters, part (c) of the theorem can be generalized to some extent:

Corollary. *With λ and s as in the theorem, suppose the adjunction morphism κ_s for $M \in \mathcal{O}_\lambda$ is injective. If $\mathrm{ch}\, M = \mathrm{ch}\, M(w \cdot \lambda)$ for some $w \in W$, then $\mathrm{ch}\,\mathrm{Sh}_s M = \mathrm{ch}\, M(ws \cdot \lambda)$.*

Proof. By assumption we have a short exact sequence

$$0 \to M \to \Theta_s M \to \mathrm{Sh}_s M \to 0.$$

Since Θ_s is an exact functor, inducing an endomorphism of $K(\mathcal{O})$, we have $\mathrm{ch}\,\mathrm{Sh}_s M = \mathrm{ch}\,\mathrm{Sh}_s M(w \cdot \lambda)$, which equals $\mathrm{ch}\, M(ws \cdot \lambda)$ thanks to part (c) of the theorem. $\qquad\square$

In what follows we look first for a convenient characterization of a family of shuffled Verma modules and then discuss several approaches to the construction which will permit verification of the injectivity of the relevant adjunction morphisms.

Example. Let $\mathfrak{g} = \mathfrak{sl}(3, \mathbb{C})$. Consider the block $\mathcal{O}_0$, where we write $M_w :=$ $M(w \cdot (-2\rho))$ and $L_w := L(w \cdot (-2\rho))$. Take s to be one of the simple reflections and t the other. If $w = st$, one can describe much of the structure of $\mathrm{Sh}_s\, M_w$ by comparing (1) and (2) above. Here the dominant Verma module $M_{w_\circ}$ has six distinct composition factors, while M_w has just four: L_w, L_s, L_t, L_1. The shuffled module has a quotient isomorphic to the embedded submodule M_w, together with a submodule isomorphic to the quotient $M_{w_\circ}/M_w$. Moreover, $\mathrm{Soc}\, \mathrm{Sh}_s\, M_w \cong L_{ts}$.

As observed above, it is less easy to say what Sh_t does to the modules produced by Sh_s.

12.3. Families of Twisted Verma Modules

Andersen–Lauritzen [5] axiomatize the formal properties required for a family $\{M_\lambda(x,y)\} \subset \mathcal{O}_\lambda$ to be called a **family of twisted Verma modules**. Here λ may be any antidominant weight (but they use the convention that λ is dominant). In the axioms, x and y run over W.

($\mathcal{S}$1) $M_\lambda(1,1) \cong M(\lambda)$.

($\mathcal{S}$2) $M_\lambda(x,y) \cong M_\lambda(xs, sy)$ if $xs > x, sy > y$.

($\mathcal{S}$3) If λ is regular and μ antidominant, then $T_\lambda^\mu M_\lambda(x,y) \cong M_\mu(x,y)$.

($\mathcal{S}$4) If λ is regular and $ys > y$ for a simple reflection s, then for all $x \in W$ the adjunction morphism is injective on $M_\lambda(x,y)$ and yields a short exact sequence

$$0 \to M_\lambda(x,y) \to \Theta_s M_\lambda(x,y) \to M_\lambda(x,ys) \to 0.$$

Thus $\mathrm{Sh}_s\, M_\lambda(x,y) \cong M_\lambda(x,ys)$.

($\mathcal{S}$5) If λ is regular and $ys > y$, then $\Theta_s M_\lambda(x,y) \cong \Theta_s M_\lambda(x,ys)$.

In case $y = 1$, we will see below that ($\mathcal{S}$1) generalizes to $M_\lambda(x,1) \cong M(x \cdot \lambda)$ for all $x \in W$. Thus ($\mathcal{S}$4) implies that $M_\lambda(x,s)$ is isomorphic to the shuffled Verma module $\mathrm{Sh}_s\, M(x \cdot \lambda)$, which by Theorem 12.2(b) has simple head $L(x \cdot \lambda)$ and formal character equal to $\mathrm{ch}\, M(xs \cdot \lambda)$.

Exercise. Let $\mathfrak{g} = \mathfrak{sl}(2, \mathbb{C})$, with $W = \{1, s\}$. Identifying Λ with $\mathbb{Z}$ as usual, assume that $\lambda \leq -1$. Using ($\mathcal{S}$1) $-$ ($\mathcal{S}$5), show how to describe all $M_\lambda(x,y)$ in terms of Verma modules and their duals.

12.4. Uniqueness of a Family of Twisted Verma Modules

Assuming the existence of a family of twisted Verma modules, we want to prove that the family is uniquely determined, up to isomorphism, while exhibiting the formal characters involved.

Theorem. *Fix an antidominant weight $\lambda \in \Lambda$. If $\{M_\lambda(x,y)\}$ with $x, y \in W$ is a collection of modules in $\mathcal{O}_\lambda$ satisfying $(\mathcal{S}1) - (\mathcal{S}5)$, then:*

(a) *$M_\lambda(x,1) \cong M(x \cdot \lambda)$ for all $w \in W$.*

(b) *If $y = s_1 \cdots s_n$ is a reduced expression, then for all $x \in W$ we have $M_\lambda(x,y) \cong \mathrm{Sh}_{s_n} \cdots \mathrm{Sh}_{s_1} M(x \cdot \lambda)$.*

(c) *$\mathrm{ch}\, M_\lambda(x,y) = \mathrm{ch}\, M(xy \cdot \lambda)$ for all $x, y \in W$.*

Proof. (a) First we use $(\mathcal{S}3)$ to reduce the proof to the case when λ is *regular*. Suppose we have already shown that $M_\lambda(x,1) \cong M(x \cdot \lambda)$ in this case, and let μ be antidominant. Combining $(\mathcal{S}3)$ with Theorem 7.6, we get

$$M_\mu(x,1) \cong T_\lambda^\mu M_\lambda(x,1) \cong T_\lambda^\mu M(x \cdot \lambda) \cong M(x \cdot \mu).$$

Fixing the regular weight λ, we now use induction on $\ell(x)$ to complete the proof. The case $x = 1$ is settled by $(\mathcal{S}1)$. Assume that $M_\lambda(x,1) \cong M(x \cdot \lambda)$. If there exists a simple reflection s with $xs > x$, we get from $(\mathcal{S}2)$ and the induction hypothesis:

$$(1) \qquad\qquad M_\lambda(xs,s) \cong M_\lambda(x,1) \cong M(x \cdot \lambda).$$

According to $(\mathcal{S}4)$, $M_\lambda(xs,1)$ is isomorphic to the kernel of the morphism $\Theta_s M_\lambda(xs,1) \to M_\lambda(xs,s)$. Replace the first module by $\Theta_s M_\lambda(xs,s)$ using $(\mathcal{S}5)$ and then replace $M_\lambda(xs,s)$ throughout by $M(x \cdot \lambda)$ using (1). The upshot is that $M_\lambda(xs,1)$ is isomorphic to the kernel of the epimorphism $\Theta_s M(x \cdot \lambda) \to M(x \cdot \lambda)$. Comparison with 12.2(1) shows that this kernel is isomorphic to $M(xs \cdot \lambda)$.

(b) As in part (a), use $(\mathcal{S}3)$ and Theorem 7.6 to reduce to the case when λ is regular.

The proof now goes by induction on $\ell(y)$. In case $y = s_1$, use $(\mathcal{S}4)$ to get

$$0 \to M_\lambda(x,1) \to \Theta_{s_1} M_\lambda(x,1) \to M_\lambda(x,s_1) \to 0.$$

After using (a) to replace $M_\lambda(x,1)$ by $M(x \cdot \lambda)$, comparison with 12.2(2) shows that $M_\lambda(x,s_1) \cong \mathrm{Sh}_{s_1} M(x \cdot \lambda)$. Now proceed inductively.

(c) If the equality of characters holds when λ is regular, apply $(\mathcal{S}2)$ on the left and Theorem 7.6 on the right to get equality in general.

Now assume that λ is regular. Using induction on $\ell(y)$, combine part (b) and $(\mathcal{S}4)$ with Theorem 12.2(c) and Corollary 12.2. $\square$

When λ is regular, the $|W|^2$ modules $M_\lambda(x, y)$ form an array whose rows and columns are indexed by the elements of W taken in some fixed order compatible with the Bruhat ordering (with the first row and column indexed by 1). In particular, the first column consists of Verma modules.

Exercise. Using Theorem 12.2, show that for any $w \in W$, a module $M_\lambda(x, y)$ with character equal to $\operatorname{ch} M(w \cdot \lambda)$ occurs in each row and each column of the array. Exhibit the array of formal characters explicitly when $\mathfrak{g} = \mathfrak{sl}(3, \mathbb{C})$.

Remark. It is natural to ask which $M_\lambda(x, y)$ are isomorphic to dual Verma modules. Experimentation suggests a systematic pattern: for all $x, y \in W$, we expect

$$(2) \qquad M_\lambda(x, y)^\vee \cong M_\lambda(xw_\circ, w_\circ y).$$

In particular, all dual Verma modules $M(x \cdot \lambda)^\vee$ should belong to the family. Here $M(x \cdot \lambda) \cong M_\lambda(x, 1)$, so (2) would force $M(x \cdot \lambda)^\vee \cong M_\lambda(xw_\circ, w_\circ)$. All of this does turn out to be true, after one has constructed a family of twisted Verma modules based on principal series Harish-Chandra modules.

It is tempting to base a proof of (2) instead on the theorem just proved. (In [5] this is stated as a corollary of the theorem, but with a significant step in the proof left unfinished.) The idea is to set $M'_\lambda(x, y) := M_\lambda(xw_\circ, w_\circ y)^\vee$ for $x, y \in W$, then verify that these modules satisfy axioms $(\mathcal{S}1)$–$(\mathcal{S}5)$. The uniqueness of a family of twisted Verma modules in $\mathcal{O}_\lambda$ will yield the asserted isomorphisms. One has to use the fact that translation functors commute with duality (Proposition 7.1). Most of the details are routine, but the verification of $(\mathcal{S}4)$ is tricky. Assuming that λ is regular and that $ys > y$ for a simple reflection s, we want to obtain the following short exact sequence involving the adjunction morphism:

$$0 \to M'_\lambda(x, y) \to \Theta_s M'_\lambda(x, y) \to M'_\lambda(x, ys) \to 0.$$

Since $\ell(ys) > \ell(y)$, we get $\ell(w_\circ ys) < \ell(w_\circ y)$ from 0.3(3); so $(w_\circ ys)s > w_\circ ys$. Therefore $(\mathcal{S}4)$ gives

$$0 \to M_\lambda(xw_\circ, w_\circ ys) \to \Theta_s M_\lambda(xw_\circ, w_\circ ys) \to M_\lambda(xw_\circ, w_\circ y) \to 0.$$

Use $(\mathcal{S}5)$ to replace the middle term by $\Theta_s M(xw_\circ, w_\circ y)$. Then apply duality, which commutes with Θ_s and reverses the order:

$$0 \to M_\lambda(xw_\circ, w_\circ y)^\vee \to \Theta_s M_\lambda(xw_\circ, w_\circ y)^\vee \to M_\lambda(xw_\circ, w_\circ ys)^\vee \to 0.$$

The question remains, however: *Is the map here identifiable with the adjunction morphism for $M_\lambda(xw_\circ, w_\circ y)^\vee$?*

12.5. Existence of Twisted Verma Modules

Andersen–Lauritzen [**5**] describe three approaches to the construction of a family of twisted Verma modules, then prove that these all give the same results:

(1) They reformulate Irving's results in [**141**] (discussed above) based on the category equivalence between a block of $\mathcal{O}$ and a category of Harish-Chandra modules (10.10).

(2) A quite different approach identifies twisted Verma modules with local cohomology modules for the action of an algebraic group G with Lie algebra $\mathfrak{g}$ on the *flag variety* (8.5). This only works directly for integral weights.

(3) A third construction (more intrinsic to category $\mathcal{O}$) is based on Arkhipov's general theory of twisting functors, which we discuss in 12.6 below. This applies to arbitrary weights.

12.6. Twisting Functors

As part of his study of "semi-infinite cohomology", Arkhipov [**8, 9**] was led to introduce **twisting functors**. These turn out to be closely related to shuffling functors, but are in some ways better behaved. On the other hand, Arkhipov's construction is rather subtle; it will be outlined in 12.7 below.

The program is to define for each $w \in W$ a functor T_w on the entire category of $U(\mathfrak{g})$-modules, then see how it restricts to $\mathcal{O}$ and its blocks. Some of the main findings can be summarized as follows, in the later formulations by Andersen–Stroppel [**7**].

Theorem. *The twisting functors T_w with $w \in W$ satisfy:*

(a) *T_w is right exact, but not left exact.*

(b) *If s is a simple reflection and $ws > w$, then $T_{ws} \cong T_w T_s$.*

(c) *T_w preserves $\mathcal{O}$ as well as each block $\mathcal{O}_\lambda$.*

(d) *Up to isomorphism, T_w commutes with the functor $M \mapsto M \otimes L$ when $\dim L < \infty$.*

(e) *T_w commutes with translation functors.*

(f) *$\operatorname{ch} T_w M(\lambda) = \operatorname{ch} M(w \cdot \lambda)$ for all $\lambda \in \mathfrak{h}^*$.*

In their further development of Arkhipov's ideas, Andersen and Stroppel show for example that the left derived functor of T_w defines an auto-equivalence of the bounded derived category $D^b(\mathcal{O}_0)$ [**7**, Cor. 4.2]. They also study how T_s acts on simple modules in $\mathcal{O}$. Moreover, they prove that twisting functors stabilize the collection of all dual Verma modules. It is shown that all twisted Verma modules are indecomposable; in fact their endomorphism algebras are one dimensional.

Remarks. (1) In [**5**, §7], Andersen and Lauritzen explain how twisting functors provide an appropriate setting for the deformation theory needed to construct precise analogues of *Jantzen filtrations* and *sum formulas* for twisted Verma modules. This is made explicit (for regular integral weights) when $\mathfrak{g}$ has type B_2. Here all composition factor multiplicities are 1, so that the sum formulas give unambiguous information about filtration layers. Already in this small case one sees that twisted Verma modules need not have a simple head or simple socle. (It is interesting to note that the same layer structures appeared earlier in the context of "weight filtrations" for related Lie group representations: see Table I in Casian–Collingwood [**62**].)

(2) Recently Abe [**1**] has worked out precise conditions for nonzero homomorphisms to exist between twisted Verma modules. (This generalizes the BGG criterion for Verma modules.) The setting here is the category equivalence involving category $\mathcal{O}$ and principal series Harish-Chandra modules described in 10.10.

12.7. Arkhipov's Construction of Twisting Functors

As promised, we explain now how Arkhipov's functors are defined, following the accounts in Andersen–Lauritzen [**5**, §6] and Andersen–Stroppel [**5**, §6]. Here the universal enveloping algebra $U = U(\mathfrak{g})$ plays an essential role.

The idea is to construct for each $w \in W$ a U-bimodule denoted S_w, then assign to a module $M \in \mathcal{O}$ the left U-module $S_w \otimes_U M$. This is not likely to lie in $\mathcal{O}$, but if we "twist" the action of U by w we get the desired module $T_w M \in \mathcal{O}$. Here we use the fact that w induces an automorphism of Φ and hence an automorphism of $\mathfrak{g}$ or U. The resulting twisted action of U on a module N takes a weight space N_μ to $N_{w\mu}$.

The definition of S_w is somewhat opaque looking at first sight. Start with the subalgebra $\mathfrak{n}_w^- := \mathfrak{n}^- \cap w^{-1}\mathfrak{n}$ of $\mathfrak{n}^-$ and call its universal enveloping algebra N_w for short. For example, if $w = s_\alpha$ is a simple reflection, $\mathfrak{n}_w^- = \mathfrak{g}_{-\alpha}$ and N_w is the polynomial algebra in y_α. The algebra N_w has a graded dual N_w^*, which is $\mathbb{Z}$-graded in nonpositive degrees. This grading starts with the standard grading of U by Λ_r, which defines a $\mathbb{Z}$-grading on U if all simple root vectors are placed in U_1. Then N_w^* is taken to be $\bigoplus_{n \geq 0}(N_w)_n^*$ (the usual dual space of $(N_w)_n$). Somewhat miraculously, it can then be shown that S_w is not just a left U-module but has a natural bimodule structure.

Theorem 12.6(b) indicates that the main focus of attention will be the special case when $w = s_\alpha$ for some $\alpha \in \Delta$ (call it s). Here the definition of T_w given above becomes much more down-to-earth. In the study of enveloping algebras (or other noncommutative noetherian rings), there is a kind of noncommutative (Ore) localization relative to certain multiplicatively closed subsets: see for example Dixmier [**84**, 3.6], Jantzen [**148**, Kap. 11]. In our situation, use the set $\{y_\alpha^n \mid n \in \mathbb{Z}^+\}$ to form an algebra $U_{(s)}$ containing U as a proper subalgebra. In effect, the negative powers of y_α are admitted (say as left factors). Then $U_{(s)}/U$ has a natural structure of bimodule for U. It turns out to be isomorphic to S_s. Thus $T_s M$ is obtained by making U act on $S_s \otimes_U M$ with a twist by s.

A basic conclusion that emerges from this construction and the proof of Theorem 12.6 is

> *The collection of modules produced by applying all T_w to all Verma modules in a given block $\mathcal{O}_\lambda$ is a family of twisted Verma modules in the sense of 12.3 above.*

The precise correlation with the earlier notation $M_\lambda(x, y)$ is a little tricky to work out. In any case, Theorem 12.6(f) shows that the resulting formal character data can be specified quite generally apart from the parametrization of a block by a dominant or antidominant weight.

Remark. Recent work has provided an alternative definition of twisting functors in terms of the completion functors defined earlier by Enright and Joseph; this will be explained in 12.12.

12.8. Twisted Versions of Standard Filtrations

Having introduced twisted versions of Verma modules, it is natural to ask whether these can be used to filter modules (for example, projective or tilting) which are known to have standard filtrations. In the case of projective modules, this was investigated by Irving [**141**, §4] (though with too few details given in the proof of his Theorem 4.1).

In this section, let $\lambda \in \Lambda$ again be regular and antidominant. For projectives there are two extremes: In the case $P(w_\circ \cdot \lambda) = M(w_\circ \cdot \lambda)$, the standard filtration by itself is the only way to filter this module using twisted Verma modules. At the other extreme, $P(\lambda)$ is self-dual and involves all $M(w \cdot \lambda)$ in a standard filtration, each with multiplicity one. So its formal character is at least compatible with the existence of many filtrations whose subquotients are twisted Verma modules.

With these examples in mind, we can formulate Irving's result as follows (under our restrictive hypotheses on λ). It says in effect that the "larger" the projective, the more ways there are to filter it by twisted Verma modules.

> *Fix $x \in W$. If y ranges over the parabolic subgroup of W generated by all simple reflections s with $xs > x$, then*

> $\mathrm{Sh}_y \, P(x \cdot \lambda) \cong P(x \cdot \lambda)$ *and* $P(x \cdot \lambda)$ *has a filtration with quotients of the form* $M_\lambda(x, y)$. *In particular,* $P(\lambda)$ *has such a filtration involving all* $M_\lambda(1, y)$ *for* $y \in W$.

Mazorchuk [**212**] goes on to consider how shuffling (or twisting) functors interact with the *tilting modules* introduced in Chapter 11. Since the self-dual projective $P(\lambda)$ is isomorphic to the tilting module with highest weight $w_\circ \cdot \lambda$, Irving's criterion applies at least here. In fact, Mazorchuk obtains a uniform result. If $x \in W$, we say that a module M has an $\mathcal{F}_x$-filtration if M has a filtration with subquotients of the form $M_\lambda(x, y)$ with $y \in W$. Define an $\mathcal{F}^y$-filtration similarly for a given y.

Theorem. *Let* $\mathcal{O}_\lambda$ *be a block parametrized by an antidominant weight* $\lambda \in \Lambda$. *Then each tilting module in* $\mathcal{O}_\lambda$ *has an* $\mathcal{F}_x$-*filtration and an* $\mathcal{F}^y$-*filtration, for all* $x, y \in W$.

The proof uses wall-crossing functors, induction on length in W, and some complicated diagram-chasing.

12.9. Complete Modules

Here we change the subject, at least temporarily. Enright [**89**] introduced a notion of "complete module" in a category of integral weight modules having substantial overlap with $\mathcal{O}$. The goal was an algebraic construction of discrete series (or more generally, "fundamental series") representations for Lie groups, permitting a natural computation of related multiplicities. Later Joseph [**158**] and Mathieu [**206**] independently broadened the construction to encompass all of $\mathcal{O}$. With the benefit of hindsight, the resulting completion functors are seen to be intimately related to the twisting functors discussed earlier; this permits an alternative approach to the definition (12.12).

To provide some insight into Enright's idea, we first sketch his original construction [**89**, §3] while sidestepping his somewhat complicated proofs (which have been superseded by the later work). Since notation varies a lot in the papers cited below, we impose our own compromise choices.

The setting is the category $\mathcal{I} = \mathcal{I}(\mathfrak{g})$ of $U(\mathfrak{g})$-modules M satisfying the conditions:

($\mathcal{I}$1) M is a weight module with weights in Λ.

($\mathcal{I}$2) M is $U(\mathfrak{n})$-finite, so each x_α with $\alpha > 0$ acts locally nilpotently on M.

($\mathcal{I}$3) M is torsion-free as a $U(\mathfrak{n}^-)$-module.

First we take $\mathfrak{g}$ to be $\mathfrak{s} = \mathfrak{sl}(2,\mathbb{C})$ with its standard basis (h,x,y) and write $\mathcal{I}(\mathfrak{s})$. Similarly, denote by $\mathcal{O}(\mathfrak{s})$ the version of category $\mathcal{O}$ for $\mathfrak{s}$. (Later $\mathfrak{s}$ will be identified with the subalgebra $\mathfrak{s}_\alpha \subset \mathfrak{g}$ for a root $\alpha > 0$.)

The axioms for $\mathcal{I}(\mathfrak{s})$ amount to the requirements: h acts on M by integral weights (identified as usual with elements of $\mathbb{Z}$); x acts locally nilpotently on M; y acts injectively on M. For Enright's application it is important that M is not required to be finitely generated. But if it is finitely generated, the third requirement translates into the statement that M is a *free* $U(\mathbb{C}y)$-module, since this polynomial algebra is a PID. Evidently the finitely generated modules in $\mathcal{I}(\mathfrak{s})$ all lie in $\mathcal{O}(\mathfrak{s})$.

For example, consider a regular (integral) block of $\mathcal{O}(\mathfrak{s})$: here the only highest weights are n and $-n-2$ with $n \in \mathbb{Z}^+$. The indecomposables in such a block were classified in 3.12:

$$L(n),\ M(n),\ M(-n-2) = L(-n-2),\ M(n)^\vee,\ P(-n-2).$$

They all satisfy $(\mathcal{I}1)$ and $(\mathcal{I}2)$, but $L(n)$ and $M(n)^\vee$ fail to satisfy $(\mathcal{I}3)$.

Now write M^x for the subspace of M annihilated by x: this is spanned by all maximal vectors. Call M **complete** if for each $n \in \mathbb{Z}^+$, the action of y^{n+1} maps M_n^x isomorphically onto M_{-n-2}^x. The reader can check that this is equivalent to requiring that for any $\mathfrak{s}$-homomorphism $\varphi : M(-n-2) \to M$, there is a unique $\mathfrak{s}$-homomorphism $\overline{\varphi} : M(n) \to M$ with $\overline{\varphi} \circ i = \varphi$, where $i : M(-n-2) \to M(n)$ is the natural embedding.

When $M \in \mathcal{I}(\mathfrak{s})$ is arbitrary, call a module $\overline{M} \in \mathcal{I}(\mathfrak{s})$ a **completion** of M if $\overline{M}$ is complete and there is an injective map $i : M \to \overline{M}$ for which $\overline{M}/M$ is a locally finite $U(\mathfrak{s})$-module. For example, $M(n)$ is a completion of $M(-n-2)$ when $n \in \mathbb{Z}^+$, whereas $P(-n-2)$ is already complete. The first nontrivial facts to be established are these:

Proposition. *Let $\mathfrak{s} \cong \mathfrak{sl}(2,\mathbb{C})$ and define $\mathcal{I}(\mathfrak{s})$ as above.*

    (a) *Each M in $\mathcal{I}(\mathfrak{s})$ has, up to isomorphism, a unique completion $C(M)$ in $\mathcal{I}(\mathfrak{s})$.*

    (b) *If $\dim L < \infty$ and $M \in \mathcal{I}(\mathfrak{s})$, then $L \otimes M$ belongs to $\mathcal{I}(\mathfrak{s})$. Moreover, $C(L \otimes M) \cong L \otimes C(M)$.*

    (c) *The assignment $M \mapsto C(M)$ defines a covariant functor on $\mathcal{I}(\mathfrak{s})$.*

To illustrate the method of proof, consider first the modules M in $\mathcal{I}(\mathfrak{s})$ which lie in $\mathcal{O}(\mathfrak{s})$. We may assume that M lies in a *regular* block of $\mathcal{O}(\mathfrak{s})$ corresponding to some fixed $n \in \mathbb{Z}^+$. As recalled above, the indecomposables in this block which lie in $\mathcal{I}(\mathfrak{s})$ are quite limited. So the given M decomposes as the direct sum of various $M(n)$ with $n \geq -1$ and various $P(-n-2)$ (these being complete) along with various $M(-n-2)$ (these not being complete).

To construct $C(M)$, one just has to replace each $M(-n-2)$ by $M(n)$ in this direct sum.

12.10. Enright's Completions

Now the problem solved by Enright is to define a similar notion of completion for any semisimple Lie algebra $\mathfrak{g}$, relative to its canonical subalgebra $\mathfrak{s} = \mathfrak{s}_\alpha \cong \mathfrak{sl}(2,\mathbb{C})$) for a fixed $\alpha > 0$. Let $\mathcal{I}_\alpha(\mathfrak{g})$ consist of $U(\mathfrak{g})$-modules M whose restriction $M_\mathfrak{s}$ to $\mathfrak{s}$ lies in $\mathcal{I}(\mathfrak{s})$. Obviously $\mathcal{I}(\mathfrak{g}) \subset \mathcal{I}_\alpha(\mathfrak{g})$. Then:

Theorem. *Fix $\alpha > 0$ and $\mathfrak{s} = \mathfrak{s}_\alpha$ as above. If $M \in \mathcal{I}_\alpha(\mathfrak{g})$, then $C(M_\mathfrak{s})$ admits a unique $\mathfrak{g}$-module structure so that the embedding $M_\mathfrak{s} \hookrightarrow C(M_\mathfrak{s})$ becomes an embedding of $\mathfrak{g}$-modules.*

Enright's proof involves a complicated "amalgamation" construction along the lines of the one described above for $\mathfrak{sl}(2,\mathbb{C})$. His theorem then makes it reasonable to write $C_\alpha M$ in place of $C(M_\mathfrak{s})$. This may be called the α-*completion* of M: clearly $C_\alpha M$ is a completion of M in the above sense for the action of $\mathfrak{s}$. If M happens to lie in $\mathcal{O}$, the construction also makes it clear that $C_\alpha M$ lies in $\mathcal{O}$. In general, $M \mapsto C_\alpha M$ defines a covariant, left exact functor on $\mathcal{I}_\alpha(\mathfrak{g})$ which is idempotent and takes Verma modules to Verma modules. Moreover, if L is finite dimensional, then $C_\alpha(L \otimes M) \cong L \otimes C_\alpha M$; so these functors commute with translation functors. In case M has a contravariant form, this form can be extended to $C_\alpha M$.

Whenever $M \in \mathcal{I}(\mathfrak{g})$, it makes sense to iterate the application of the functors C_α: taking a fixed reduced expression $w = s_1 \cdots s_n$ in W with all reflections $s_i = s_{\alpha_i}$ simple, just set $C_w M := \big(C_{\alpha_1} \circ \cdots \circ C_{\alpha_n}\big)(M)$. But then it remains a problem to decide whether a different reduced expression will yield the same functor C_w. For his application, Enright is able to bypass this question; but it is taken up in later constructions (12.11).

Remark. Motivation for the introduction of completion functors came from the search for an algebraic construction of *discrete series* representations of a real semisimple Lie group G (with finite center) relative to a maximal compact subgroup K. The representations in question occur in $L^2(G)$ and are not acccssible by standard induction methods, though their "characters" were determined by Harish-Chandra. Denoting by $\mathfrak{k}$ and $\mathfrak{g}$ the complexifications of the respective Lie algebras, the discrete series exists when $\mathfrak{g}$ has a "compact" Cartan subalgebra $\mathfrak{h}$ (one which lies in $\mathfrak{k}$). A further problem is to derive algebraically the known multiplicities of (finite dimensional!) simple $\mathfrak{k}$-modules in a discrete series module, as specified by Blattner's Conjecture.

The basic idea in the algebraic construction is to start with a known simple module M (such as an antidominant Verma module in $\mathcal{O}$) and use a sequence of completions to build up a "lattice" of more complicated modules above M indexed by the Weyl group of $\mathfrak{k}$; the elusive discrete series module D is then the simple module at the "top" (analogous to $L(n)$ if the process for $\mathfrak{sl}(2,\mathbb{C})$ begins with $M(-n-2) = L(-n-2)$). As a byproduct of this construction, Enright obtains a resolution of D involving the modules in the lattice, using the maps in the BGG resolutions (6.1) of finite dimensional simple $\mathfrak{k}$-modules. Then the $\mathfrak{k}$-multiplicities in the discrete series module can easily be read off.

The paper by Enright [**89**], which covers more generally the "fundamental series", followed his joint work with Varadarajan (see also Wallach [**258**]). His lecture notes [**90**] provide further exposition, while a brief sketch is given by Humphreys [**128**, §4].

12.11. Completion Functors

Apart from the laborious and somewhat *ad hoc* construction employed by Enright, some natural questions remained open: Do his functors satisfy braid relations? Can the functors be extended in a reasonable way to all of $\mathcal{O}$, and if so, which features will generalize? Two rather different approaches to these questions have emerged:

(1) Deodhar [**77**] uses a localization procedure to streamline the construction of C_α, where we always take α to be a simple root. This leads to a proof of the braid relations and other refinements of Enright's results, but still at the cost of doing complicated direct calculations. (An independent proof of the braid relations is due to Bouaziz [**44**].) Moreover, the functors are still limited to the category $\mathcal{I}$.

Briefly put, Deodhar realizes C_α as a subfunctor of a functor D_α defined concretely as follows. Setting $y = y_\alpha$, one defines for $M \in \mathcal{I}$ an equivalence relation on the set of symbols $\{y^{-n} \cdot v \mid n \in \mathbb{Z}^+, v \in M\}$ by $y^{-n} \cdot v \sim y^{-m} \cdot v'$ if and only if $y^m \cdot v = y^n \cdot v'$. The set $D_\alpha M$ of equivalence classes admits a natural vector space structure as well as a natural action of $U(\mathfrak{g})$ (via the commutation relations); moreover, M embeds in $D_\alpha M$. Then $C_\alpha M$ can be defined to consist of all $v \in D_\alpha M$ on which every root vector x_β with $\beta > 0$ acts nilpotently. (In effect, Deodhar is using an *Ore localization* process; this also occurs in Arkhipov's construction of twisting functors outlined in 12.7.) The intuition behind the use of localization here can be seen in the rank one case, where the process of completing $M(-n-2)$ to $M(n)$ amounts to applying negative powers of y to a maximal vector.

Deodhar's functors can be identified with Enright's. Besides yielding the braid relations, his construction shows conceptually that completions exist and are unique up to isomorphism.

In later work on the classification of simple weight modules, Mathieu [**206**, Appendix] generalizes this localization technique in such a way as to extend Deodhar's construction to a wider range of modules including all of $\mathcal{O}$, while recovering the earlier results such as braid relations. The methods are then refined by König–Mazorchuk [**193**], using a totally different approach: they translate the problem into the category of modules over a finite dimensional algebra attached to a block of $\mathcal{O}$. This streamlines the proof of the braid relations, in particular.

(2) Joseph [**158**, **159**] provides a much less computational approach, defining completion functors on all of $\mathcal{O}$ which satisfy the main properties found earlier including the braid relations. His construction shows in addition that a contravariant form on a module M lifts to its completion. This motivates the work, since it suggests an algebraic approach to understanding the Jantzen Conjecture.

Joseph's method relies heavily on the category equivalence relating $\mathcal{O}$ and a category of Harish-Chandra modules, which we outlined briefly at the end of Chapter 10. In particular, the Harish-Chandra bimodules denoted $\mathcal{L}(M, N)$ there come into play. Take $\lambda \in \Lambda$ to be dominant and regular. Then for a fixed simple root α and any $M \in \mathcal{O}$, set

$$C_\alpha M := \mathcal{L}\big(M(s_\alpha \cdot \lambda), M\big) \otimes_{U(\mathfrak{g})} M(\lambda).$$

From this it is possible to compute all $C_\alpha L(\mu)$ and to show that $C_\alpha M(w \cdot \lambda)$ is isomorphic to either $M(w \cdot \lambda)$ or $M(s_\alpha w \cdot \lambda)$. Moreover, this C_α agrees with the Enright/Deodhar functor on $\mathcal{I} \cap \mathcal{O}$. But C_α is no longer idempotent in general. And it sometimes differs from Mathieu's functor on $\mathcal{O}$, as illustrated in the table below.

12.12. Comparison of Functors

While the twisting and completion functors introduced in this chapter have diverse motivations and histories, they turn out to be close relatives. Mazorchuk and Stroppel draw together the relevant literature in [**219**], starting with a broad overview of endofunctors on $\mathcal{O}$ defined relative to a simple root (or reflection): for example, wall-crossing functors, shuffling functors, twisting functors, or completion functors as defined by Enright, Joseph, Matthieu. In tables and diagrams they summarize concisely the relationships among all of these. Among the issues addressed are: left or right exactness, existence of left or right adjoints, derived functors, morphisms between functors.

The rank one case is easy to illustrate: for a regular integral block of $\mathfrak{sl}(2, \mathbb{C})$ with $n \in \mathbb{Z}^+$, the table below shows the effect on standard modules of Arkhipov's twisting functor as well as the completion functors in variants defined by Joseph and Mathieu.

Table 1. Functors acting on a regular integral block, $\mathfrak{g} = \mathfrak{sl}(2, \mathbb{C})$

	Arkhipov	Joseph	Mathieu
$L(n)$	0	0	0
$M(-n-2)$	$M(n)^\vee$	$M(n)$	$M(n)$
$M(n)$	$M(-n-2)$	$M(n)$	$M(n)$
$M(n)^\vee$	$M(n)^\vee$	$M(-n-2)$	$M(n)$
$P(-n-2)$	$P(-n-2)$	$P(-n-2)$	$P(-n-2)$

In their study of Arkhipov's twisting functor T_w, Andersen–Stroppel [7, §4] show that its right adjoint functor is isomorphic to $D \circ T_{w^{-1}} \circ D$, where D is the duality functor on $\mathcal{O}$; see also their Remark 5.5. Then in [185, §5] Khomenko–Mazorchuk identify this adjoint functor with Joseph's completion functor:

Theorem. *For $w \in W$, denote by T_w Arkhipov's twisting functor and by G_w Joseph's completion functor. On a regular integral block of $\mathcal{O}$, G_w is right adjoint to T_w. If D denotes the duality on $\mathcal{O}$, then moreover*

$$(*) \qquad\qquad G_w \cong D \circ T_{w^{-1}} \circ D.$$

The reader can easily check the isomorphism $(*)$ on standard modules for $\mathfrak{sl}(2, \mathbb{C})$ using the table; but the full isomorphism of functors requires more work. In general, w^{-1} occurs here because a reduced expression $w = s_1 \cdots s_n$ yields $T_w = T_{s_1} \cdots T_{s_n}$ (Theorem 12.6(b)), whereas $G_w = G_{s_n} \cdots G_{s_1}$.

The methods of Khomenko–Mazorchuk [185] were inspired by the representation theory of finite dimensional algebras, enabling them to go further in several directions. In particular, they derive the braid relations for twisting functors from the known braid relations satisfied by completion functors.

Remark. Mazorchuk and Stroppel [219, Thm. 1] develop an interesting application to the study of knot invariants: when $\mathfrak{g} = \mathfrak{sl}(n, \mathbb{C})$, well-chosen endofunctors of $\mathcal{O}_0$ yield a categorification of the Baez–Birman "singular braid monoid". This results from assigning generators of the braid monoid to well-chosen functors.

Chapter 13

Complements

This final chapter is structured more loosely than the previous ones, but has two main themes:

(1) First we sketch some of the interactions between category $\mathcal{O}$ and other parts of representation theory. Here we can cite only a small sample of the extensive literature. We have already pointed out close ties between category $\mathcal{O}$ and Lie group representations. To illustrate the widespread influence of $\mathcal{O}$ in other areas, we discuss both older and newer work, under several headings:

- *Universal enveloping algebras:* classification of primitive ideals (13.1–13.3); Kostant's problem (13.4).

- *Parallels to category $\mathcal{O}$:* Kac–Moody algebras (13.5–13.6); analogues of highest weight modules in various categories, leading to axiomatic approaches (13.7).

- *Blocks:* their quivers and representation types (13.8–13.10).

(2) To round out the chapter, we then look at more advanced viewpoints on $\mathcal{O}$ which have been in the forefront of recent research: Soergel's theorem on endomorphism algebras of self-dual projective modules and the coinvariant algebra of W (13.11–13.13); endomorphism algebras of arbitrary projectives (13.14); Koszul duality (13.15).

Concerning notation, the basic conventions used earlier remain in effect except as indicated. But when the work discussed ranges more broadly, we sometimes follow the mildly conflicting notation used in the cited papers.

13.1. Primitive Ideals in $U(\mathfrak{g})$

We start by looking back at a subject which reached maturity during the decade 1972–1982 but still has unsolved problems. The universal enveloping algebra $U(\mathfrak{g})$ is an interesting object of study in its own right. It is a noncommutative noetherian ring, generated by any basis $x_1, \ldots, x_n$ of $\mathfrak{g}$. A resulting PBW basis induces a filtration of $U(\mathfrak{g})$, whose associated graded algebra is isomorphic to the polynomial algebra $S(\mathfrak{g})$ in n indeterminates (as seen in the proof of the PBW Theorem). The center $Z(\mathfrak{g})$ of $U(\mathfrak{g})$ is itself a polynomial algebra in ℓ variables, isomorphic to the W-invariants for the dot-action in $U(\mathfrak{h}) \cong S(\mathfrak{h})$; its characters or their kernels naturally parametrize the W-linkage classes in $\mathfrak{h}^*$.

The general theory of noetherian rings provides a range of tools for the study of $U(\mathfrak{g})$, while at the same time $U(\mathfrak{g})$ serves as an illuminating example for this theory. More to the point for us, the study of $U(\mathfrak{g})$ has a symbiotic relationship with the representation theory of $\mathfrak{g}$ and especially with category $\mathcal{O}$. Here we focus on *primitive ideals* of $U(\mathfrak{g})$: the kernels of irreducible representations. Although there is no reasonable algebraic classification of simple $U(\mathfrak{g})$-modules, we can still ask for a classification of primitive ideals. These form a partially ordered set $\operatorname{Prim} U(\mathfrak{g})$, which has a Jacobson topology (analogous to the Zariski topology in the commutative case). While the topology on $\operatorname{Prim} U(\mathfrak{g})$ is still incompletely understood, its structure as an ordered set has been well studied.

The results quoted below can be found (with references to primary sources) in the books by Dixmier [84] and Jantzen [148] together with the surveys by Borho [40, 41], Jantzen [149, 150], Joseph [160]. Jantzen's book documents especially the extensive contributions of Joseph. We have also listed in the references a small sample of related papers, including Borho–Jantzen [42], Conze–Dixmier [72], Conze–Duflo [73], Duflo [86, 87, 88], Joseph [154, 155, 157, 162]. The subject is unexpectedly rich, with links to Kazhdan–Lusztig theory, Springer's Weyl group representations, and the geometry of nilpotent orbits in $\mathfrak{g}$.

As pointed out in the Notes to Chapter 1, every simple $U(\mathfrak{g})$-module M has a central character $\chi_M : Z(\mathfrak{g}) \to \mathbb{C}$. If $J = \operatorname{Ann} M$, then $J \cap Z(\mathfrak{g}) = \operatorname{Ker} \chi_M$. This yields a natural map

$$\pi : \operatorname{Prim} U(\mathfrak{g}) \to \operatorname{Max} Z(\mathfrak{g}),$$

which plays a fundamental role in the study of primitive ideals. Since $Z(\mathfrak{g})$ is a polynomial ring, its maximal ideal spectrum may be identified with the affine space $\mathbb{C}^\ell$.

13.2. Classification of Primitive Ideals

Even though $\mathcal{O}$ contains "few" of the simple $U(\mathfrak{g})$-modules, the classical work of Dixmier, Duflo, and others shows that $\mathcal{O}$ does encode indirectly a great deal of information about $\operatorname{Prim} U(\mathfrak{g})$:

Theorem. *Let $\mathfrak{g}$ be a semisimple Lie algebra over $\mathbb{C}$. Then:*

   (a) *If $\lambda \in \mathfrak{h}^*$, the annihilator of $M(\lambda)$ is the two-sided ideal of $U(\mathfrak{g})$ generated by $\operatorname{Ker}\chi_\lambda$; thus it is the same for all weights linked to λ and may be denoted J_χ with $\chi = \chi_\lambda$.*

   (b) *For any central character $\chi : Z(\mathfrak{g}) \to \mathbb{C}$, the ideal J_χ is primitive.*

   (c) *Every primitive ideal of $U(\mathfrak{g})$ is of the form $\operatorname{Ann} L(\lambda)$ for some simple highest weight module $L(\lambda), \lambda \in \mathfrak{h}^*$; call this ideal J_λ.*

   (d) *As χ ranges over the central characters, the ideals J_χ are precisely the minimal primitive ideals of $U(\mathfrak{g})$.*

We already quoted parts (a) and (b) in connection with projective functors in 10.6. While (a) is a fairly deep result, part (b) is then an easy consequence: take $\chi = \chi_\lambda$ with $\lambda \in \mathfrak{h}^*$ *antidominant*, so that $M(\lambda) = L(\lambda)$.

Part (c) is a later theorem of Duflo [**88**].

To deduce (d) from the other statements, observe that $\operatorname{Ann} M(\mu) \subset \operatorname{Ann} L(\mu)$ for all $\mu \in \mathfrak{h}^*$. In particular, when λ is antidominant and $\mu \in W \cdot \lambda$, it follows from (a) and (b) that $J_\lambda = J_\chi = \operatorname{Ann} M(\mu) \subset J_\mu$. Combined with (c), this shows that such ideals J_χ are the minimal ones in $\operatorname{Prim} U(\mathfrak{g})$.

The emerging picture can be summarized in a commutative diagram:

$$
\begin{array}{ccc}
\mathfrak{h}^* & \longrightarrow & \operatorname{Prim} U(\mathfrak{g}) \\
\downarrow & & \downarrow{\scriptstyle\pi} \\
\mathfrak{h}^*/W & \xrightarrow{\ \chi\ } & \operatorname{Max} Z(\mathfrak{g})
\end{array}
$$

The top horizontal map sends $\lambda \in \mathfrak{h}^*$ to J_λ, which π then sends to $J_\lambda \cap Z(\mathfrak{g}) = \operatorname{Ker}\chi_\lambda$. Thanks to the theorem, both of these maps are surjective. The left vertical map sends λ to the W-linkage class $\{w \cdot \lambda \mid w \in W\}$; this too is surjective. Using the fact that $J_\chi = J_\lambda$ when $\chi = \chi_\lambda$ and λ is antidominant, the lower horizontal map sends the linkage class to $\operatorname{Ker}\chi$. This last map is actually bijective, since the natural correspondence between linkage classes and central characters (or their kernels) is bijective.

It is customary to write briefly $\mathcal{X} := \operatorname{Prim} U(\mathfrak{g})$. The fibers $\mathcal{X}_\chi$ of the map π are the finite sets of primitive ideals J_λ with λ running over a single linkage class (where $\chi = \chi_\lambda$). Moreover, no inclusions can exist between primitive ideals lying in distinct fibers. Part (d) of the above theorem shows that $\mathcal{X}_\chi$ has a unique minimal element J_χ. Conze–Dixmier [**72**, Lemme 2] show that the fiber also has a unique maximal element: any ideal J_λ with λ dominant and $\chi = \chi_\lambda$. To describe $\mathcal{X}$ as a partially ordered set reduces now to describing the order structure of each fiber. Along the way one also has to determine the sizes of the fibers.

Borho and Jantzen [**42**] apply translation functors systematically to reduce the problem further to the case when λ is regular and integral. They show that *all fibers of this type are isomorphic as partially ordered sets*, while the singular cases exhibit predictable degeneracies. Understanding a single case such as $\lambda = 0$ is then crucial.

13.3. Structure of a Fiber

In view of the Borho–Jantzen results, we can focus now on the fiber $\mathcal{X}_0$, writing J_w in place of $J_{w\cdot 0}$. The fiber has at most $|W|$ elements, including the unique minimal element $J_{w_\circ}$ and the unique maximal element J_1. Beyond this the structure of $\mathcal{X}_0$ is mysterious. But the work of Borho–Jantzen suggests strongly that it should somehow depend just on W, a situation reminiscent of the multiplicity problem for Verma modules. Part of the problem here is to describe the equivalence relation on W which corresponds to $J_w = J_{w'}$.

For the finer classification of primitive ideals it is useful to consider two invariants, which can be defined for an arbitrary primitive ideal J in terms of the ring-theoretic properties of $U(\mathfrak{g})/J$.

(1) In the classical setting, a simple $U(\mathfrak{g})$-module M with $\dim M < \infty$ corresponds to a matrix representation $U(\mathfrak{g}) \to M_n(\mathbb{C})$ having kernel J; moreover, this map is surjective (Burnside). If $\dim M = \infty$, we still have an injective map $U(\mathfrak{g})/J \to \operatorname{End} M$; but it need not be surjective (see 13.4 below). In general one can assign to a $U(\mathfrak{g})$-module M a measure of "growth" called the *Gelfand–Kirillov dimension* (which was referred to earlier in the discussion of Theorem 9.14). This is denoted $\operatorname{Dim} M$ and is 0 precisely when $\dim M < \infty$. In our situation it takes values in $\mathbb{Z}^+$, because it has a well-behaved commutative counterpart when one passes from $U(\mathfrak{g})$ to $S(\mathfrak{g})$ and replaces J by a graded ideal in the polynomial ring.

(2) By Goldie's Theorem, the ring $U(\mathfrak{g})/J$ embeds into its complete ring of fractions (Ore) which in turn is isomorphic to the ring of $r \times r$ matrices over a division algebra. Call r the *Goldie rank*, denoted $\operatorname{rk}(U(\mathfrak{g})/J)$. Now Joseph goes on to define *Goldie rank polynomials*: Fix $J_w \in \mathcal{X}_0$ and let λ be a regular integral weight in the w-chamber; then set $p_w(\lambda) := \operatorname{rk}(U(\mathfrak{g})/J_\lambda)$. The Zariski density in $\mathfrak{h}^*$ of the regular elements of Λ then allows one to extend p_w to a polynomial function on $\mathfrak{h}^*$. Its degree is given explicitly by

$$\deg p_w = |\Phi^+| - \frac{1}{2}\operatorname{Dim}(U(\mathfrak{g})/J_w).$$

For example, when $w = 1$, this is $|\Phi^+|$ and the polynomial p_1 is just the classical Weyl dimension polynomial. By combining these and related ideas, one concludes that $J_w = J_{w'}$ *if and only if $p_w = p_{w'}$*. Similarly there is a precise algorithm for the inclusion of one ideal in another.

The various J_w in $\mathcal{X}_0$ can be further organized into "clans". In a clan, all the Goldie rank polynomials have the same degree; moreover, they form a basis for a simple $\mathbb{C}W$-module.

Another essential ingredient is supplied by Borho. He associates to a primitive ideal J a nilpotent orbit in $\mathfrak{g}$ under the action of the adjoint group G and proves that $\mathrm{Dim}(U(\mathfrak{g})/J)$ is just the dimension of this orbit. In particular, the orbit dimensions are all *even* and range from 0 to $|\Phi|$, consistent with the above formula for $\deg p_w$.

Kazhdan–Lusztig theory [**170**] draws these diverse strands together and relates them to Springer's Weyl group representations as well. The equivalence classes described above in W are the *left cells* in W, and the $\mathbb{C}W$-modules are those involved in Springer's theory, where the correspondence with nilpotent orbits also plays a natural role. In type A_ℓ one gets all irreducible representations of $W = S_{\ell+1}$, which are parametrized by partitions of $\ell + 1$ (as are the nilpotent orbits, in terms of sizes of Jordan blocks). But in general the picture is much more subtle.

Example. Borho and Jantzen [**42**, 4.17] work out a number of examples to illustrate the ideas in the classification of primitive ideals, including the full order diagram for a regular integral fiber when $\mathfrak{g} = \mathfrak{sl}(4, \mathbb{C})$ or $\mathfrak{sl}(5, \mathbb{C})$. The data for the first of these can be summarized as follows. The fiber $\mathcal{X}_0$ has 10 elements, corresponding to 10 left cells in $W = S_4$. The Gelfand–Kirillov dimensions are $0, 6, 8, 10, 12$ (the dimensions of the five nilpotent orbits), while the Goldie ranks are respectively $6, 3, 2, 1, 0$. The corresponding clans have (respectively) $1, 3, 2, 3, 1$ elements, these being the degrees of irreducible representations of S_4 (whose squares sum to $24 = |W|$). Here 10 is the sum of these degrees and in this case counts the number of involutions in S_4.

13.4. Kostant's Problem

Techniques developed in category $\mathcal{O}$ have a bearing on a classical (still unsolved) problem arising in Kostant's early work on Harish-Chandra modules. This concerns a $U(\mathfrak{g})$-module M and the related bimodule $\mathcal{L}(M, M)$. Recall from 10.9 the construction, which starts with the linear space $\mathrm{Hom}(M, M)$ viewed as a (U, U)-bimodule (where $U = U(\mathfrak{g})$). When $\mathfrak{g} = \mathfrak{k}$ is embedded suitably in $\mathfrak{g} \times \mathfrak{g}$, the space of $\mathfrak{k}$-finite vectors in $\mathrm{Hom}(M, M)$ is a Harish-Chandra module denoted $\mathcal{L}(M, M)$. As observed in Remark 10.9, the bimodule U and its quotient $U/\mathrm{Ann}\, M$ are themselves locally finite under this action of $\mathfrak{g}$. So the natural homomorphism $U \to \operatorname{End} M$ induces an injection $U/\operatorname{Ann} M \to \mathcal{L}(M, M)$. This raises the natural question:

Problem (Kostant). *For which simple $U(\mathfrak{g})$-modules M is the natural injective homomorphism $U(\mathfrak{g})/\operatorname{Ann} M \to \mathcal{L}(M, M)$ surjective?*

While the main interest is focused on simple modules, one can of course pose the question more generally. For example, it is known for a dominant weight λ that $M(\lambda)$ and all its quotients such as $L(\lambda)$ yield positive answers: see Jantzen [**148**, 6.9]. (In case $\dim L(\lambda) < \infty$, this already follows from the classical theorem of Burnside as noted in 13.3(1) above.)

The problem is still unsettled even for the simple highest weight modules $L(\lambda)$, where both positive and negative answers have been found in special cases. Joseph [**156**] studied the problem systematically in terms of Goldie rank, finding a negative answer for type B_2 but a positive answer for some $L(\lambda)$. (See [**156**, 9.5] and [**220**, 11.5].) Further cases where the answer is positive also emerged from the work of Gabber–Joseph [**105**].

In recent years the problem has been revisited by Mazorchuk and his collaborators. Here is a quick overview. In [**213**] Kostant's Problem is studied from the standpoint of the Arkhipov twisting functors discussed in Chapter 12. Their close connection with completion functors (an important ingredient in Joseph's work) makes twisting functors a natural tool here. On the one hand, Mazorchuk solves the problem affirmatively in some new cases, while on the other hand he recovers known results in a more unified framework. For example, let $\lambda \in \Lambda$ be dominant and regular. Take a subset $I \subset \Delta$, with corresponding parabolic subgroup W_I of W and longest element w_I. For a fixed $\alpha \in I$, set $w := s_\alpha w_I w_\circ$. Then *Kostant's problem has a positive answer for $L(w \cdot \lambda)$.*

In the special case $\mathfrak{g} = \mathfrak{sl}(n, \mathbb{C})$, where W is the symmetric group S_n, Mazorchuk–Stroppel [**220**, §11] approach the problem in terms of the Kazhdan–Lusztig (one-sided) cells of W and find positive answers in new as well as known cases. While this encourages optimism about finding only positive answers in type A, counterexamples soon emerge even in this case [**221**]. Definitive results for $\mathfrak{sl}(n, \mathbb{C})$ when $n \leq 5$ are then worked out by Kåhrström and Mazorchuk [**168**] for regular integral blocks.

13.5. Kac–Moody Algebras

We turn now to the role played by category $\mathcal{O}$ ideas in diverse module categories related to Lie theory. *To avoid excessive and often conflicting notation, we temporarily suspend in this and the next section our standard conventions about the symbols $\mathfrak{g}, \mathfrak{h}, W, \ldots$.*

The nearest relatives of a finite dimensional semisimple Lie algebra are the infinite dimensional *Kac–Moody algebras*, which arose independently in the mid-1960s (with different motivations) in the thesis work of Victor Kac and Robert Moody. The starting point for their construction is the presentation of a finite dimensional algebra by generators and relations perfected by Serre: see Bourbaki [**46**, VIII, §4], Carter [**60**, 7.5], Humphreys [**125**, 18.3]. The generators are indexed by simple roots and would be denoted x_i, y_i, h_i in our set-up (but are usually labelled e_i, f_i, h_i in the Kac–Moody literature). The relations depend only on the *Cartan matrix* with integer entries $\langle \alpha_i, \alpha_j^\vee \rangle$, equivalent to the data in the Dynkin diagram.

By relaxing the conditions on matrix entries one gets a *generalized Cartan matrix* $A = (A_{ij})$: its diagonal entries are all 2, its off-diagonal entries are nonpositive integers, with $A_{ij} = 0$ just when $A_{ji} = 0$. The analogue of Serre's presentation then typically yields an infinite dimensional Lie algebra which we again call $\mathfrak{g}$. Somewhat surprisingly, $\mathfrak{g}$ retains many features of the "finite type" $\mathfrak{g}$ such as a triangular decomposition of the form $\mathfrak{g} = \mathfrak{n}^- \oplus \mathfrak{h} \oplus \mathfrak{n}$, with $\mathfrak{h}$ abelian and finite dimensional. There is also an infinite "root system" and "Weyl group" W. Here W is a "crystallographic" Coxeter group in the sense that the orders of products of distinct involutive generators lie in $\{2, 3, 4, 6, \infty\}$. One novelty is that the root spaces $\mathfrak{g}_\alpha$ spanning $\mathfrak{n}, \mathfrak{n}^-$ need not be one dimensional; write $m_\alpha := \dim \mathfrak{g}_\alpha$.

The closest analogue of a finite dimensional $\mathfrak{g}$ is the related *affine Lie algebra*, which we denote by $\widehat{\mathfrak{g}}$. Here the matrix A builds on the Cartan matrix for $\mathfrak{g}$ by adding an extra row and column, corresponding to the extended Dynkin diagram obtained by using as an extra root the negative of the highest root. Unlike general Kac–Moody algebras, $\widehat{\mathfrak{g}}$ admits an explicit model: first construct the *loop algebra* $\mathbb{C}[t, t^{-1}] \otimes \mathfrak{g}$ using the Laurent polynomial ring and define an obvious Lie bracket, then pass to a one dimensional central extension (this is the subtle but essential part). (For technical reasons one then often extends the algebra by a derivation.) In this situation there are two kinds of roots: *real* (conjugate under W to the given "simple" roots) and *imaginary* (with root spaces all of dimension ℓ). The latter are nonzero integral multiples of a single root.

Affine Lie algebras have been well studied and have many links to other parts of mathematics and physics. Beyond these the best behaved Kac–Moody algebras are the *symmetrizable* ones: here A can be made symmetric by multiplying it by a diagonal matrix with positive integer entries on the diagonal. This includes both finite and affine types.

There is by now a vast literature devoted to Kac–Moody algebras as well as their generalizations: vertex operator algebras, Borcherds algebras, etc. Three books which cover the basic material, though from different perspectives, are Kac [**165**], Moody–Pianzola [**223**] (emphasizing the triangular decomposition), Carter [**60**] (integrating the finite and affine types). To convey the flavor of the published research dealing with topics parallel to those in the original category $\mathcal{O}$, we mention here just a tiny sample of papers published over three decades: Kac [**164**], Garland–Lepowsky [**108**], Kac–Kazhdan [**166**], Deodhar–Gabber–Kac [**82**], Rocha–Wallach [**227, 228**], Kumar [**197**], Kashiwara–Tanisaki [**169**], Tanisaki [**249**], Fiebig [**94, 95**].

13.6. Category $\mathcal{O}$ for Kac–Moody Algebras

Any Kac–Moody algebra admits a category of "highest weight" modules analogous to $\mathcal{O}$. As before a weight λ is a linear function on $\mathfrak{h}$. But one has to relax the finite generation assumption, allowing for infinitely many "composition factors". Moreover, the center of the universal enveloping algebra plays no significant organizing role. As a result it is highly nontrivial to generalize proofs even though a surprising number of theorems for the finite dimensional case have good analogues here.

One starts with natural analogues of Verma modules, again written $M(\lambda)$; these have easily computed formal characters and unique simple quotients $L(\lambda)$. There is a reasonable analogue here of finite dimensional modules: a module M in $\mathcal{O}$ is called *integrable* if all root vectors e_i, f_i act locally nilpotently on it; then $\dim M_\mu = \dim M_{w\mu}$ for all μ and all $w \in W$. For a simple module $L(\lambda)$ this is equivalent to requiring that λ be *dominant integral*: all $\lambda(h_i) \in \mathbb{Z}^+$. Now ρ is chosen (non-uniquely) to satisfy $\rho(h_i) = 1$ for all i. In 1974 Kac [**164**] was able to prove:

Theorem (Kac). *Let $\mathfrak{g}$ be a symmetrizable Kac–Moody algebra. If λ is dominant and integral, then $L(\lambda)$ is integrable, with*

$$Q * \operatorname{ch} L(\lambda) = \sum_{w \in W} (-1)^{\ell(w)} e\big(w(\lambda + \rho)\big).$$

Here

$$Q = e(\rho) * \prod_{\alpha > 0} \big(1 - e(-\alpha)\big)^{m_\alpha} = \sum_{w \in W} (-1)^{\ell(w)} e(w\rho).$$

All of this closely resembles the classical Weyl formula (Theorem 2.4) except for the root multiplicities $m_\alpha \geq 1$ appearing in the denominator formula. But in general the sums and products here are all infinite. The proof by Kac is similar to the BGG proof we followed in the finite dimensional case, but relies essentially on a *Casimir operator* (a formal infinite sum which acts on a module just as a finite sum) as a substitute for the full center of the enveloping algebra.

While at first the theorem appears to be a purely formal generalization of classical results, it was motivated in large part by a spectacular interpretation of the above formula for Q. This expresses an infinite product over positive roots in terms of an infinite sum over the Weyl group. Special cases of this were discovered earlier in an *ad hoc* way by Macdonald; this recaptures in particular the classical Jacobi Triple Product Identity related to Dedekind's η-function. The Lie algebra approach reveals many new "sum equals product" identities, within a conceptually unified framework.

As in the finite dimensional case, the character formula can also be realized by a BGG-type resolution. Associated with this are Lie algebra homology computations. (See for example Kumar [197].)

Generalizing the Weyl–Kac character formula when $L(\lambda)$ is integrable, one can formulate a precise analogue of the KL Conjecture for any $L(\lambda)$: since W is a Coxeter group, all of the combinatorial machinery is already in place. But the proof in the Kac–Moody case requires substantial new ideas, supplied by the work of Casian and Kashiwara–Tanisaki [169, 249].

Besides the study of multiplicities in highest weight modules, some of the literature cited above delves into other more categorical aspects of $\mathcal{O}$ for a Kac–Moody algebra. But nothing is quite so straightforward as in the case when $\mathfrak{g}$ is finite dimensional.

13.7. Highest Weight Categories

Beyond the case of Kac–Moody algebras, there are quite a few module categories which resemble $\mathcal{O}$ in some ways but differ significantly in others. Each has its own extensive literature (and local notation), defying even the type of brief sketch we gave in the case of Kac–Moody algebras. Here we just list some of the active areas of study in recent decades, with reference to books, surveys, and a somewhat arbitrary sample of research papers.

- simple weight modules: Mathieu [206].

- Virasoro algebra: Rocha–Wallach [229], Boe–Nakano–Wiesner [38].

- Lie superalgebras: Brundan [49, 50].

- finite W-algebras: Brundan–Goodwin–Kleshchev [53].

- quantum enveloping algebras (and quantized function algebras): Andersen [4], Jantzen [151], Joseph [161].

- symplectic reflection algebras, including as a special case the rational Cherednik algebras: Gordon [113], Ginzburg–Guay–Opdam–Rouquier [111], Guay [116], Khare [172].

- representations in prime characteristic of algebraic groups, their Lie algebras, finite groups of Lie type: Jantzen [**152**], Humphreys [**131**].

- Schur algebras and q-Schur algebras: Donkin [**85**].

- BGG Reciprocity for perverse sheaves: Mirollo–Vilonen [**222**].

Is it possible to extract from the diversity here some common core in axiomatic form? This could be modelled for example on the features of a single block of the original category $\mathcal{O}$. A number of axiomatic approaches have been proposed: see Cline–Parshall–Scott [**63, 64**], Irving [**137**], Futorny–Mazorchuk [**103**], Gomez–Mazorchuk [**112**], Khare [**173**].

The set-up formulated by Cline–Parshall–Scott is rather general and has been especially influential. In [**63**, §3] an abelian category $\mathcal{C}$ is dubbed a *highest weight category* if it is "locally artinian", with enough injectives, and satisfies a number of axioms. One can define the composition factors of an object $A \in \mathcal{C}$ to be composition factors of a subobject of finite length, leading to multiplicities $[A : S]$ which may be infinite. Now the simple objects $S(\lambda)$ are assumed to be indexed by an interval-finite partially ordered set (thought of as "weights"), as are objects $A(\lambda)$ modelled on dual Verma modules, so that $S(\lambda) \subset A(\lambda)$ and all composition factors $S(\mu)$ of $A(\lambda)/S(\lambda)$ satisfy $\mu < \lambda$. Moreover, $S(\lambda)$ has an injective envelope $I(\lambda)$ in $\mathcal{C}$ which is filtered by various $A(\mu)$ (each of finite multiplicity).

This framework can be seen to encompass many of the above examples both in characteristic 0 and in characteristic p. Moreover, there turns out to be a close connection with the *quasi-hereditary* algebras introduced by Scott: for a finite dimensional algebra A over a field, its module category is a highest weight category precisely when A is quasi-hereditary [**63**, Thm. 3.6].

Irving [**137**] studies a narrower class of highest weight categories attached to *BGG algebras*. Here the category is required to have a duality functor, leading to BGG Reciprocity.

13.8. Blocks and Finite Dimensional Algebras

By decomposing $\mathcal{O}$ (or $\mathcal{O}^{\mathfrak{p}}$) into a direct sum of subcategories, each having only finitely many simple objects, one is able (in principle) to invoke the well-developed techniques of the representation theory of finite dimensional algebras. This is usually not easy to do in practice, but some new approaches and insights outside the usual framework of Lie theory are suggested. Here we look briefly at two related topics:

(1) Quiver attached to a block.

(2) Representation type of a block.

The easiest decomposition of $\mathcal{O}$ involves the subcategories $\mathcal{O}_\chi$ attached to central characters χ (or W-linkage classes of weights). It was shown in Proposition 3.13 that each $\mathcal{O}_\chi$ is equivalent to the category of (right) A-modules for some finite dimensional algebra A; a convenient choice is the endomorphism algebra of a projective generator.

If the weights involved are required to be integral, the subcategories $\mathcal{O}_\chi$ are actually the *blocks* of $\mathcal{O}$: see 1.13. (As before the reader is cautioned to be aware of different usages for the term "block" in the literature cited.) Otherwise $\mathcal{O}_\chi$ has a refined block decomposition, using the nonempty intersections of a linkage class with cosets of Λ_r in $\mathfrak{h}^*$. Here the blocks $\mathcal{O}_\lambda$ are indexed by their unique antidominant weights λ: see 4.9 and Remark 3.5. The blocks in $\mathcal{O}_\chi$ are all isomorphic as partially ordered sets (even equivalent as categories, thanks to Mathieu [**206**, A.3]).

We remark that Soergel's work, discussed in 13.13 below, shows when $\lambda \notin \Lambda$ that $\mathcal{O}_\lambda$ is the subcategory corresponding to a linkage class of $W_{[\lambda]}$ in the analogue of category $\mathcal{O}$ for a semisimple Lie algebra having Weyl group $W_{[\lambda]}$. This makes it possible to limit attention just to *integral* weights, which we now do. The main ideas can be illustrated conveniently for the principal block, using the previous convention that $L_w = L\big(w \cdot (-2\rho)\big)$, etc.

13.9. Quiver Attached to a Block

The use of quivers in Lie theory goes back to early work of I. M. Gelfand and his collaborators. But the first systematic study of quivers for $\mathcal{O}$ seems to have been carried out by Irving, starting with his unpublished notes [**132**] (see [**133**]). Recent work by Stroppel [**246**] and Vybornov [**257**] goes much further. The quiver viewpoint has the advantage of stripping away all but the most essential features of the module category in question.

Here is a quick review of the construction. Given a finite dimensional algebra A, its quiver is a directed graph having vertices in bijection with the simple modules S_i. There are n_{ij} arrows $i \to j$ if $\dim \mathrm{Ext}_A(S_j, S_i) = n_{ij}$ (the multiplicity of S_i as a summand of the top radical layer of the projective cover P_j of S_j). Each arrow represents a nonsplit extension having length 2 (equivalently, a nonzero homomorphism $P_i \to P_j$ whose image meets the top radical layer of P_j in a copy of S_i).

Up to Morita equivalence, A may be assumed here to be a *basic* algebra: all $\dim S_i = 1$; so $A \cong \bigoplus_i P_i$. Then A is isomorphic to the quotient of its *path algebra* by the ideal of relations, where the path algebra has a basis consisting of all directed paths in the quiver and an obvious product ($= 0$ when two paths cannot be composed). Thus one wants to describe the quiver with relations in order to recover the full category of A-modules. These modules correspond to *representations* of the quiver, defined by assigning a vector space to each vertex and a linear map to each arrow.

In the case of $\mathcal{O}$, Theorem 3.2(e) ensures that

$$\dim \mathrm{Ext}_{\mathcal{O}}\big(L(\lambda), L(\mu)\big) = \dim \mathrm{Ext}_{\mathcal{O}}\big(L(\mu), L(\lambda)\big) \text{ for all } \lambda, \mu.$$

Rather than use pairs of arrows, we can just connect vertices with a double arrow when $\mathrm{Ext} \neq 0$. The simplest example is the principal block for $\mathfrak{sl}(2, \mathbb{C})$ [**246**, 5.1.1]. Labelling the vertices (left to right) by the weights 0 and -2, we get the quiver

$$\bullet \longleftarrow \bullet.$$

Observe how to identify the five nonisomorphic indecomposable modules in $\mathcal{O}_0$ specified in 3.12 with five (indecomposable) representations of the quiver. In this case one needs only the single relation

$$(\bullet \longleftarrow \bullet) \circ (\bullet \longrightarrow \bullet) = 0.$$

But beyond rank one the number of relations required grows very fast.

In general the KL Conjecture implies for the block $\mathcal{O}_0$ that if $x < w$ in W, then $\dim \mathrm{Ext}_{\mathcal{O}}(L_x, L_w) = \mu(x, w)$; this is the lead coefficient of the KL polynomial $P_{x,w}(q)$ if its degree is as large as possible, or 0 otherwise. (See Theorem 8.15(c).) But it is not obvious how to describe all the relations involved in the quiver.

Going beyond Irving's direct methods, Stroppel [**246**] takes advantage of Soergel's identification of a regular block of $\mathcal{O}$ with a category of modules over the coinvariant algebra of W (discussed in 13.12 below). She formulates an algorithm for computing the quiver with relations of any such block and carries out the details for examples in rank 2 as well as for $\mathfrak{sl}(4, \mathbb{C})$. In this last case the resulting quiver diagram and list of relations (given in the Appendix) are impressive; as she observes, there are "a lot of arrows and very many relations". Indecomposable modules can then be pictured in detail as representations of the quiver, for example the dominant Verma module in a regular integral block of $\mathfrak{sl}(4, \mathbb{C})$. Here some multiple composition factors occur, leading to submodules which are not sums of Verma submodules and which can be located precisely using the quiver.

Vybornov [**257**] refines Stroppel's quiver algorithm for an arbitrary simple $\mathfrak{g}$, leading to a computer executable algorithm. His approach is entirely different, based on modeling categories of perverse sheaves on various spaces including the flag variety by "IC-modules" (without use of derived categories).

13.10. Representation Type of a Block

In studying the module category of a finite dimensional algebra A, one has to look at all possible indecomposable modules, including the finitely many simple modules and their projective covers. There are three possibilities for the "representation type" of A: *finite type*, meaning that up to isomorphism A has only finitely many indecomposables; *tame type*, meaning that A has infinitely many nonisomorphic indecomposables, which can nonetheless be parametrized in some reasonable way; *wild type*, when neither of these alternatives is true. In the literature, there are precise definitions of tame type (differing somewhat in the details) which we won't pursue here.

When A is the group algebra of a finite group, much work has been done to sort out the possibilities here in terms of the structure of the group. In the case of a finite group of Lie type, the outcome has a somewhat negative flavor: in only a few small rank cases does one have finite or tame representation type. (These are summarized in Humphreys [**131**, 8.9].) Recent work on blocks of category $\mathcal{O}$ or $\mathcal{O}^{\mathfrak{p}}$ leads to a somewhat similar outcome. Here we have already seen in 3.12 a small rank case of finite representation type: a regular integral block of $\mathfrak{g} = \mathfrak{sl}(2, \mathbb{C})$, which contains just five indecomposable modules. Of course, the singular block involving just the weight $-\rho$ is also of finite type (and is moreover semisimple).

Roughly speaking, the classification of blocks by representation type is done indirectly. Typically one shows (using quivers) that a known wild type of module category embeds in the given block, ensuring that it too is of wild type. Then a lot of case-by-case work is needed to sort out the relatively few remaining cases. We can summarize the end result for $\mathcal{O}$ in the case when $\mathfrak{g}$ is simple. Consider a block $\mathcal{O}_\lambda$ with $\lambda \in \Lambda$ antidominant. The isotropy group W_λ° of λ is then the Weyl group of a root subsystem $\Psi \subset \Phi$.

Theorem. *Fix an antidominant weight $\lambda \in \Lambda$, with isotropy group W_λ° and root system $\Psi \subset \Phi$. Then the representation type of $\mathcal{O}_\lambda$ depends just on the pair (Φ, Ψ), as follows.*

(a) *Finite type:* (Φ, Φ), $(\mathsf{A}_1, \emptyset)$, $(\mathsf{A}_2, \mathsf{A}_1)$.

(b) *Tame type:* $(\mathsf{A}_3, \mathsf{A}_2)$, $(\mathsf{B}_2, \mathsf{A}_1)$.

(c) *Wild type: all others.*

This theorem is the main result of Futorny–Nakano–Pollack [**104**], who rely mainly on techniques such as quivers from the representation theory of finite dimensional algebras together with a comparison of blocks for Levi subalgebras of $\mathfrak{g}$. The theorem is recovered a bit later in a completely different way by Brüstle–König–Mazorchuk [**54**], building in part on a paper by the last two authors [**193**]. They exploit heavily the work of Soergel on projectives in $\mathcal{O}$ and the coinvariant algebra of W (13.12 below) along with Schubert calculus for the cohomology algebra of the related flag variety.

The story for blocks in parabolic categories $\mathcal{O}^{\mathfrak{p}}$ is considerably more complicated and less complete, the blocks themselves not yet being fully determined. Boe–Nakano [**37**] have worked out the representation type of certain blocks, while further cases have been treated by Boe and his student K. Platt. As above, most blocks are wild. When the weights involved are *regular* (and integral), the blocks of each representation type are known. Very few of them are of finite type: for the root systems of types A_ℓ, B_ℓ, C_ℓ, the subset $I \subset \Delta$ must be respectively of type $A_{\ell-1}, B_{\ell-1}, C_{\ell-1}$; the only other case is G_2 with I of type A_1. Among the blocks involving the linkage class of a singular weight λ, detailed results are obtained for those which depend on two *disjoint* subsets of Δ: one is the given set I with $\mathfrak{p} = \mathfrak{p}_I$ while the other defines the subsystem consisting of all roots α for which $\langle \lambda + \rho, \alpha^\vee \rangle = 0$.

13.11. Soergel's Functor $\mathbb{V}$

Now we change gears once again and introduce some influential work of Soergel [**237**] and Beilinson–Ginzburg–Soergel [**22**], which was stimulated by earlier unpublished conjectures of Bernstein and these authors. They look at category $\mathcal{O}$ from a more advanced homological viewpoint, making further connections to perverse sheaves and the geometry of flag varieties. One striking theme that emerges is the role of *graded* (not just filtered) categories. Eventually the geometric side of the subject is captured under the rubric "geometric representation theory", as in Gaitsgory's lecture notes [**107**].

The first goal is to formulate precisely some of the main results in [**237**, §2], where the methods are still algebraic; but we follow our own notational conventions when conflicts occur. Here it is important to allow arbitrary weights $\lambda \in \mathfrak{h}^*$ and blocks $\mathcal{O}_\lambda$, with the corresponding reflection subgroups $W_{[\lambda]}$ of W. In particular, let λ be *antidominant* and consider the projective module $P = P(\lambda)$. When λ is integral, we saw in 4.10 that P is self-dual, hence also injective, while each Verma module in its block $\mathcal{O}_\lambda$ occurs precisely once as a quotient in any standard filtration. With more effort, the same result was later proved in general using translation functors (7.16).

Write $E = \operatorname{End}_{\mathcal{O}} P$ and denote by $\mathcal{M}$ the category of finite dimensional E-modules. For any $M \in \mathcal{O}$, observe that the vector space $\operatorname{Hom}_{\mathcal{O}}(P, M)$ has a natural structure of E-module: simply compose an endomorphism $P \to P$ with a given homomorphism $P \to M$. Now Soergel defines a functor $\mathbb{V} = \mathbb{V}_P$ from $\mathcal{O}$ to $\mathcal{M}$, which sends $M \mapsto \operatorname{Hom}_{\mathcal{O}}(P, M)$ and behaves in the obvious way on morphisms. For example, when $M = P$ we just get $\mathbb{V}P = E$. With this set-up, a fundamental structure theorem states:

Struktursatz. *Let* $\lambda \in \mathfrak{h}^*$ *be antidominant and write* $P = P(\lambda)$, $E = \mathrm{End}_{\mathcal{O}}\, P$. *If* Q *is any projective module in the block* $\mathcal{O}_\lambda$ *and* $M \in \mathcal{O}$, *then the functor* $\mathbb{V}$ *defined above induces a vector space isomorphism*

$$\mathrm{Hom}_{\mathcal{O}}(M, Q) \xrightarrow{\sim} \mathrm{Hom}_E(\mathbb{V}M, \mathbb{V}Q).$$

Remarks. (1) For an easy example, take $M = Q = P$: then the theorem just expresses the natural vector space isomorphism from E onto $\mathrm{Hom}_E(E, E)$.

(2) More recently Beilinson–Bezrukavnikov–Mirković [**19**] have recovered the Struktursatz in a geometric setting.

To make the functor $\mathbb{V}$ effective, one of course has to understand the structure of the finite dimensional algebra E. Indeed, the proof of the Struktursatz given in [**237**, 2.3] relies on a concrete description of E. This turns out to have a natural connection with W. The computation of $\dim E$ is already suggestive. According to Theorem 3.9(c), $\dim E = [P : L(\lambda)]$. This equals the index in $W_{[\lambda]}$ of the isotropy group W_λ° thanks to the way P is filtered by Verma modules in $\mathcal{O}_\lambda$ and the fact that $L(\lambda)$ occurs just once as a composition factor of each.

13.12. Coinvariant Algebra of W

In general it is not obvious how to determine the endomorphism algebra of an indecomposable projective in $\mathcal{O}$. There is one obvious step in this direction: left multiplication by an element of $Z(\mathfrak{g})$ is always a $U(\mathfrak{g})$-endomorphism, so there is a natural map from $Z(\mathfrak{g})$ to the endomorphism algebra. When $E = \mathrm{End}\, P$ with $P = P(\lambda)$ for an antidominant $\lambda \in \mathfrak{h}^*$, it follows from Soergel's description of E in the theorem below that this natural map is surjective. In particular, E is *commutative*. The key result is that E depends only on $W_{[\lambda]}$ and W_λ°, and is in fact isomorphic to the *coinvariant algebra* C of $W_{[\lambda]}$ if λ is regular.

We have to recall how C is defined, referring for details to Bourbaki [**45**, V, §5], especially 5.2, Thm. 1 and Thm. 2 (see also Humphreys [**129**, 3.6]). The definition makes sense for an arbitrary finite reflection group, which we just call W for convenience.

Viewing W as a subgroup of a suitable $\mathrm{GL}(V)$ with $\dim V = n$, its action on W extends naturally to the (graded) symmetric algebra $S(V)$. By Chevalley's theorem, the subalgebra $S(V)^W$ of invariants is itself a polynomial algebra in n indeterminates and is generated by homogeneous elements of $S(V)$; their degrees are uniquely determined and have product $|W|$. (When

W is a symmetric group, this is just the theorem on elementary symmetric polynomials.) In turn, the ideal I generated by homogeneous invariants of positive degree is homogeneous and the quotient $S(V)/I$ is a graded finite dimensional algebra called the *coinvariant algebra of* W. Denote it by C. As a W-module it affords the regular representation of W; thus $\dim C = |W|$.

It is well known from the classical work of Borel that the coinvariant algebra of the Weyl group W of $\mathfrak{g}$ is isomorphic as a graded algebra to the cohomology algebra of the flag variety G/B (as in 8.5). This fact is used in the proof of the theorem below; it also suggests links between $\mathcal{O}$ and the geometry of Schubert varieties implicated in the KL Conjecture.

Endomorphismensatz. *Let λ and $P = P(\lambda)$ be as in the Struktursatz, and recall that W_λ° denotes the isotropy group of λ in $W_{[\lambda]}$. Let C be the coinvariant algebra of $W_{[\lambda]}$ as above, with subalgebra $C^{W_\lambda^\circ}$ of invariants under W_λ°. Then there is a surjective homomorphism $Z(\mathfrak{g}) \to C^{W_\lambda^\circ}$, having as kernel $\mathrm{Ann}_{U(\mathfrak{g})} P$. As a result,*

$$\mathrm{End}_\mathcal{O} P \cong C^{W_\lambda^\circ},$$

which is just C if λ is regular.

For the proof of the theorem in [**237**, 2.2], Soergel relies heavily on translation functors (especially wall-crossing functors), together with a close study of the interaction between C and $Z(\mathfrak{g})$.

Once the Endomorphismensatz is in place, the proof of the Struktursatz is relatively short but requires a number of techniques. In one step, the copy of the dominant Verma module embedded in P is characterized as the subspace of P annihilated by the unique maximal ideal of C. In another step, the modules in $\mathcal{O}$ killed by $\mathbb{V}$ are characterized as those having Gelfand–Kirillov dimension $< \dim \mathfrak{n}$; then projective functors come into play.

13.13. Application: Category Equivalence

The results of Soergel discussed above make more precise the earlier observation that key features of $\mathcal{O}$ depend solely on the Weyl group. For example, the multiplicities of composition factors of Verma modules with integral weights are given by the values at 1 of KL polynomials attached to W and therefore coincide for Lie algebras of types B_ℓ and C_ℓ.

Here we get insight into the way in which a block of $\mathcal{O}_\lambda$ associated with a *nonintegral* weight λ, together with the subgroup $W_{[\lambda]}$ of W, can be related to an *integral* block of another Lie algebra having $W_{[\lambda]}$ as its Weyl group. (As mentioned in earlier chapters, this reduces many problems such as the proof of the KL Conjecture to the integral case.) Soergel's theorem [**237**, 2.5, Thm. 11] can be stated formally as follows.

Theorem. *Let* $\mathfrak{g}$ *and* $\mathfrak{g}'$ *be semisimple Lie algebras, with respective Weyl groups* W *and* W'. *Fix weights* λ *for* $\mathfrak{g}$ *and* λ' *for* $\mathfrak{g}'$, *with corresponding blocks* $\mathcal{O}_\lambda$ *and* $\mathcal{O}'_{\lambda'}$ *and reflection subgroups* $W_{[\lambda]}$ *and* $W'_{[\lambda']}$. *If there is an isomorphism between these Coxeter groups which takes the isotropy group of* λ *to the isotropy group of* λ', *then* $\mathcal{O}_\lambda$ *is equivalent to* $\mathcal{O}'_{\lambda'}$, *with* $L(\lambda)$ *sent to* $L(\lambda')$ *and* $M(\lambda)$ *sent to* $M(\lambda')$.

For the application just indicated, take λ to be nonintegral and then use the fact that $W_{[\lambda]}$ is the Weyl group of the root system $\Phi_{[\lambda]}$ (Theorem 3.4) to find a semisimple Lie algebra $\mathfrak{g}'$ whose Weyl group is isomorphic to $W_{[\lambda]}$. Then λ' can be taken to be any integral weight whose isotropy group agrees with that of λ. (Note a subtle point in this category equivalence: since there is typically a different weight ρ for each Lie algebra, one cannot directly compare a linkage class of weights for $\mathfrak{g}$ with one for $\mathfrak{g}'$. So the proof of the theorem is necessarily somewhat indirect.)

To prove the theorem, a preliminary reduction is made (using translation functors) to the case of regular weights. Now the two blocks in question are determined by the endomorphism algebras of projective generators, which under the functor $\mathbb{V}$ can be compared in terms of C-modules for the shared reflection group. Here one builds up the images under $\mathbb{V}$ of indecomposable projectives inductively in the C-module setting using the counterpart of wall-crossing functors.

The simple modules have categorical meaning and therefore correspond naturally. This also happens for the Verma modules: first characterize them categorically as projective objects in appropriate subcategories (this was remarked after the proof of Proposition 3.8).

13.14. Endomorphisms and Socles of Projectives

Building on Soergel's work, Stroppel [**246**] studies the endomorphism algebras of arbitrary projective modules in an integral block of $\mathcal{O}$ (along with related quivers, discussed in 13.9 above). For convenience we state the main result [**246**, Thm. 7.1] just for the block $\mathcal{O}_0$, with standard modules L_w, M_w, P_w labelled by the weight $w \cdot (-2\rho)$.

Theorem. *If* $x \in W$, *the following three statements are equivalent:*

(a) $\mathrm{End}_{\mathcal{O}}\, P_x$ *is commutative.*

(b) *The filtration multiplicity* $(P_x : M_{w_\circ}) = 1$.

(c) *There exists an epimorphism* $Z(\mathfrak{g}) \to \mathrm{End}_{\mathcal{O}}\, P_x$.

Although this is stated without explicit reference to Soergel's functor $\mathbb{V}$, the proof uses his results in an essential way; indeed, a different proof of the implication (a) $\Rightarrow$ (b) was given by Soergel. In this regular situation Stroppel observes that (b) implies the more precise conclusion that the endomorphism ring is a quotient of the coinvariant algebra C.

By BGG Reciprocity, (b) is the same as $[M_{w_\circ} : L_x] = 1$. This can be checked using Jantzen's multiplicity one criterion (8.7), which predates the KL Conjecture.

Soergel's work already covers the extreme case when $x = 1$; here P_x is the self-dual projective module in the block. At the other extreme, $P_{w_\circ} = M_{w_\circ}$ has the trivial endomorphism ring $\mathbb{C}$ (4.2). In case $\mathfrak{g} = \mathfrak{sl}(3, \mathbb{C})$, the multiplicity in (b) is always 1. Here Stroppel exhibits concretely the way each endomorphism ring can be written as a quotient of C.

Under the hypothesis of the theorem, it is also shown in [**246**, Thm. 8.1] that $\operatorname{Soc} P_x$ *is the direct sum of* $(P_x : M_{w_\circ})$ *copies of the simple Verma module* L_1. This number is one just when the endomorphism ring is commutative, by the theorem. The result is of course consistent with the two extreme cases. Again its statement does not involve Soergel's functor $\mathbb{V}$. The proof uses the fact that $\mathbb{V}$ kills all simple modules not of maximal Gelfand–Kirillov dimension: this leaves only copies of the simple Verma module in the socle.

Even when the conditions of the theorem are satisfied, it is not obvious in general how to describe the structure of $\operatorname{End}_{\mathcal{O}} P_x$ precisely. However, in all cases the *dimension* of the endomorphism ring is computable (in principle) from Kazhdan–Lusztig theory: thanks to Theorem 3.9(c), it is equal to $[P_x : L_x]$. After filtering P_x by Verma modules M_w with $x \le w$ and using BGG Reciprocity, version 8.4(2) of the KL Conjecture yields

$$\dim \operatorname{End}_{\mathcal{O}} P_x = \sum_{x \le w} [M_w : L_x] = \sum_{x \le w} P_{w_\circ w, w_\circ x}(1).$$

This raises the question of whether the structure of each $\operatorname{End}_{\mathcal{O}} P_x$ depends only on W and its Bruhat ordering.

13.15. Koszul Duality

By exploiting further the geometric methods used in the proof of the KL Conjecture (especially perverse sheaves), Soergel [**237**] and then Beilinson–Ginzburg–Soergel [**22**] refine significantly the algebraic results described in 13.11 and 13.12. In particular, this work brings *graded* modules into the picture, by comparing a graded Ext algebra with a not obviously graded End algebra. Important precursors include the preprints of Beilinson–Ginzburg [**20**] and Soergel [**236**].

In [**237**] the coinvariant algebra C plays a major role in this transition to geometric thinking. In one of his main theorems (Zerlegungssatz), Soergel essentially categorifies the Hecke algebra $\mathcal{H}$ of W (8.2) using the category of finitely generated graded C-modules. This leads to certain graded C-modules (in bijection with the Kazhdan–Lusztig self-dual basis of $\mathcal{H}$) which are graded versions of the $\mathbb{V}P_w$ ($w \in W$). The price paid for the use of deeper geometric methods here is the limitation to Weyl groups, even though the C-module framework makes sense for arbitrary finite reflection groups.

A high point in [**22**] is a surprising *Koszul duality* theorem, which expresses a hidden "parabolic–singular duality". In general, a ring A is called a *Koszul ring* if $A = \bigoplus_{n \geq 0} A_n$ is a graded ring, where A_0 is a semisimple ring and the A-module $A/A_+ \cong A_0$ (with $A_+ := \bigoplus_{n > 0} A_n$) has a projective resolution by graded modules P_n with P_n generated by its nth graded component. Under a mild finiteness condition, there is a well-defined *Koszul dual ring* $A^!$: the (opposite ring of the) graded ring $\mathrm{Ext}_A^\bullet(A_0, A_0)$. Then $(A^!)^! \cong A$. For example, the symmetric algebra of a finite dimensional vector space and the exterior algebra of its dual space are Koszul algebras and in fact Koszul duals.

These ideas come into play in $\mathcal{O}_\lambda$ when it is viewed as the module category over the endomorphism ring of a projective generator. (Here the duality functor can be used to pass to the opposite algebra.) In what follows all weights are assumed to be *integral* but possibly singular. Each antidominant weight $\lambda \in \Lambda$ determines a block $\mathcal{O}_\lambda$ as well as a subset $\mathrm{I} = \mathrm{I}^\lambda$ of Δ:

$$\mathrm{I}^\lambda := \{\alpha \in \Delta \mid \langle \lambda + \rho, \alpha^\vee \rangle = 0\}.$$

As before we denote by W_I the resulting subgroup of W and by W^I the set of minimal length right coset representatives in $W_I \backslash W$. Here W_I is just the isotropy group W_λ°; thus W^I parametrizes the simple modules in the block $\mathcal{O}_\lambda$ to which $L(\lambda)$ belongs. Now set

$$L^\lambda := \bigoplus_{w \in W^\mathrm{I}} L(w \cdot \lambda) \quad \text{and} \quad P^\lambda := \bigoplus_{w \in W^\mathrm{I}} P(w \cdot \lambda).$$

Here P^λ is a projective generator for the block.

On the other hand, I determines a parabolic subalgebra $\mathfrak{p} = \mathfrak{p}_\mathrm{I}$ and the *regular* block $\mathcal{O}_0^\mathfrak{p}$ of $\mathcal{O}^\mathfrak{p}$ containing the trivial module. The simple modules $L(x \cdot 0)$ lying in this block are parametrized by left coset representatives in W/W_I, which can be chosen to be those of maximal length. Here we denote by $L^\mathfrak{p}$ the direct sum of simples in the block and by $P^\mathfrak{p}$ the direct sum of their projective covers in $\mathcal{O}^\mathfrak{p}$.

In these two blocks, the respective simple modules $L(w \cdot \lambda)$ and $L(x \cdot 0)$ are in natural bijection, if we send $w \mapsto w_\circ x^{-1}$. Although the simple modules in the two situations are otherwise unrelated, a duality emerges (somewhat miraculously) in [**22**].

Koszul Duality. *With the above notation, there are algebra isomorphisms*

$$\operatorname{End}_{\mathcal{O}} P^{\lambda} \;\cong\; \operatorname{Ext}^{\bullet}_{\mathcal{O}^{\mathfrak{p}}}(L^{\mathfrak{p}}, L^{\mathfrak{p}}),$$
$$\operatorname{End}_{\mathcal{O}} P^{\mathfrak{p}} \;\cong\; \operatorname{Ext}^{\bullet}_{\mathcal{O}}(L^{\lambda}, L^{\lambda}).$$

The two Ext *algebras are Koszul algebras and are in fact Koszul duals.*

When λ is *regular* (essentially the case of $\mathcal{O}_0$) this restates the earlier result of Soergel [**237**, 3.5]; here the Ext algebra is self-dual in the Koszul sense. In general the isomorphisms in the theorem reveal an underlying graded structure in each of the End algebras, which does not appear to be accessible by algebraic means alone. In particular, category $\mathcal{O}$ is seen to be a "graded semisimple" category. All of this lies fairly deep: for example, the fact that the algebra defining $\mathcal{O}_0$ is Koszul and self-dual is actually enough to imply the KL Conjecture for $\mathcal{O}_0$, though at present its proof depends on the truth of that conjecture.

Remark. Further work has brought out more strongly the ubiquity of Koszul duality as a theme in the study of $\mathcal{O}$ and related geometry. For example, Backelin [**11**] applies the methods of [**22**] to the parabolic category $\mathcal{O}^{\mathfrak{p}}$. He shows that the algebra which defines an integral (but possibly singular) block of $\mathcal{O}^{\mathfrak{p}}$ is also Koszul and works out the Koszul dual. In another direction, Ryom–Hansen [**230**] shows that the category equivalence of [**22**] interchanges the derived functors of translation functors for the singular category and Zuckerman functors for the parabolic category. This result is recovered by Mazorchuk, Ovsienko, and Stroppel [**216**], where the theme of Koszul duality for pairs of functors is carried much further. In [**215**] Mazorchuk extends and reformulates more algebraically the treatment of Koszul dual functors in [**11**].

A concluding word: Chapters 10–13 have introduced a number of loosely related further topics to supplement the more unified treatment of highest weight modules in Part I. The methods of geometric representation theory offer a new perspective on much of this material, as illustrated by the paper of Beilinson–Ginzburg [**21**]. This part of their work was not incorporated into the joint paper [**22**] with Soergel, but provides an alternative pathway to his Endomorphismensatz and Struktursatz. It also brings into the picture translation functors and wall-crossing functors, projective functors, and tilting modules.

Bibliography

1. N. Abe, *On the existence of homomorphisms between principal series of complex semisimple Lie groups*, arXiv:0712.2122 [math.RT]

2. K. Akin, *On complexes relating the Jacobi–Trudy identity with the Bernstein–Gelfand–Gelfand resolution*, J. Algebra **117** (1988), 494–503.

3. H. H. Andersen, *Schubert varieties and Demazure's character formula*, Invent. Math. **79** (1985), 611–618.

4. ———, *Twisted Verma modules and their quantized analogues*, pp. 1–10, *Combinatorial and Geometric Representation Theory (Seoul, 2001)*, Contemp. Math., 325, Amer. Math. Soc., Providence, RI, 2003.

5. H. H. Andersen and N. Lauritzen, *Twisted Verma modules*, pp. 1–26, *Studies in Memory of Issai Schur*, Progr. Math., 210, Birkhäuser, Boston, 2003.

6. H. H. Andersen and J. Paradowski, *Fusion categories arising from semisimple Lie algebras*, Comm. Math. Phys. **169** (1995), 563–588.

7. H. H. Andersen and C. Stroppel, *Twisting functors on $\mathcal{O}_0$*, Represent. Theory **7** (2003), 681–699.

8. S. Arkhipov, *A new construction of the semi-infinite BGG resolution*, unpublished preprint, arXiv:q-alg/9605043.

9. ———, *Algebraic construction of contragredient quasi-Verma modules in positive characteristic*, pp. 27–68, *Representation Theory of Algebraic Groups and Quantum Groups*, Adv. Stud. Pure Math., 40, Math. Soc. Japan, Tokyo, 2004.

10. E. Backelin, *Representation of the category $\mathcal{O}$ in Whittaker categories*, Internat. Math. Res. Notices **1997**, no. 4, 153–172.

11. ———, *Koszul duality for parabolic and singular category $\mathcal{O}$*, Represent. Theory **3** (1999), 139–152.

12. ———, *The Hom-spaces between projective functors*, Represent. Theory **5** (2001), 267–283.

13. D. Barbasch, *Filtrations on Verma modules*, Ann. Sci. École Norm. Sup. (4) **16** (1983), 489–494.

14. H. Bass, *Algebraic K-Theory*, W. A. Benjamin, New York, 1968.

15. A. Beilinson, *Localization of representations of reductive Lie algebras*, pp. 699–710, Proc. Intern. Congr. Math. (Warsaw, 1983), PWN, Warsaw, 1984.

16. A. Beilinson and J. Bernstein, *Localisation de $\mathfrak{g}$-modules*, C.R. Math. Acad. Sci. Paris **292** (1981), 15–18.

17. A. Beilinson, J. Bernstein, and P. Deligne, *Faisceaux pervers*, Analyse et topologie sur les espaces singuliers, I (Luminy, 1981), 5–171, *Astérisque*, **100**, Soc. Math. France, Paris, 1982.

18. _______, *A proof of Jantzen conjectures*, I. M. Gelfand Seminar, 1–50, Adv. Soviet Math., 16, Part 1, Amer. Math. Soc., Providence, RI, 1993.

19. A. Beilinson, R. Bezrukavnikov, and I. Mirković, *Tilting exercises*, Mosc. Math. J. **4** (2004), 547–557.

20. A. Beilinson and V. Ginzburg, *Mixed categories, Ext-duality and representations (results and conjectures)*, preprint, 1986.

21. _______, *Wall-crossing functors and $\mathcal{D}$-modules*, Represent. Theory **3** (1999), 1–31.

22. A. Beilinson, V. Ginzburg, and W. Soergel, *Koszul duality patterns in representation theory*, J. Amer. Math. Soc. **9** (1996), 473–527.

23. J. Bernstein, I. Frenkel, and M. Khovanov, *A categorification of the Temperley–Lieb algebra and Schur quotients of $U(\mathfrak{sl}_2)$ via projective and Zuckerman functors*, Selecta Math. (N.S.) **5** (1999), 199–241.

24. J. Bernstein and S. Gelfand, *Tensor products of finite and infinite dimensional representations of semisimple Lie algebras*, Compositio Math. **41** (1980), 245–285.

25. I. N. Bernstein, I. M. Gelfand, and S. I. Gelfand, *Structure of representations generated by highest weights*, Funktsional. Anal. i Prilozhen. **5** (1971), no. 1. 1–9; English transl., Funct. Anal. Appl. **5** (1971), 1–8.

26. _______, *Differential operators on the base affine space and a study of $\mathfrak{g}$-modules*, pp. 21–64, Lie Groups and their Representations (Proc. Summer School, Bolyai János Math. Soc., Budapest, 1971), Halsted, New York, 1975.

27. _______, *On a category of $\mathfrak{g}$-modules*, Funktsional. Anal. i Prilozhen. **10** (1976), no. 2, 1–8; English transl., Funct. Anal. Appl. **10** (1976), 87–92.

28. R. Biagioli, *Closed product formulas for extensions of generalized Verma modules*, Trans. Amer. Math. Soc. **356** (2004), 159–184.

29. A. Björner and F. Brenti, *Combinatorics of Coxeter groups*, Springer-Verlag, New York, 2005.

30. B. D. Boe, *Homomorphisms between generalized Verma modules*, Trans. Amer. Math. Soc. **288** (1985), 791–799.

31. _______, *A counterexample to the Gabber–Joseph conjecture*, pp. 1–3, *Kazhdan–Lusztig Theory and Related Topics (Chicago, IL, 1989)*, Contemp. Math., 139, Amer. Math. Soc., Providence, RI, 1992.

32. B. D. Boe and D. H. Collingwood, *A comparison theory for the structure of induced representations*, J. Algebra **94** (1985), 511–545.

33. _______, *A comparison theory for the structure of induced representations II*, Math. Z. **190** (1985), 1–11.

34. _______, *A multiplicity one theorem for holomorphically induced representations*, Math. Z. **192** (1986), 265–282.

35. _______, *Multiplicity free categories of highest weight representations*, Comm. Algebra **18** (1990), 947–1032; part II, 1033–1070.

36. B. D. Boe and M. Hunziker, *Kostant modules in blocks of category $\mathcal{O}_S$*, arXiv:math.RT/0604336.

37. B. D. Boe and D. K. Nakano, *Representation type of the blocks of category $\mathcal{O}_S$*, Adv. Math. **196** (2005), 193–256.

38. B. D. Boe, D. K. Nakano, and Emilie Wiesner, *Category $\mathcal{O}$ for the Virasoro algebra: cohomology and Koszulity*, Pacific J. Math. **234** (2007), 1–22.

39. A. Borel and N. Wallach, *Continuous cohomology, discrete subgroups, and representations of reductive groups*, 2nd ed., Math. Surveys Monogr., 67, Amer. Math. Soc., Providence, RI, 2000.

40. W. Borho, *A survey on enveloping algebras of semisimple Lie algebras. I*, pp. 19–50, *Lie algebras and related topics (Windsor, Ont., 1984)*, CMS Conf. Proc., 5, Amer. Math. Soc., Providence, RI, 1986.

41. ———, *Nilpotent orbits, primitive ideals, and characteristic classes (a survey)*, pp. 350–359, Proc. Intern. Congr. Math. (Berkeley, Calif., 1986), Amer. Math. Soc., Providence, RI, 1987.

42. W. Borho and J. C. Jantzen, *Über primitive Ideale in der Einhüllenden einer halbeinfachen Lie-Algebra*, Invent. Math. **39** (1977), 1–53.

43. R. Bott, *Homogeneous vector bundles*, Ann. of Math. **66** (1957), 203–248.

44. A. Bouaziz, *Sur les représentations des algèbres de Lie semi-simples construites par T. Enright*, pp. 57–68, *Noncommutative harmonic analysis and Lie groups (Marseille, 1980)*, Lecture Notes in Math., 880, Springer-Verlag, Berlin, 1981.

45. N. Bourbaki, *Groupes et algèbres de Lie*, Chapters 4–6, Hermann, Paris, 1968; 2nd ed., Masson, Paris, 1981; English transl., Springer-Verlag, Berlin, 2002.

46. ———, *Groupes et algèbres de Lie*, Chapters 7–8, Hermann, Paris, 1975; English transl. (with Chapter 9), Springer-Verlag, Berlin, 2005.

47. F. Brenti, *Kazhdan–Lusztig polynomials: history, problems, and combinatorial invariance*, Sém. Lothar. Combin., **49** (2002/04), Art. B49b, 30 pp.

48. K. S. Brown, *Extensions of "thickened" modules of the Virasoro algebra*, J. Algebra **269** (2003), 160–188.

49. J. Brundan, *Kazhdan–Lusztig polynomials and character formulae for the Lie superalgebra $\mathfrak{gl}(m \mid n)$*, J. Amer. Math. Soc. **16** (2003), 185–231.

50. ———, *Tilting modules for Lie superalgebras*, Comm. Algebra **32** (2004), 2251–2268.

51. ———, *Centers of degenerate cyclotomic Hecke algebras and parabolic category $\mathcal{O}$*, arXiv:math.RT/0607717.

52. ———, *Symmetric functions, parabolic category $\mathcal{O}$ and the Springer fiber*, arXiv:math/0608235 [math.RT], to appear in Duke Math. J.

53. J. Brundan, S. M. Goodwin, and A. Kleshchev, *Highest weight theory for finite W-algebras*, arXiv:0801.1337 [math.RT].

54. Th. Brüstle, S. König, and V. Mazorchuk, *The coinvariant algebra and representation types of blocks of category $\mathcal{O}$*, Bull. London Math. Soc. **33** (2001), 669–681.

55. J. L. Brylinski and M. Kashiwara, *Kazhdan–Lusztig conjecture and holonomic systems*, Invent. Math. **64** (1981), 387–410.

56. K. Carlin, *Extensions of Verma modules*, Trans. Amer. Math. Soc. **294** (1986), 29–43.

57. ———, *Completion and translation in $\mathcal{O}$*, Comm. Algebra **16** (1988), 1921–1932.

58. ———, *Local systems of Shapovalov elements*, Comm. Algebra **23** (1995), 3039–3049.

59. H. Cartan and S. Eilenberg, *Homological algebra*, Princeton Univ. Press, Princeton, 1956.

60. R. W. Carter, *Lie algebras of finite and affine type*, Cambridge Univ. Press, Cambridge, 2005.

61. L. Casian and D. H. Collingwood, *The Kazhdan–Lusztig conjecture for generalized Verma modules*, Math. Z. **195** (1987), 581–600.

62. ______, *Weight filtrations for induced representations of real reductive Lie groups*, Adv. in Math. **73** (1989), 79–146.

63. E. Cline, B. Parshall, and L. Scott, *Finite dimensional algebras and highest weight modules*, J. Reine Angew. Math. **391** (1988), 85–99.

64. ______, *Abstract Kazhdan–Lusztig theories*, Tohoku Math. J. **45** (1993), 511–534.

65. D. H. Collingwood, *Category $\mathcal{O}'$, $\mathfrak{n}$-homology and the reducibility of generalized principal series representations*, Duke Math. J. **50** (1983), 1201–1224.

66. ______, *Representations of rank one Lie groups*, Research Notes in Mathematics, 137, Pitman (Advanced Publishing Program), Boston, MA, 1985.

67. ______, *The $\mathfrak{n}$-homology of Harish-Chandra modules: generalizing a theorem of Kostant*, Math. Ann. **272** (1985), 161–187.

68. ______, *Jacquet modules for semisimple Lie groups having Verma module filtrations*, J. Algebra **136** (1991), 353–375.

69. D. H. Collingwood and R. S. Irving, *A decomposition theorem for certain self-dual modules in the category $\mathcal{O}$*, Duke Math. J. **58** (1989), 89–102.

70. D. H. Collingwood, R. S. Irving, and B. Shelton, *Filtrations on generalized Verma modules for Hermitian symmetric pairs*, J. Reine Angew. Math. **383** (1988), 54–86.

71. D. H. Collingwood and B. Shelton, *A duality theorem for extensions of induced highest weight modules*, Pacific J. Math. **146** (1990), 227–237.

72. N. Conze and J. Dixmier, *Idéaux primitifs de l'algèbre enveloppante d'une algèbre de Lie semi-simple*, Bull. Sci. Math. **96** (1972), 339–351.

73. N. Conze-Berline and M. Duflo, *Sur les représentations induites des groupes semi-simples complexes*, Compositio Math. **34** (1977), 307–336.

74. C. W. Curtis and I. Reiner, *Methods of representation theory, I*, Wiley, New York, 1981.

75. P. Delorme, *Extensions dans la categorie $\mathcal{O}$ de Bernstein–Gelfand–Gelfand. Applications*, preprint, 1978.

76. ______, *Extensions in the Bernstein–Gelfand–Gelfand category $\mathcal{O}$. Applications*, Funktsional. Anal. i Prilozhen. **14** (1980), no 3, 77–78; English transl., Funct. Anal. Appl. **14** (1980), 228–229.

77. V. V. Deodhar, *On a construction of representations and a problem of Enright*, Invent. Math. **57** (1980), 101–118.

78. ______, *On the Kazhdan–Lusztig conjectures*, Indag. Math. **44** (1982), 1–17.

79. ______, *On some geometric aspects of Bruhat orderings, II. The parabolic analogue of Kazhdan–Lusztig polynomials*, J. Algebra **111** (1987), 483–506.

80. ______, *Duality in parabolic set up for questions in Kazhdan–Lusztig theory*, J. Algebra **142** (1991), 201–209.

81. ______, ed., *Kazhdan–Lusztig Theory and Related Topics (Chicago, IL, 1989)*, Contemp. Math., 139, Amer. Math. Soc., Providence, RI, 1992.

82. V. V. Deodhar, O. Gabber, and V. Kac, *Structure of some categories of representations of infinite-dimensional Lie algebra*, Adv. in Math. **45** (1982), 92–116.

83. V. V. Deodhar and J. Lepowsky, *On multiplicity in the Jordan–Hölder series of Verma modules*, J. Algebra **49** (1977), 512–524.

84. J. Dixmier, *Algèbres enveloppantes*, Gauthier–Villars, Paris, 1974; reprint of English translation, *Enveloping algebras*, Amer. Math. Soc., Providence, RI, 1996.

85. S. Donkin, *The q-Schur algebra*, London Math. Soc. Lecture Note Series, 253, Cambridge Univ. Press, Cambridge, 1998.

86. M. Duflo, *Construction of primitive ideals in an enveloping algebra*, pp. 77–93, Lie Groups and their Representations (Proc. Summer School, Bolyai János Math. Soc., Budapest, 1971), Halsted, New York, 1975.

87. ______, *Représentations irréductibles des groupes semi-simples complexes*, pp. 26–88, *Analyse harmonique sur les groupes de Lie*, Lecture Notes in Math., 497, Springer-Verlag, Berlin, 1975.

88. ______, *Sur la classification des idéaux primitifs dans l'algèbre enveloppante d'une algèbre de Lie semi-simple*, Ann. of Math. **105** (1977), 107–120.

89. T. J. Enright, *On the fundamental series of a real semisimple Lie algebra: their irreducibility, resolutions and multiplicity formulae*, Ann. of Math. **110** (1979), 1–82.

90. ______, *Lectures on representations of complex semisimple Lie groups*, Tata Inst. Lectures on Mathematics and Physics, 66, Springer-Verlag, Berlin, 1981.

91. T. J. Enright and B. Shelton, *Decompositions in categories of highest weight modules*, J. Algebra **100** (1986), 380–402.

92. ______, *Categories of highest weight modules: Applications to Hermitian symmetric spaces*, Mem. Amer. Math. Soc. **67** (1987), no. 367.

93. T. J. Enright and N. R. Wallach, *Notes on homological algebra and representations of Lie algebras*, Duke Math. J. **47** (1980), 1–15.

94. P. Fiebig, *Centers and translation functors for the category $\mathcal{O}$ over Kac–Moody algebras*, Math. Z. **243** (2003), 689–717.

95. ______, *The combinatorics of category $\mathcal{O}$ over symmetrizable Kac–Moody algebras*, Transform. Groups **11** (2006), 29–49.

96. J. Franklin, *Homomorphisms between Verma modules in characteristic p*, J. Algebra **112** (1988), 58–85.

97. W. Fulton and J. Harris, *Representation theory*, Springer-Verlag, New York, 1991.

98. V. Futorny, S. König, and V. Mazorchuk, *$\mathcal{S}$-subcategories in $\mathcal{O}$*, Manuscripta Math. **102** (2000), 487–503.

99. ______, *A combinatorial description of blocks in $\mathcal{O}(\mathcal{P}, \Lambda)$ associated with $\mathfrak{sl}(2)$-induction*, J. Algebra **231** (2000), 86–103.

100. ______, *Categories of induced modules for Lie algebras with triangular decomposition*, Forum Math. **13** (2001), 641–661.

101. V. Futorny and V. Mazorchuk, *Structure of α-stratified modules for finite-dimensional Lie algebras. I*, J. Algebra **183** (1996), 456–482.

102. ______, *BGG-resolution for α-stratified modules over simply-laced finite-dimensional Lie algebra*, J. Math. Kyoto Univ. **38** (1998), 229–240.

103. ______, *Highest weight categories of Lie algebra modules*, J. Pure Appl. Math. **138** (1999), 107–118.

104. V. Futorny, D. K. Nakano, and R. D. Pollack, *Representation type of the blocks of category $\mathcal{O}$*, Q. J. Math. **52** (2001), 285–305.

105. O. Gabber and A. Joseph, *On the Bernstein–Gelfand–Gelfand resolution and the Duflo sum formula*, Compositio Math. **43** (1981), 107–131.

106. ———, *Towards the Kazhdan–Lusztig conjecture*, Ann. Sci. École Norm. Sup. (4) **14** (1981), 261–302.

107. D. Gaitsgory, *Geometric representation theory*, lecture notes, Harvard Univ., 2005.

108. H. Garland and J. Lepowsky, *Lie algebra homology and the Macdonald–Kac formulas*, Invent. Math. **34** (1976), 37–76.

109. S. Gelfand and R. MacPherson, *Verma modules and Schubert cells: a dictionary*, pp. 1–50, *P. Dubreil and M.-P. Malliavin Algebra Seminar (Paris, 1981)*, Lecture Notes in Math., 924, Springer-Verlag, Berlin, 1982.

110. S. I. Gelfand and Yu. I. Manin, *Methods of homological algebra*, 2nd ed., Springer-Verlag, Berlin, 2003.

111. V. Ginzburg, N. Guay, E. Opdam, and R. Rouquier, *On the category $\mathcal{O}$ for rational Cherednik algebras*, Invent. Math. **154** (2003), 617–651.

112. X. Gomez and V. Mazorchuk, *On an analogue of BGG-reciprocity*, Comm. Algebra **29** (2001), 5329–5334.

113. I. G. Gordon, *Symplectic reflection algebras*, arXiv:0712.1568 [math.RT].

114. M. Goresky, *Tables of Kazhdan–Lusztig polynomials*, currently available online at http://www.math.ias.edu/~goresky/tables.html.

115. W. A. de Graaf, *Constructing homomorphisms between Verma modules*, J. Lie Theory **15** (2005), 415–428.

116. N. Guay, *Projective modules in the category $\mathcal{O}$ for the Cherednik algebra*, J. Pure Appl. Algebra **182** (2003), 209–221.

117. K. Günzl, *The fine structure of translation functors*, Represent. Theory **3** (1999), 223–249.

118. A. Gyoja, *Further generalization of generalized Verma modules*, Publ. Res. Inst. Math. Sci. **29** (1993), 349–395.

119. ———, *A remark on homomorphisms between generalized Verma modules*, J. Math. Kyoto Univ. **34** (1994), 695–697.

120. ———, *A duality theorem for homomorphisms between generalized Verma modules*, J. Math. Kyoto Univ. **40** (2000), 437–450.

121. Harish-Chandra, *On some applications of the universal enveloping algebra of a semisimple Lie algebra*, Trans. Amer. Math. Soc. **70** (1951), 28–96.

122. P. J. Hilton and U. Stammbach, *A course in homological algebra*, Springer-Verlag, New York, 1971.

123. A. van den Hombergh, *Note on a paper by Bernstein, Gelfand and Gelfand on Verma modules*, Indag. Math. **36** (1974), 352–356.

124. R. Hotta, K. Takeuchi, and T. Tanisaki, *D-modules, perverse sheaves, and representation theory*, Progr. Math., 236, Birkhäuser, Boston, 2008.

125. J. E. Humphreys, *Introduction to Lie algebras and representation theory*, Springer-Verlag, New York, 1972.

126. ———, *A construction of projective modules in the category $\mathcal{O}$ of Bernstein–Gelfand–Gelfand*, Indag. Math. **39** (1977), 301–303.

127. ———, *Finite and infinite dimensional modules for semisimple Lie algebras*, pp. 1–64, *Lie Theories and their Applications*, Queen's Papers in Pure and Appl. Math. No. 48, Kingston, Ont., 1978.

128. ______, *Highest weight modules for semisimple Lie algebras*, pp. 72–103, *Representation Theory I*, Lecture Notes in Math., 831, Springer-Verlag, Berlin, 1980.

129. ______, *Reflection groups and Coxeter groups* (Cambridge Studies in Advanced Mathematics, 29), Cambridge Univ. Press, Cambridge, 1990.

130. ______, *Conjugacy classes in semisimple algebraic groups*, Math. Surveys Monographs, 43, Amer. Math. Soc., Providence, RI, 1995.

131. ______, *Modular representations of finite groups of Lie type*, London Math. Soc. Lecture Note Series, 326, Cambridge Univ. Press, Cambridge, 2006.

132. R. S. Irving, *Projective modules in the category $\mathcal{O}$*, unpublished manuscript, 1982.

133. ______, *Projective modules in the category $\mathcal{O}_S$: self-duality*, Trans. Amer. Math. Soc. **291** (1985), 701–732.

134. ______, *Projective modules in the category $\mathcal{O}_S$: Loewy series*, Trans. Amer. Math. Soc. **291** (1985), 733–754.

135. ______, *The socle filtration of a Verma module*, Ann. Sci. École Norm. Sup. (4) **21** (1988), 47–65.

136. ______, *Singular blocks of the category $\mathcal{O}$*, Math. Z. **204** (1990), 209–224.

137. ______, *BGG algebras and the BGG reciprocity principle*, J. Algebra **135** (1990), 363–380.

138. ______, *A filtered category $\mathcal{O}_S$ and applications*, Mem. Amer. Math. Soc. **83** (1990), no. 419.

139. ______, *Graded BGG algebras*, pp. 181–200, *Abelian Groups and Nonabelian Rings*, Contemp. Math., 130, Amer. Math. Soc., Providence, RI, 1992.

140. ______, *Singular blocks of the category $\mathcal{O}$, II*, pp. 237–248, *Kazhdan–Lusztig Theory and Related Topics (Chicago, IL, 1989)*, Contemp. Math., 139, Amer. Math. Soc., Providence, RI, 1992.

141. ______, *Shuffled Verma modules and principal series modules over complex semisimple Lie algebras*, J. London Math. Soc. **48** (1993), 263–277.

142. R. S. Irving and B. Shelton, *Loewy series and simple projective modules in the category $\mathcal{O}_S$*, Pacific J. Math. **132** (1988), 319–342; correction, ibid. **135** (1988), 395–396.

143. N. Jacobson, *Lie algebras*, Wiley Interscience, New York, 1962; Dover reprint, 1979.

144. J. C. Jantzen, *Darstellungen halbeinfacher algebraischer Gruppen und zugeordnete kontravariante Formen*, Bonner Math. Schriften, No. 67, 1973.

145. ______, *Zur Charakterformel gewisser Darstellungen halbeinfacher Gruppen und Lie-Algebren*, Math. Z. **140** (1974), 127–149.

146. ______, *Kontravariante Formen auf induzierten Darstellungen halbeinfacher Lie-Algebren*, Math. Ann. **226** (1977), 53–65.

147. ______, *Moduln mit einem höchsten Gewicht*, Lecture Notes in Math., 750, Springer-Verlag, Berlin, 1979.

148. ______, *Einhüllende Algebren halbeinfacher Lie-Algebren*, Springer-Verlag, Berlin, 1983 (reviewed by D. A. Vogan, Jr. in Bull. Amer. Math. Soc. **12** (1985), 279–283).

149. ______, *Einhüllende Algebren halbeinfacher Lie-Algebren*, pp. 393–401, *Proc. Intern. Congr. Math. (Warsaw, 1983)*, PWN, Warsaw, 1984.

150. ______, *Primitive ideals in the enveloping algebra of a semisimple Lie algebra*, pp. 29–36, *Noetherian Rings and their Applications*, Math. Surv. Monogr., 24, Amer. Math. Soc., Providence, RI, 1987.

151. ______, *Lectures on quantum groups*, Amer. Math. Soc., Providence, RI, 1996.

152. ______, *Representations of algebraic groups*, Academic Press, Orlando, 1987; 2nd ed., Amer. Math. Soc., Providence, RI, 2003.

153. ______, *Character formulae from Hermann Weyl to the present*, to appear.

154. A. Joseph, *Gelfand–Kirillov dimension for the annihilators of simple quotients of Verma modules*, J. London Math. Soc. **18** (1978), 50–60.

155. ______, *Dixmier's problem for Verma and principal series submodules*, J. London Math. Soc. **20** (1979), 193–204.

156. ______, *Kostant's problem, Goldie rank and the Gelfand–Kirillov conjecture*, Invent. Math. **56** (1980), 191–213.

157. ______, *Towards the Jantzen conjecture*, Compositio Math. **40** (1980), 35–67; II **40**, 69–78; III **41** (1981), 23–30.

158. ______, *The Enright functor and the Bernstein–Gelfand–Gelfand category $\mathcal{O}$*, Invent. Math. **67** (1982), 423–445.

159. ______, *Completion functors in the $\mathcal{O}$ category*, pp. 80–105, *Noncommutative harmonic analysis and Lie groups (Marseille, 1982)*, Lecture Notes in Math., 1020, Springer-Verlag, Berlin, 1983.

160. ______, *Primitive ideals in enveloping algebras*, pp. 403–414, *Proc. Intern. Congr. Math. (Warsaw, 1983)*, PWN, Warsaw, 1984.

161. ______, *Quantum groups and their primitive ideals*, Springer-Verlag, Berlin, 1995.

162. ______, *Sur l'annulateur d'un module de Verma*, pp. 237–300, *Representation Theories and Algebraic Geometry (Montreal, PQ, 1997)*, NATO Adv. Sci. Inst. Ser. C Math. Phys. Sci., 514, Kluwer Acad. Publ., Dordrecht, 1998.

163. A. Joseph, G. S. Perets, and P. Polo, *Sur l'équivalence de catégories de Beilinson et Bernstein*, C.R. Math. Acad. Sci. Paris **313** (1991), 705–709.

164. V. G. Kac, *Infinite-dimensional Lie algebras and Dedekind's η-function*, Functional Anal. Appl. **8** (1974), 68–70.

165. ______, *Infinite dimensional Lie algebras*, 3rd ed., Cambridge Univ. Press, Cambridge, 1990.

166. V. G. Kac and D. A. Kazhdan, *Structure of representations with highest weight of infinite-dimensional Lie algebras*, Adv. Math. **34** (1979), 97–108.

167. J. Kåhrström, *Tensoring with infinite-dimensional modules in $\mathcal{O}_0$*, arXiv:0708/2218 [math.RT].

168. J. Kåhrström and V. Mazorchuk, *A new approach to Kostant's problem*, arXiv:0712.3117 [math.RT].

169. M. Kashiwara and T. Tanisaki, *Characters of irreducible modules with non-critical highest weights over affine Lie algebras*, pp. 275–296, *Representations and Quantizations (Shanghai, 1998)*, China High. Educ. Press, Shanghai, 2000.

170. D. Kazhdan and G. Lusztig, *Representations of Coxeter groups and Hecke algebras*, Invent. Math. **53** (1979), 165–184.

171. ______, *Schubert varieties and Poincaré duality*, pp. 185–203, *Geometry of the Laplace operator*, Proc. Sympos. Pure Math., XXXVI, Amer. Math. Soc., Providence, R.I., 1980.

172. A. Khare, *Category $\mathcal{O}$ over a deformation of the symplectic oscillator algebra*, J. Pure Appl. Algebra **195** (2005), 131–166; erratum **199** (2005), 319–320.

173. ______, *Axiomatic framework for the BGG category $\mathcal{O}$*, arXiv:math.RT/0502227.

174. O. Khomenko, *Categories with projective functors*, Proc. London Math. Soc. **90** (2005), 711–737.

175. O. Khomenko and V. Mazorchuk, *The Shapovalov from for generalized Verma modules*, Dopov. Nats. Akad. Nauk Ukr. Mat. Prirodozn. Tekh. Nauki **1999**, no. 6, 28–32.

176. ______ , *A note on simplicity of generalized Verma modules*, Comment. Math. Univ. St. Paul. **48** (1999), 145–148.

177. ______ , *On the determinant of Shapovalov form for generalized Verma modules*, J. Algebra **215** (1999), 318–329.

178. ______ , *Generalized Verma modules over the Lie algebra of type G_2*, Comm. Algebra **27** (1999), 777–783.

179. ______ , *Schubert filtration for simple quotients of generalized Verma modules*, Ark. Mat. **38** (2000), 319–326.

180. ______ , *An irreducibility criterion for generalized Verma modules*, Dopov. Nats. Akad. Nauk Ukr. Mat. Prirodozn. Tekh. Nauki **2001**, no. 5, 22–24.

181. ______ , *Generalized Verma modules induced from $\mathfrak{sl}(2,\mathbb{C})$ and associated Verma modules*, J. Algebra **242** (2001), 561–576.

182. ______ , *On multiplicities of simple subquotients in generalized Verma modules*, Czechoslovak Math. J. **52 (127)** (2002), 337–343.

183. ______ , *Rigidity of generalized Verma modules*, Colloq. Math. **92** (2002), 45–57.

184. ______ , *Structure of modules induced from simple modules with minimal annihilator*, Canad. J. Math. **56** (2004), 293–309.

185. ______ , *On Arkhipov's and Enright's functors*, Math. Z. **249** (2005), 357–386.

186. S. Khoroshkin, *Projective functors and the restriction of a Verma module to a subalgebra of Levi type*, Ann. Global Anal. Geom. **10** (1992), 81–86.

187. S. L. Kleiman, *The development of intersection homology theory*, Pure Appl. Math. Q. **3** (2007), 225–282.

188. A. W. Knapp, *Lie groups, Lie algebras, and cohomology*, Math. Notes **34**, Princeton Univ. Press, Princeton, NJ, 1988.

189. S. König, *The global dimension of the regular blocks of the BGG-category $\mathcal{O}$ of a semisimple complex Lie algebra*, unpublished preprint.

190. ______ , *Cartan decompositions and BGG-resolutions*, Manuscripta Math. **86** (1995), 103–111.

191. ______ , *Blocks of category $\mathcal{O}$, double centralizer properties, and Enright's completions*, pp. 113–134 in *Algebra—Representation Theory*, K. W. Roggenkamp and M. Ştefănescu, eds., Kluwer, Dordrecht, 2001.

192. ______ , *Ringel duality and Kazhdan–Lusztig theory*, Pacific J. Math. **203** (2002), 415–428.

193. S. König and V. Mazorchuk, *Enright's completions and injectively copresented modules*, Trans. Amer. Math. Soc. **354** (2002), 2725–2743.

194. ______ , *An equivalence of two categories of $\mathfrak{sl}(n,\mathbb{C})$-modules*, Algebr. Represent. Theory **5** (2002), 319–329.

195. B. Kostant, *A formula for the multiplicity of a weight*, Trans. Amer. Math. Soc. **93** (1959), 53–73.

196. ______ , *Lie algebra cohomology and the generalized Borel–Weil–Bott theorem*, Ann. of Math. **74** (1961), 329–387.

197. S. Kumar, *Bernstein–Gelfand–Gelfand resolution for arbitrary Kac–Moody algebras*, Math. Ann. **286** (1990), 709–729.

198. J. Lepowsky, *Conical vectors in induced modules*, Trans. Amer. Math. Soc. **208** (1975), 219–272.

199. ______, *Existence of conical vectors in induced modules*, Ann. of Math. **102** (1975), 17–40.

200. ______, *Uniqueness of embeddings of certain induced modules*, Proc. Amer. Math. Soc. **56** (1976), 55–58.

201. ______, *Generalized Verma modules, the Cartan–Helgason theorem, and the Harish-Chandra homomorphism*, J. Algebra **49** (1977), 470–495.

202. ______, *A generalization of the Bernstein–Gelfand–Gelfand resolution*, J. Algebra **49** (1977), 496–511.

203. G. Lusztig and D. A. Vogan, Jr., *Singularities of closures of K-orbits on flag manifolds*, Invent. Math. **71** (1983), 365–379.

204. A. V. Lutsyuk, *Homomorphisms of the modules M_χ*, Funktsional. Anal. i Prilozhen. **8** (1974), no. 4, 91–92; English transl., Funct. Anal. Appl. **8** (1975), 351–352.

205. C. Marastoni, *Generalized Verma modules, b-functions of semi-invariants and duality for twisted $\mathcal{D}$-modules on generalized flag manifolds*, C.R. Math. Acad. Sci. Paris **335** (2002), 111–116.

206. O. Mathieu, *Classification of irreducible weight modules*, Ann. Inst. Fourier (Grenoble) **50** (2000), 537–592.

207. H. Matumoto, *On the existence of homomorphisms between scalar generalized Verma modules*, pp. 259–274, *Representation Theory of Groups and Algebras*, Contemp. Math., 145, Amer. Math. Soc., Providence, RI, 1993.

208. ______, *The homomorphisms between scalar generalized Verma modules associated to maximal parabolic subalgebras*, Duke Math. J. **131** (2006), 75–118.

209. V. Mazorchuk, *α-stratified modules over the Lie algebra $\mathfrak{sl}(n,\mathbb{C})$*, Ukrain. Mat. Zh. **45** (1993), no. 9, 1215–1224; English transl., Ukrainian Math. J. **45** (1993), no. 9, 1360–1371.

210. ______, *On the structure of an α-stratified generalized Verma module over Lie algebra $\mathfrak{sl}(n,\mathbb{C})$*, Manuscripta Math. **88** (1995), 59–72.

211. ______, *Generalized Verma modules*, Mathematical Studies Monograph Series, 8. VNTL Publishers, L'viv, 2000.

212. ______, *Twisted and shuffled filtrations on tilting modules*, C.R. Math. Acad. Sci. Soc. R. Can. **25** (2003), no. 1, 26–32.

213. ______, *A twisted approach to Kostant's problem*, Glasg. Math. J. **47** (2005), 549–561.

214. ______, *Some homological properties of the category $\mathcal{O}$*, Pacific J. Math. **232** (2007), 313–341.

215. ______, *Applications of the category of linear complexes of tilting modules associated with the category $\mathcal{O}$*, to appear in Algebr. Represent. Theory.

216. V. Mazorchuk, S. Ovsienko, and C. Stroppel, *Quadratic duals, Koszul dual functors, and applications*, arXiv:math.RT/0603475, to appear in Trans. Amer. Math. Soc.

217. V. Mazorchuk and C. Stroppel, *Translation and shuffling of projectively presentable modules and a categorification of a parabolic Hecke module*, Trans. Amer. Math. Soc. **357** (2005), 2939–2973.

218. ______, *Projective-injective modules, Serre functors and symmetric algebras*, arXiv:math.RT/0508119, to appear in J. Reine Angew. Math.

219. ______, *On functors associated to a simple root*, J. Algebra **314** (2007), 97–128.

220. ______, *Categorification of (induced) cell modules and the rough structure of generalized Verma modules*, arXiv:math.RT/070281.

221. ______, *Categorification of Wedderburn's basis for* $\mathbb{C}[S_n]$, arXiv:0708.3949 [math.RT], to appear in Arch. Math. (Basel).

222. R. Mirollo and K. Vilonen, *Bernstein–Gelfand–Gelfand reciprocity on perverse sheaves*, Ann. Sci. École Norm. Sup. (4) **20** (1987), 311–324.

223. R. V. Moody and A. Pianzola, *Lie algebras with triangular decompositions*, Wiley Interscience, New York–Toronto, 1995.

224. H. Oda and T. Oshima, *Minimal polynomials and annihilators of generalized Verma modules of the scalar type*, J. Lie Theory **16** (2006), 155–219.

225. P. Ostapenko, *Inverting the Shapovalov form*, J. Algebra **147** (1992), 90–95.

226. A. Rocha-Caridi, *Splitting criteria for $\mathfrak{g}$-modules induced from a parabolic and the Bernstein–Gelfand–Gelfand resolution of a finite-dimensional, irreducible $\mathfrak{g}$-module*, Trans. Amer. Math. Soc. **262** (1980), 335–366.

227. A. Rocha-Caridi and N. R. Wallach, *Projective modules over graded Lie algebras. I*, Math. Z. **180** (1982), 151–177.

228. ______, *Highest weight modules over graded Lie algebras: resolutions, filtrations and character formulas*, Trans. Amer. Math. Soc. **277** (1983), 133–162.

229. ______, *Characters of irreducible representations of the Virasoro algebra*, Math. Z. **185** (1984), 1–21.

230. S. Ryom-Hansen, *Koszul duality of translation and Zuckerman functors*, J. Lie Theory **14** (2004), 151–163.

231. N. N. Shapovalov, *On a bilinear form on the universal enveloping algebra of a complex semisimple Lie algebra*, Funktsional. Anal. i Prilozhen. **6** (1972), no. 4, 65–70; English transl., Funct. Anal. Appl. **6** (1972), 307–312.

232. B. Shelton, *Extensions between generalized Verma modules: the Hermitian symmetric cases*, Math. Z. **197** (1988), 305–318.

233. W. Soergel, *$\mathcal{D}$-modules et équivalence de Enright–Shelton*, C.R. Math. Acad. Sci. Paris **307** (1988), 19–22.

234. ______, *Universelle versus relative Einhüllende: Eine geometrische Untersuchung von Quotienten von universellen Einhüllenden halbeinfacher Lie-Algebren*, Math. Ann. **284** (1989), 177–198.

235. ______, *$\mathfrak{n}$-cohomology of simple highest weight modules on walls and purity*, Invent. Math. **98** (1989), 565–680.

236. ______, *Parabolisch-singuläre Dualität für Kategorie $\mathcal{O}$*, Max-Planck-Institut, Bonn, MPI/89–68.

237. ______, *Kategorie $\mathcal{O}$, perverse Garben und Moduln über den Koinvarianten zur Weylgruppe*, J. Amer. Math. Soc. **3** (1990), 421–445.

238. ______, *Gradings on representation categories*, pp. 800–806, Proc. Intern. Congr. Math. (Zürich, 1994), Birkhäuser Verlag, Basel, 1995.

239. ______, *Kazhdan–Lusztig polynomials and a combinatoric for tilting modules*, Represent. Theory **1** (1997), 83–114.

240. ______, *Character formulas for tilting modules over Kac–Moody algebras*, Represent. Theory **2** (1998), 432–448.

241. ______, *Andersen filtration and Hard Lefschetz*, Geom. Funct. Anal. **17** (2008), 2066–2089.

242. T. A. Springer, *Quelques applications de la cohomologie d'intersection*, Bourbaki Seminar, Vol. 1981/1982, Exp. 589, *Astérisque*, **92–93**, Soc. Math. France, Paris, 1982.

243. R. Steinberg, *Lectures on Chevalley groups*, Yale Univ. Math. Dept., 1967–68.

244. C. Stroppel, *Homomorphisms and extensions of principal series*, J. Lie Theory **13** (2003), 193–212.

245. ______, *Category $\mathcal{O}$: gradings and translation functors*, J. Algebra **268** (2003), 301–326.

246. ______, *Category $\mathcal{O}$: quivers and endomorphism rings of projectives*, Represent. Theory **7** (2003), 322–345.

247. ______, *Composition factors of quotients of the universal enveloping algebra by primitive ideals*, J. London Math. Soc. **70** (2004), 643–658.

248. ______, *Categorification of the Temperley–Lieb category, tangles, and cobordisms via projective functors*, Duke Math. J. **126** (2005), 547–596.

249. T. Tanisaki, *Character formulas of Kazhdan–Lusztig type*, pp. 261–276, *Representations of finite dimensional algebras and related topics in Lie theory and geometry*, Fields Inst. Commun., 40, Amer. Math. Soc., Providence, RI, 2004.

250. V. S. Varadarajan, *Lie groups, Lie algebras, and their representations*, Prentice–Hall, Englewood Cliffs, NJ, 1974; reprinted by Springer-Verlag, New York, 1984.

251. D. N. Verma, *Structure of certain induced representations of complex semisimple Lie algebras*, Ph.D. thesis, Yale Univ., 1966.

252. ______, *Structure of certain induced representations of complex semisimple Lie algebras*, Bull. Amer. Math. Soc. **74** (1968), 160–166; errata, 628.

253. D. A. Vogan, Jr., *Irreducible characters of semisimple Lie groups I*, Duke Math. J. **46** (1979), 61–108.

254. ______, *Irreducible characters of semisimple Lie groups II. The Kazhdan–Lusztig conjectures*, Duke Math. J. **46** (1979), 805–859.

255. ______, *Irreducible characters of semisimple Lie groups III. Proof of the Kazhdan–Lusztig conjecture in the integral case*, Invent. Math. **71** (1983), 381–417.

256. ______, *The character table for* E_8, Notices Amer. Math. Soc. **54** (2007), 1122–1134.

257. M. Vybornov, *Perverse sheaves, Koszul IC-modules, and the quiver for the category $\mathcal{O}$*, Invent. Math. **167** (2007), 19–46.

258. N. R. Wallach, *On the Enright–Varadarajan modules: a construction of the discrete series*, Ann. Sci. École Norm. Sup. (4) **9** (1976), 81–101.

259. C. A. Weibel, *An introduction to homological algebra*, Cambridge Univ. Press, Cambridge, 1994.

260. Wai Ling Yee, *The signature of the Shapovalov form on irreducible Verma modules*, Represent. Theory **9** (2005), 638–677.

261. A. V. Zelevinsky, *Resolutions, dual pairs, and character formulas*, Funktsional. Anal. i Prilozhen. **21** (1987), no. 2, 74–75; English transl., Funct. Anal. Appl. **21** (1987), 152–154.

262. G. Zuckerman, *Tensor products of finite and infinite dimensional representations of semisimple Lie groups*, Ann. of Math. **106** (1977), 295–308.

Frequently Used Symbols

Symbol	Description	Section
$\mathfrak{g}$	semisimple Lie algebra	0.1
$\mathfrak{h}$	Cartan subalgebra	0.1
ℓ	rank of $\mathfrak{g}$ $\ (= \dim \mathfrak{h})$	0.1
Φ	root system of $\mathfrak{g}$ relative to $\mathfrak{h}$	0.1
Δ	simple system in Φ	0.1
Φ^+	positive roots relative to Δ	0.1
$\mathfrak{g}_\alpha$	root space	0.1
$\mathfrak{n}$	sum of positive root spaces	0.1
$\mathfrak{n}^-$	sum of negative root spaces	0.1
$\mathfrak{b}$	standard Borel subalgebra $\mathfrak{h} \oplus \mathfrak{n}$	0.1
$\mathfrak{p} = \mathfrak{p}_I$	parabolic subalgebra, $I \subset \Delta$	0.1
$\mathfrak{l} = \mathfrak{l}_I$	Levi subalgebra	0.1
$\mathfrak{u} = \mathfrak{u}_I$	nilradical of $\mathfrak{p}$	0.1
$(h_\alpha, x_\alpha, y_\alpha)$	standard basis elements, $\alpha \in \Phi^+$	0.1
E	euclidean space spanned by Φ	0.2
$\alpha^\vee$	coroot $2\alpha/(\alpha, \alpha)$	0.2
s_α	reflection relative to $\alpha \in \Phi$	0.2
Λ_r	root lattice in E	0.2

Symbol	Description	Section
W	Weyl group	0.3
$\ell(w)$	length of $w \in W$	0.3
$w_\circ$	longest element of W	0.3
W_I	parabolic subgroup of W $(\mathrm{I} \subset \Delta)$	0.3
$w' \leq w$	Bruhat ordering of W	0.4
$U(\mathfrak{g})$	universal enveloping algebra of $\mathfrak{g}$	0.5
$Z(\mathfrak{g})$	center of $U(\mathfrak{g})$	0.5
τ	transpose map on $U(\mathfrak{g})$	0.5
Λ	integral weight lattice in E	0.6
Λ^+	dominant integral weights	0.6
$\mu \leq \lambda$	partial ordering of weights	0.6
Γ	$\mathbb{Z}^+$-linear combinations of Δ	0.6
ϖ_α	fundamental weight, $\alpha \in \Delta$	0.6
ρ	sum of fundamental weights	0.6
C	Weyl chamber in E	0.6
$\mathrm{Mod}\,U(\mathfrak{g})$	category of all $U(\mathfrak{g})$-modules	0.7
M_λ	weight space in M for $\lambda \in \mathfrak{h}^*$	0.7
$\Pi(M)$	set of weights of M	0.7
$\mathcal{O}$	subcategory of $\mathrm{Mod}\,U(\mathfrak{g})$	1.1
$\mathcal{O}_{\mathrm{int}}$	modules in $\mathcal{O}$ with integral weights	1.1
$M(\lambda)$	Verma module	1.3
$L(\lambda)$	simple quotient of $M(\lambda)$	1.3
$N(\lambda)$	maximal submodule of $M(\lambda)$	1.3
χ_λ	central character	1.7
$w \cdot \lambda$	$w(\lambda + \rho) - \rho$ $(w \in W, \lambda \in \mathfrak{h}^*)$	1.8
$[M : L(\lambda)]$	composition factor multiplicity	1.11
$K(\mathcal{O})$	Grothendieck group of $\mathcal{O}$	1.11
$\mathrm{Rad}\,M$	radical of module M	1.11
$\mathrm{Soc}\,M$	socle of module M	1.11
$\mathcal{O}_\chi$	subcategory of $\mathcal{O}$	1.12
M^χ	summand of M in $\mathcal{O}_\chi$	1.12
$\mathcal{O}_0$	principal block of $\mathcal{O}$	1.13
$\mathcal{X}_0$	additive group of formal characters	1.14
$\mathrm{ch}\,M$	formal character of M	1.14
p	Kostant function in $\mathcal{X}_0$	1.16
$\mathcal{P}$	original Kostant partition function	1.16

Symbol	Description	Section
$M^\vee$	dual of module $M \in \mathcal{O}$	3.2
$\Phi_{[\lambda]}$	integral root system of $\lambda \in \mathfrak{h}^*$	3.4
$\Delta_{[\lambda]}$	simple system in $\Phi_{[\lambda]}$	3.4
$W_{[\lambda]}$	integral Weyl group of $\lambda \in \mathfrak{h}^*$	3.4
$\mathcal{O}_\lambda$	block in $\mathcal{O}$	3.5
$\big(M : M(\lambda)\big)$	standard filtration multiplicity	3.7
$P(\lambda)$	projective cover of $L(\lambda)$	3.9
$Q(\lambda)$	injective hull of $L(\lambda)$	3.9
$\mu \uparrow \lambda$	strong linkage of weights	5.1
$M(\lambda)^i$	ith submodule in Jantzen filtration	5.3
$M(\lambda)_i$	ith layer in Jantzen filtration	5.3
Φ_λ^+	index set in Jantzen Sum Formula	5.3
T_λ^μ	translation functor	7.1
H_α	α-hyperplane shifted by $-\rho$ in E	7.3
$\widehat{F}$	upper closure of facet F	7.3
$\widehat{C}$	upper closure of Weyl chamber C	7.3
$E(\lambda)$	span of $\Phi_{[\lambda]}$ in E	7.4
$\lambda^\natural$	integral part of $\lambda \in \mathfrak{h}^*$	7.4
W_λ°	stabilizer of λ	7.4
w_λ	longest element in $W_{[\lambda]}$	7.13
Θ_s	wall-crossing functor	7.14
$\ell(x, w)$	$\ell(w) - \ell(x)$ when $x \le w$	8.0
$P_{x,w}(q)$	Kazhdan–Lusztig polynomial, $x \le w$	8.2
G/B	flag variety	8.5
X_w	Bruhat cell in G/B	8.5
$\mathcal{O}^{\mathfrak{p}}$	parabolic subcategory of $\mathcal{O}$	9.3
$M_{\mathrm{I}}(\lambda)$	parabolic Verma module	9.4
W^{I}	minimal coset representatives in $W_{\mathrm{I}} \backslash W$	9.4

Index